Nongonococcal Urethritis and Related Infections

Nongonococcal Urethritis and Related Infections

EDITORS:

Derek Hobson
New Medical School
University of Liverpool
Liverpool, England

King K. Holmes
University of Washington School of Medicine and
U.S. Public Health Service Hospital
Seattle, Washington

American Society for Microbiology
Washington, D.C.
1977

1913 I Street, NW

Washington, DC 20006

Library of Congress Cataloging in Publication Data

Workshop on Nongonococcal Urethritis and Related Infections, Lake Placid, N.Y., 1976.

Nongonococcal urethritis and related infections.

Includes indexes.

1. Chlamydia infections—Congresses. 2. Urethritis—Congresses. 3. Conjunctivitis—Congresses. 4. Generative organs—Diseases—Congresses. 5. Chlamydia trachomatis—Congresses. I. Hobson, Derek. II. Holmes, King, 1937– III. Title. [DNLM: 1. Chlamydia trachomatis—Congresses. 2. Chlamydia infections—Congresses. 3. Gynecology—Congresses. 4. Ophthalmology—Congresses. 5. Genital diseases, Male—Congresses. 6. Urethritis—Congresses. WJ600 W926p 1976]

RC124.5.W67 1976 616.9′518 77-24329

ISBN 0-914826-11-5

Printed in the United States of America

List of Contributors

E. Russell Alexander
School of Medicine, University of Washington, Seattle, WA 98114

J. Allami
School of Public Health, Teheran, Iran

Charles R. Ashley
Public Health Laboratory, Kingsdown, Bristol BS2 8EL, England

R. Palmer Beasley
School of Public Health and Community Medicine, Seattle, WA 98195, and United States Naval Medical Research Unit no. 2, Taipei, Taiwan

Marc O. Beem
Department of Pediatrics, University of Chicago, Chicago, IL 60637

William R. Bowie
Department of Medicine, University of Washington, Seattle, WA 98114

Harlan D. Caldwell
Department of Pathobiology, University of Washington School of Public Health, Seattle, WA 98195

E. Owen Caul
Public Health Laboratory, Kingsdown, Bristol BS2 8EL, England

John Chandler
Department of Ophthalmology, University of Washington School of Medicine, Seattle, WA 98195

Suzanne K. R. Clarke
Public Health Laboratory, Kingsdown, Bristol BS2 8EL, England

S. Cubitt
Institute of Ophthalmology, University of London, London WC1H 9QS, England

S. Darougar
Institute of Ophthalmology, University of London, London WC1H 9QS, England

Chandler R. Dawson
G. W. Hooper Foundation, University of California, San Francisco, CA 94112

Jill Day
Institute of Ophthalmology, University of London, London WC1H 9QS, England

Ronald F. DiGiacomo
Department of Epidemiology and International Health, University of Washington School of Public Health, Seattle, WA 98195

R. J. Dines
Institute of Ophthalmology, University of London, London WC1H 9QS, England

H. El-Sheikh
Institute of Ophthalmology, University of London, London WC1H 9QS, England

Marilyn English
Department of Epidemiology and International Health, University of Washington School of Public Health, Seattle, WA 98195

William R. Fair
Department of Surgery, Washington University, St. Louis, MO 63110

Victor S. C. Fan
Department of Microbiology, University of Louisville Medical School, Louisville, KY 40202

Denys K. Ford
Department of Medicine, Vancouver General Hospital, Vancouver, British Columbia, Canada

T. Forsey
Institute of Ophthalmology, University of London, London WC1H 9QS, England

James L. Gale
Department of Epidemiology, University of Washington, Seattle, WA 98195

J. Thomas Grayston
School of Public Health and Community Medicine, University of Washington, Seattle, WA 98195

Lavelle Hanna
Department of Microbiology, University of California Medical Center, San Francisco, CA 94122

Leonard Hayflick
Department of Medical Microbiology, Stanford University, Palo Alto, CA 92605

Derek Hobson
Department of Medical Microbiology, New Medical School, University of Liverpool, Liverpool L69 3BX, England

King K. Holmes
Department of Medicine, University of Washington School of Medicine, Seattle, WA 98114

Howard M. Jenkin
The Hormel Institute, University of Minnesota, Austin, MN 55912

F. William A. Johnson
Department of Medical Microbiology, New Medical School, University of Liverpool, Liverpool L69 3BX, England

Barrie R. Jones
Institute of Ophthalmology, University of London, London WC1H 9QS, England

George E. Kenny
Department of Pathobiology, University of Washington, Seattle, WA 98114

Mark D. Kiviat
Department of Urology, University of Washington School of Medicine, Seattle, WA 98195

Cho-Chou Kuo
Department of Pathobiology, University of Washington School of Public Health, Seattle, WA 98195

Juey-Shin L. Lin
Channing Laboratories, Boston, MA 01288

C. Lyon
F. I. Proctor Foundation, University of California, San Francisco, CA 94143

William M. McCormack
Department of Microbiology, Harvard University, Boston, MA 02115
G. Philip Manire
Department of Bacteriology and Immunology, University of North Carolina School of Medicine, Chapel Hill, NC 27514
Per-Anders Mårdh
Institute of Medical Microbiology, University of Lund, Solvegatan 23, Sweden
Gerald K. Masover
Department of Surgery, Stanford University, Palo Alto, CA 92605
M. A. Monnickendam
Institute of Ophthalmology, University of London, London WC1H 9QS, England
Carl H. Mordhorst
Statens Serum Institute, Copenhagen, Denmark
Edward S. Murray
Department of Microbiology, Harvard University School of Public Health, Boston, MA 02115
J. David Oriel
University College Hospital, Department of Genito-Urinary Medicine, London WC1E 6AU, England
Marina Palant
Department of Medical Microbiology, Stanford University, Palo Alto, CA 92605
Terrence A. Pheifer
Department of Obstetrics and Gynecology, University of Washington School of Medicine, Seattle, WA 98195
Shmuel Razin
Department of Medical Microbiology, Stanford University, Palo Alto, CA 92605
Elisabeth Rees
Department of Medical Microbiology, New Medical School, University of Liverpool, Liverpool L69 3BX, England
Shirley J. Richmond
Public Health Laboratory, Kingsdown, Bristol BS2 8EL, England
K. Torvald Ripa
Institute of Medical Microbiology, University of Lund, Solvegatan 23, Sweden
Teresa R. Rota
Harvard School of Public Health, Boston, MA 02115
Evelyn M. Saxon
University of Chicago, Chicago, IL 60637
Julius Schachter
G. W. Hooper Foundation, University of California, San Francisco, CA 94112
Maurice C. Shepard
Navy Medical Center, Communicable Disease Laboratory, Camp Lejeune, NC 28542
Patricia Skahen
Department of Ophthalmology, University of Washington School of Medicine, Seattle, WA 98195
S. Squires
May and Baker Ltd., Dagenham, Essex, England

Stephen L. Swartz
Bureau of Laboratories, Center for Disease Control, Atlanta, GA 30333

I. Anne Tait
Department of Medical Microbiology, New Medical School, University of Liverpool, Liverpool L69 3BX, England

David Taylor-Robinson
Clinical Research Center, Harrow, Middlesex HA1 3UJ, England

J. D. Treharne
Institute of Ophthalmology, University of London, London WC1H 9QS, England

San-Pin Wang
School of Public Health and Community Medicine, University of Washington, Seattle, WA 98195

Berttina B. Wentworth
State of Michigan Department of Public Health, Bureau of Disease Control and Laboratory Service, Lansing, MI 48914

Lars Weström
Department of Obstetrics and Gynecology, University Hospital, S-221 85 Lund, Sweden

J. P. Whitcher
G. W. Hooper Foundation, University of California, San Francisco, CA 94112

Paul J. Wiesner
Venereal Disease Control Division, Center for Disease Control, Atlanta, GA 30333

R. M. Woodland
Institute of Ophthalmology, University of London, London WC1H 9QS, England

Lung-Jui Yeh
Trachoma Control Program, Provincial Health Department, Taiwan

C. K. Yeo
Institute of Ophthalmology, University of London, London WC1H 9QS, Endland

Zenaida Zerrudo
Department of Medical Microbiology, Stanford University, Palo Alto, CA 92605

List of Participants

Susan Alpert
Channing Laboratories, Boston, MA 02118

Michael F. Barile
Bureau of Biologics, Bethesda, MD 20014

Almen L. Barron
Department of Microbiology and Immunology, University of Arkansas, Medical Sciences, Little Rock, AR 72201

Marc Beem
Department of Pediatrics, University of Chicago, Chicago, IL 60637

Billie Bird
Center for Disease Control, Atlanta, GA 30333

Abraham I. Braude
University of California, University Hospital, San Diego, CA 92103

Merlin L. Brubaker
World Health Organization/Pan American Health Organization, Washington, DC 20037

Gerald I. Byrne
Department of Microbiology, University of Chicago, Chicago, IL 60637

Harlan D. Caldwell
University of Washington, Seattle, WA 98195

Georges Y. Causse
VDT Unit, World Health Organization, 1211 Geneva 27, Switzerland

Dorland J. Davis
American Social Health Association, Bethesda, MD 20034

G. J. Delisle
Department of Microbiology, Queen's University, Kingston, Ontario, Canada K76 3N6

J. A. Embil
Infectious Diseases Unit, The Izaak Walton Killam Hospital for Children, Halifax, Nova Scotia, Canada B3J 3G6

George T. Frommell
University of Colorado Medical Center, Denver, CO 80220

Morey Gardner
950 E. 59th Street, Box 415, Chicago, IL 60637

Milton Gordon
Division of Research Grants, National Institutes of Health, Bethesda, MD 20014

Doris Graham
Department of Microbiology, University of Melbourne, Parkville, Victoria 3052, Australia

William W. Gregory
Department of Microbiology, University of Chicago, Chicago, IL 60637

Dieter W. Gump
University of Vermont, Given Medical Building, Burlington, VT 05401

Margaret Hamerslag
Channing Laboratories, Boston, MA 02118

Thomas P. Hatch
Department of Microbiology, University of Texas, Austin, TX 78712
James C. Hill
National Institute of Allergy and Infectious Diseases, Bethesda, MD 20014
Bruce E. Ivins
Department of Bacteriology and Immunology, University of North Carolina, Chapel Hill, NC 27514
Mary Kendrick
Channing Laboratories, Boston, MA 02118
Sam Knox
American Social Health Association, New York, NY 10019
Katalin Ladnay
Channing Laboratories, Boston, MA 02118
Hayes Lamont
Department of Microbiology, Harvard School of Public Health, Boston, MA 02115
David A. Lennette
Department of Microbiology, Hahnemann Medical College and Hospital, Philadelphia, PA 19102
Harvey Liebhaber
Washington University School of Medicine, The Jewish Hospital of St. Louis, St. Louis, MO 63110
Dorothy E. McComb
Department of Microbiology, Harvard School of Public Health, Boston, MA 02115
David Madden
National Institutes of Health, Bethesda, MD 20014
Per-Anders Mårdh
Institute of Medical Microbiology, University of Lund, Solvegatan 23, Lund, Sweden
Janet M. Mattson
National Institutes of Health, Bethesda, MD 20014
Tomasz Mroczkowski
Venereal Disease Control Division, Center for Disease Control, Atlanta, GA 30333
J. Orfila
Service de Bacteriologie-Virologie, Immunologie Generale, Centre Hospitalo-Universitaire, 80030 Amiens Cedex, France
Torvald Ripa
Institute of Medical Microbiology, University of Lund, Solvegatan 23, Lund, Sweden
Janet A. Robertson
Department of Medical Bacteriology, The University of Alberta, Edmonton, Alberta T6G 4S6, Canada
J. L. Schuller
Department of Dermatology, Erasmus University, Rotterdam, Holland
Deborah Semine
Department of Microbiology, Harvard School of Public Health, Boston, MA 02115

Thomas F. Smith
Mayo Clinic, Rochester, MN 55901
William Todd
Department of Microbiology, Hahnemann Medical College and Hospital, Philadelphia, PA 19102
Joseph G. Tully
National Institutes of Health, Bethesda, MD 20014
J. R. Waters
Johnston Labs Inc., Cockeysville, MD 21030
Douglas Wigent
The Hormel Institute, University of Minnesota, Austin, MN 55912
K. H. Wong
Center for Disease Control, Atlanta, GA 30333
Priscilla B. Wyrick
Department Bacteriology, University of North Carolina, School of Medicine, Chapel Hill, NC 27514

CONTENTS

Session IV

Chlamydia: Biology and Experimental Pathology

Session V
Laboratory Diagnosis of *Chlamydia trachomatis* Infections

Session VI
Practical Problems in Isolation and Growth of *Chlamydia trachomatis* in Tissue Culture

Session VII
Ureaplasma urealyticum

Introduction

This symposium represents an amalgamation of two symposia which were being planned independently early in 1976. The National Institutes of Health was planning a meeting dealing with nongonococcal urethritis (NGU) and related infections, while the W. Alton Jones Cell Science Center of the Tissue Culture Association was organizing a workshop on *Chlamydia trachomatis*. A decision was made to combine the meetings under the auspices of the Cell Science Center and the National Institutes of Health, to allow participation of the maximum possible number of clinical and laboratory investigators in the United States and abroad who are active in these fields. Nearly all of the invitees were able to participate, thus ensuring a comprehensive and authoritative monograph. Enthusiasm for this symposium, which was held 28–30 April 1976 at Lake Placid, N.Y., arose from a series of related developments. These included the rapidly increasing incidence of NGU on a worldwide basis; the increasing evidence for a causal role of *C. trachomatis* and perhaps of ureaplasmas in NGU; the very recent studies suggesting that *C. trachomatis* causes several other serious infections, such as pelvic inflammatory disease, epididymitis, and infant pneumonia; and recent advances in the cultivation and antigenic characterization of *C. trachomatis*.

The investigation of these infective conditions has been multidisciplinary, and thus inevitably somewhat fragmented, shared in the clinical field among venereologists, gynecologists and obstetricians, ophthalmologists, especially trachomatologists, and pediatricians; and in the laboratory among bacteriologists, virologists, mycoplasmatologists, chlamydiologists, immunologists, and experimental pathologists, who have rarely had the opportunity to meet jointly to discuss the interplay of their various findings or to establish mutual grounds for future research.

It seemed timely to draw together experts in all these varied fields of endeavor to reassess the clinical spectrum of chlamydial and ureaplasmal infections; to bring the experience of trachomatologists to bear on the newer problems of genital chlamydial infection; to review recent advances (or non-advances) in the cultivation, in vitro propagation, and serological study of chlamydiae and ureaplasmas; and to present recent progress in studies of the basic biology of *Chlamydia trachomatis* and *Ureaplasma urealyticum*, and of the pathogenesis and immunology of infections caused by these organisms. It was further intended, at the meeting and through this monograph, to provide a continuing-education program for the many younger clinicians and laboratory workers now becoming involved in this expanding field. Finally, and not least importantly, it was hoped that a broad exchange of achievements, problems, and opinions on clinical and laboratory experiences gained throughout the United States and in Europe, the Middle East, Africa, and Australia would constitute a useful body of information for national and international health authorities who are and will increasingly be concerned with halting the spread and reducing the present high socioeconomic cost of these conditions at the population level.

Just how much concern about NGU and related infections is warranted? The "classical" venereal diseases are gonorrhea, syphilis, chancroid, granuloma inguinale, and lymphogranuloma venereum. In modern clinical practice, all but gonorrhea and syphilis have been largely replaced by a host of new sexually transmissible diseases. NGU in the male is now the commonest sexually transmissible disease syndrome seen in developed countries. This condition is increasing at an epidemic rate, faster even than gonorrhea. However, there has been relatively little diffusion of information about NGU and related infections from the clinical specialist to non-specialists. Even among venereologists, there is a prevailing opinion that NGU is a trivial infection. Thus, in North America, some public venereal disease clinics and many physicians in private practice treat all cases of urethritis in men with penicillin G or a semisynthetic penicillin, effective for gonorrhea but not for NGU. Many public clinics do perform diagnostic tests to differentiate gonorrhea from NGU, but only offer treatment for those cases of urethritis caused by the gonococcus. Men with NGU may be told to avoid alcohol, spicy foods, or repeated self-examination—as if the condition were irritative or imaginary, rather than infective. Even when antimicrobial treatment is offered for both diseases in men, treatment of female sex partners is usually limited to those exposed to gonorrhea, whereas those exposed to NGU go untreated. Most public authorities still regard penicillin G as the treatment-of-choice for gonorrhea, despite the fact that *C. trachomatis* can be isolated from 20% or more of patients with gonorrhea and is not eradicated by penicillin therapy.

One goal of this symposium, then, is to define from the clinical and epidemiological standpoints the nature and extent of morbidity of NGU and related infections. The term "nongonococcal urethritis" was chosen rather than "nonspecific urethritis" because of increasing evidence for specific microbial etiologies of NGU. "Related infections" refers to those conditions that are either etiologically related, being caused by the putative agents of NGU, or are often clinically associated with NGU. To assess the morbidity attributable to NGU and related infection, several questions must be answered. How common is NGU today, and who is affected? How effective is treatment? In men, is the disease self-limiting, or are epididymitis, prostatitis, or Reiter's syndrome actually complications of NGU? Are vaginitis, urethritis, cervicitis, pelvic inflammatory disease, or other conditions in women etiologically related to NGU? Is significant neonatal or infant morbidity etiologically related to NGU? In developed countries, ocular infections caused by *C. trachomatis* are often regarded as uncommon, trivial, and self-limited.

Nongonococcal urethritis and related ophthalmic diseases are certainly of multiple etiologies. It has been clearly recognized since the beginning of this century that the triad of abacterial conjunctivitis in the neonate, cervical infection in the mother, and urethritis in the father was frequently associated with an agent that produced intracellular inclusions closely resembling those characteristically found in the eyes of tropical patients with hyperendemic trachoma. Hence it has been a major target of modern research to see how much present-day disease in developed temperate countries can be explained by *C. trachomatis*. Other obvious candidates as etiological agents for at least some nongonococcal genital infections include the ureaplasmas, which are sexually transmissible and in some studies have been associated with NGU and with pelvic inflammatory disease. In all centers in the U.S. and abroad where NGU and nongonococcal cervicitis and related infections have been extensively investigated, *C. trachomatis* is consistently isolated from the urethras of 30 to 50% of men with NGU and from the cervixes of many of their sex partners. These findings raise several important questions concerning etiology which form some of the main motifs throughout the symposium:

1. Is *C. trachomatis* merely a "normal" resident of the genital tract of sexually active adults, or, at the most, an opportunistic secondary invader in lesions caused by other agents? Can it ever be regarded as a primary pathogen in the genital tract, fulfilling Koch's postulates?

2. Is recurrent genital chlamydial infection due to repeated reinfection, or to reactivation of latent genital chlamydial infection?

3. If a substantial proportion of NGU can be defined as nonchlamydial, what are the other possible etiological agents? Do the ureaplasmas play a significant role in this respect?

The "new" problems now posed by *C. trachomatis* in its "new" role in the genital tract may well be solved most readily by reference to the agent's behavior in its classical role in the eye. Controlled clinical and epidemiological surveys into hyperendemic trachoma in Asia, Asia Minor, and Africa have provided a wealth of data which are now highly relevant to the genitourinary field, and are represented in the symposium papers, for example: (i) the transmissibility of *C. trachomatis* and the influence of previous infection with this agent, or of superinfection with other organisms, or of socioeconomic circumstances on the outcome of infection; (ii) the effect of chemotherapy and immunological mechanisms in controlling the disease; and (iii) the development of rapid diagnostic methods.

Have the status of chlamydiae or ureaplasmas as pathogens, and the diagnostic technology, advanced to the point that the clinical microbiology laboratory should begin to offer diagnostic tests for chlamydial and ureaplasmal infection on a routine basis? A variety of tissue culture and serological innovations have been made during the past few years. Few of the advances in knowledge of chlamydial and ureaplasmal infections would have been possible without the recent rapid development of methods of rapid isolation and identification of these agents in the laboratory, and reliably passaging such strains for further detailed investigation. Modern trachomatology (of ophthalmic or genital varieties) owes a large debt of gratitude to the many workers from the early 1930s onward who pioneered the laboratory study of *C. psittaci*, first by growth in the yolk sac of chick embryo, but particularly by showing that this agent could be grown in tissue cultures similar to those now conventionally used in virus research. Work with *C. trachomatis* has introduced many new procedural problems, which are described in the following papers, but it is now apparent that the isolation of this organism from genital or acute ophthalmic infections can be achieved in tissue culture with a degree of precision, sensitivity, and reproducibility that is quite comparable to that already accepted in clinical practice for gonorrhea or for mycoplasmal or viral diseases of the genital tract. The high correlation between chlamydial isolation rates and specific seroconversion in cases of "first-episode" urethritis suggests that, if specimens are well taken and the present laboratory methods are used efficiently, very few acutely infected cases will fail to be diagnosed. It is now important to simplify these methods sufficiently to bring them into use on a routine worldwide basis.

In the past, the selection of antibiotics to control chlamydial infections has perforce been somewhat empirical, but the advent of precise quantitative tissue culture procedures for propagating patients' strains of *C. trachomatis* in the laboratory has also permitted the measurement of minimum inhibitory concentrations of antibiotics, analogous to those routinely used in the guidance and control of chemotherapy for bacteriological infections. Although this has not yet led to routine testing of isolates as a guide to individual patient management, it has become possible to use susceptibility testing to determine which antibiotics are consistently most effective against *C. trachomatis*, as a rational guide to future therapeutic trials.

Genital infection with *C. trachomatis*, both in men with acute NGU and in women with cervicitis, has often been considered to be a noncausal relationship; i.e., it is assumed that some other infective agent is primarily responsible for the clinical condition, with the chlamydial infection representing only a reactivation of a long-standing resident organism, or a superimposed secondary invasion. This situation can still not be entirely resolved, but it is already becoming clear that demonstrable chlamydial infections of patients experiencing their first clinical attack of urethritis induce seroconversion of specific immunoglobulin M antibody, and that specific antibiotic treatment to eradicate the chlamydial infection rapidly terminates the signs and symptoms of urethritis or cervicitis, whereas such treatment is usually not effective in cases of NGU where *C. trachomatis* is not isolated. Furthermore, animal experimental infections with the inoculation of various strains of *C. psittaci* in the eyes and genital tracts of guinea pigs and kittens and experimental infections with *C. trachomatis* itself, inoculated into the genital tracts of primates, have now convincingly shown that chlamydiae can behave as primary pathogens in the genital tract, as has long been accepted for their role in the eye. These models, described in the following papers, may permit definition of the events involved in the pathogenesis of mucosal inflammation produced by these agents.

Recent advances in the immunology and molecular biology of chlamydiae, discussed here, have significantly increased our knowledge of the relationship of the TRIC strains of temperate countries to strains isolated from classical cases of trachoma and also to the chlamydial agent of lymphogranuloma venereum, which is, at least on taxonomic grounds, also regarded as *C. trachomatis*. The chlamydial agents found in natural genital diseases of many animals, but collectively classified with *C. psittaci*, are immunologically distinct from *C. trachomatis*, except for the common group antigen. The fundamental biological relationship of these agents to each other is, however, close enough for research findings with any one of them to have direct relevance to workers with other chlamydiae and chlamydial infections. A major question which must be addressed concerning the immunology of both *C. trachomatis* and *U. urealyticum* concerns whether natural or acquired immunity to either agent occurs. If so, is immunity type specific, and how does it work?

Writers of introductions to conference proceedings should adopt what is fashionably known as a "low profile" and let the speakers and their results speak for themselves. This we now propose to do, pausing only to thank most warmly all those who contributed in formal communications and in vigorous unscripted discussions to a wide-ranging interchange of views, and particularly to express our gratitude to the joint organizers, Dr. Milton Puziss of the National Institutes of Health and Dr. Paul J. Chapple of the W. Alton Jones Cell Science Center of the Tissue Culture

Association, for initiating and sponsoring this conference, and to the Royal Society of Medicine, London, for additional support.

This Conference was supported by the National Institute of Allergy and Infectious Diseases. Publication of the Proceedings was aided by a grant from Pfizer, Inc., and the Venereal Disease Research Fund of the American Social Health Association.

King K. Holmes
Derek Hobson

Session I

NONGONOCOCCAL URETHRITIS

Selected Aspects of the Epidemiology of Nongonococcal Urethritis

PAUL J. WIESNER

The epidemiology of nongonococcal urethritis (NGU) is difficult to discuss because of (i) the use of noncomparable case definitions, (ii) ignorance about the prevalence and incidence of this disease, resulting from the worldwide lack of systematic recording, and (iii) the paucity of information about risk factors. The most critical reservation about such a discussion is exemplified by the content of this conference—multiple etiological agents are likely to be involved, each with a potentially unique epidemiology and very probably of different significance from one geographic area to the next.

We shall focus on areas where data exist. Therefore, my discussion will be limited to the following: (i) estimates of NGU cases among American naval personnel stationed throughout the world, (ii) what is known about the epidemiology of NGU in the United States and the United Kingdom, and (iii) the results of a study of individual characterisitics associated with the risk of urethritis.

Melton has analyzed the outpatient cases of gonorrhea and nongonococcal urethritis reported to the U.S. Navy Medical Data Center from 1966 to 1974 (1). Although diagnostic criteria and reporting have varied, the two diseases have increased dramatically in these men in the past 8 years (averaging a 21% annual increase for gonorrhea and 27% annual increase for NGU). Highest attack rates for both diseases were in the Pacific area. Whereas the ratio of gonococcal urethritis (GCU) to NGU for all cases was 1.3:1, this ratio ranged from a low of 0.3:1 in Europe to a high of 1.7:1 in the Western Pacific. The ratio of GCU to NGU was 0.9:1 in the naval personnel based in the United States.

The geographic distribution of case rates of GCU and NGU per 1,000 personnel vary with Naval District within the United States. For gonorrhea, the highest rates are in southern coastal districts. For NGU, the highest rates are on the west coast and in the northeastern area.

Two additional sources of information are available for examining civilian cases in the United States. One source, the IMS System, is a national random sample consisting of 1,500 office-based physicians who report on 12,000 physician-days per year. All diagnoses associated with every patient seen on the sample days are recorded and subsequently coded according to the International Classification of Disease. Figure 1 depicts the estimated number of first visits for NGU to office-based private physicians in the United States from 1960 to 1974. In 1974, private physicians saw approximately 397,000 cases of NGU among men. In this survey, these patients were seen most frequently by urologists, followed by general or family physicians, internists, osteopaths, and all others. Men between the ages of 10 and 29 years accounted for 78% of NGU cases, which were most numerous in the South and least common in the Midwest.

Our second data source is public clinic records. The prevalence of NGU among consecutive urethritis cases seen in clinics throughout the United States is depicted in Table 1. The clinics were selected simply because data were available, either in

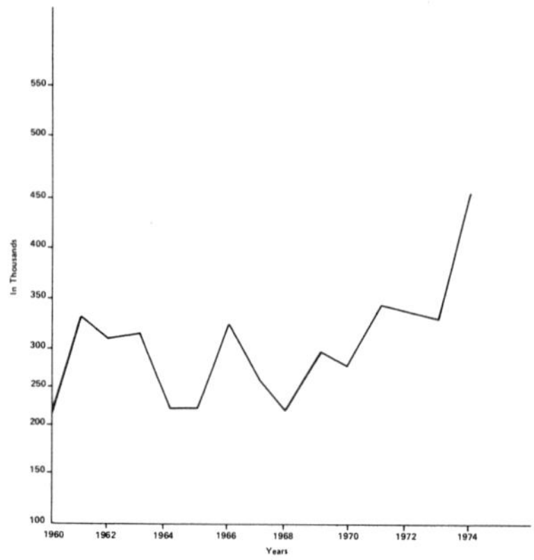

FIG. 1. *Estimated number of first visits to office-based private physicians for nongonococcal urethritis in the United States, 1960–1974.*

published form or through a search of unpublished records at the Venereal Disease Control Division. The range of percentages of urethritis cases which are attributed to NGU is dramatic. For venereal disease clinics, the percentage ranges from a low of 18.8% in Memphis, Tenn., to a high of 78% in Rochester, Minn. University student health clinics had the highest percentages: 86.9% of urethritis cases in Providence, R. I., and 92% in Charlottesville, Va., were attributed to NGU. Since this is not a systematic sample, and since populations attending clinics are not uniform, I hesitate to speak of specific regional differences, but the variability is impressive.

TABLE 1. *Prevalence of nongonococcal urethritis (NGU) among urethritis cases in selected United States clinics*

City[a]	Year[b]	Percent NGU
Veneral Disease Clinics		
Memphis, Tenn.	1972	18.8
Albuquerque, N.M.	1975	20.9
Montgomery, Ala.	1975	26.5
Boston, Mass.	1975	29.1
Columbus, Ohio	1975	31.3
Atlanta, Ga. (2)	1974	45.4
Denver, Colo. (3)	1973	51.8
New York, N.Y.	1974	57.0
Seattle, Wash. (4)	1974	62.0
Missoula, Mont.	1973	67.0
Rochester, Minn. (5)	1973	78.0
University Student Health Centers		
Providence, R.I. (6)	1974	86.9
Charlottesville, Va. (7)	1972	92.0

[a] Reference numbers are given in parentheses.
[b] Year observation was made.

The influence of race upon the prevalence of NGU among urethritis cases in three separate clinics (the first two being in the same city) is shown in Table 2 (W. McCormack, personal communication). In all three clinics, black men with urethritis were more likely to have gonorrhea than to have NGU, whereas in clinic C white men with urethritis were more likely to have NGU than gonorrhea (2).

Great Britain is the only country in the world that has systematically recorded NGU for long periods of time—indeed, since 1951. Table 3 shows the most recent figures (kindly supplied by J. Berrie of the Department of Health and Social Security). Since 1965, the incidence of NGU exceeded the incidence of GCU in men. Diagnostic criteria undoubtedly vary greatly from clinic to clinic and over time, but some trends can be discerned. NGU increased 133% from 1966 to 1974. Table 4 shows the yearly percent change of incidence of NGU and of GCU. From 1965 to 1970, NGU cases rose relatively consistently. I am reluctant to interpret the years after 1970 mainly because of the extremely erratic nature of the data dealing with gonorrhea incidence (0, −5, +8, and −1% for 1971–1974). Our counterparts in Great Britain may have opinions on this subject, but I would not conclude that NGU cases are decreasing.

Now I would like to briefly review the results of a special study reported by a group of investigators from the Department of Navy, the University of Washington, and the Center for Disease Control (O. G. Jones et al., Am. Public Health Assoc. Meet., 1975, Chicago, Ill.)

En route to the Western Pacific, the men aboard a large naval vessel were given the opportunity to participate in a study which had the primary purpose of testing the efficacy of a single-dose antibiotic in preventing urethritis after sex exposure.

During a 57-day period, of which 28 days were in port and 29 days were at sea, all diagnoses of venereal disease made in sickbay were recorded on study forms. In a portion of the report, the investigators attempted to measure and correlate the prospective risk of acquiring urethritis with the personal characteristics of the study participants assessed at the beginning of the Western Pacific cruise by use of a personal history questionnaire.

The personal history questionnaire included four sections. The first contained personal identifiers such as age and marital status; the second, sex history and

TABLE 2. *Race of men with gonococcal urethritis (GCU) and nongonococcal urethritis (NGU)*

Clinic	Infection	No. of men: White	No. of men: Black
A	GCU	34 (52%)	124 (79%)
	NGU	31 (48%)	34 (21%)
	Total	65	158
B	GCU	66 (52%)	38 (79%)
	NGU	62 (48%)	10 (21%)
	Total	128	48
C	GCU	27 (27%)	71 (71%)
	NGU	73 (73%)	29 (29%)
	Total	100	100

TABLE 3. *Occurrence of gonorrhea, nongonococcal urethritis (NGU), and nonspecific genital infections in England and Wales, 1951–1974*

Year	Gonorrhea			NGU in males[a]	Nonspecific genital infection[b]		
	Male	Female	Total		Male	Female	Total
1951	14,975	3,089	18,064	10,794			
1952	15,510	3,585	19,095	11,552			
1953	15,242	4,021	19,263	13,157			
1954	13,962	3,574	17,536	13,279			
1955	14,079	3,766	17,845	14,269			
1956	16,377	4,011	20,388	14,825			
1957	19,620	4,761	24,381	16,066			
1958	22,398	5,489	27,887	17,606			
1959	24,964	6,380	31,344	20,227			
1960	26,618	7,152	33,770	22,004			
1961	29,519	7,588	37,107	24,472			
1962	28,329	7,109	35,438	24,494			
1963	27,895	8,154	36,049	25,289			
1964	29,050	8,615	37,665	27,521			
1965	27,886	8,805	36,691	29,141			
1966	27,921	9,562	37,483	30,462			
1967	30,645	11,184	41,829	32,318			
1968	32,595	12,367	44,962	35,721			
1969	36,379	14,881	51,260	41,189			
1970	37,811	16,953	54,764	47,292			
1971	37,929	18,059	55,988		59,023	13,397	72,420
1972	36,095	18,879	54,974		63,813	14,562	78,375
1973	38,901	21,357	60,258		69,993	15,022	85,015
1974	38,466	21,288	59,754		70,988	15,069	86,057

[a] Including cases with arthritis.
[b] Excluding cases with arthritis.

history of condom use; the third, history of venereal infection; and the fourth, knowledge about the symptoms of the venereal diseases.

Diagnoses of gonorrhea were based on positive smears and/or cultures; diagnoses of NGU were based on signs and symptoms, the finding of 20 or more white blood cells in the first-voided urine, and the absence of positive laboratory findings for gonorrhea.

Overall, 15.5% of the men were infected with gonorrhea during the 57-day study period; by race, 12.6% of whites and 39.1% of blacks became infected with gonorrhea.

For NGU, the overall infection rate was 11%, with 10.9% of whites and 15.5% of blacks developing infection. Multiple infections of NGU were not observed, primarily as a result of arbitrary diagnostic criteria in which continuation or reappearance of urethritis within 6 weeks after treatment for gonorrhea was called postgonococcal urethritis. Whereas gonorrhea occurred about 3 times more frequently in blacks than in whites (39.1 versus 12.6% infected), NGU occurred only 1.5 times as frequently in blacks as in whites. Since the purpose of this section is to describe risk factors associated with the prospective occurrence of gonorrhea and NGU, and because of the differences in relative rates of occurrence of the two diseases within races, the authors decided to look at the risk factors separately for each race in anticipation that such an analysis might indicate reasons for these

TABLE 4. *Yearly percent change[a] in incidence of nongonococcal urethritis (NGU) and gonococcal urethritis (GCU) in males in England and Wales*

Year	NGU	GCU
1965	+6	−4
1966	+4	0
1967	+6	+10
1968	+11	+6
1969	+15	+12
1970	+15	+4
1971	+25	0
1972	+8	−5
1973	+10	+8
1974	+1	−1

[a] Percent change = (number of cases in index year − number of cases in the prior year)/ number of cases in index year.

differences. In a multistep regression analysis, each risk factor, plus some interactions, was entered into the analysis separately for each of the two diseases, GCU and NGU, within each of two races, whites and blacks. The results were as follows:

1. Age, marital status, and lifetime number of gonorrhea cases were significantly correlated with risk of infection in both races for both diseases.
2. The lifetime number of sex partners was significantly related to risk of both GCU and NGU infections among whites, but these risk factors were not significant among blacks.
3. A history of mononucleosis infection was significantly related to risk of both GCU and NGU infections among whites. Mononucleosis as a prospective risk factor among blacks could not be evaluated, since no blacks indicated a history of mononucleosis.
4. Among blacks, a history of usual condom use before military service was significantly related to lower risk of infection with both GCU and NGU. History of condom use was not a significant factor among whites for either disease.
5. Two other factors, by multiple regression analysis, were significant for only one race and disease: lack of circumcision was a significant high-risk factor only for gonorrhea among whites, and lifetime history of NGU infection was a significant risk factor only for NGU among whites.

In general, the men at greatest risk of acquiring urethritis in this study were young, not married, had less than 2 years of military service, had the least amount of education, were of low rank, and were on their first cruise; they began sex at a very early age, had experienced a large number of sex partners, and had a prior history of venereal disease. Correct knowledge about the signs and symptoms of venereal disease did not appear to be associated with risk.

This study was performed among personnel on liberty in a foreign port with easily available female consorts among whom the prevalence of disease is very high. The factors associated with higher and lower prospective risks of acquiring a venereal disease cannot be generalized to the civilian population because of the wide differences in circumstances.

Earlier, I mentioned that one reason for examining GCU and NGU separately within each race was to look for factors which might account for the higher incidence of GCU than NGU in blacks than in whites. In several instances among blacks, the risk of NGU was opposite to the risk of gonorrhea; that is, if a group of blacks was at high risk for gonorrhea it was at low risk for NGU and vice versa.

Gonorrhea and NGU are caused by two different etiological agents, so an individual may acquire both diseases at the same time. Because of the diagnostic methods and criteria used in this study, if both diseases existed simultaneously in a patient, gonorrhea would be the only immediate diagnosis made. In fact, since urethritis continuing for as long as 6 weeks after treatment for gonorrhea was called postgonococcal urethritis, a diagnosis of NGU was precluded for 6 weeks after a diagnosis of gonorrhea. Since the study ran for a period of only 57 days, the diagnosis of NGU was essentially precluded in those patients who had a diagnosis of gonorrhea.

If GCU and NGU occur independently so that an individual may acquire either of the diseases separately or both at the same time, the chance of acquiring both at the same time will be higher if his risk for each disease is high and lower if his risk of acquiring each disease is low.

Since an error favoring deficit diagnosis of NGU occurs in coexistent disease, and since coexistent disease will occur more frequently in high-risk groups than in low-risk groups, then relatively more GCU than NGU will be diagnosed in high-risk groups than in low-risk groups.

This hypothesis suggests that NGU was diagnosed relatively less frequently than gonorrhea in blacks than in whites because of diagnostic bias which occurred more frequently in the high-risk black group than in the lower-risk whites. It has also been reported that gonorrhea occurs relatively more frequently than NGU among venereal disease clinic patients than among private physician patients. If the patients served by venereal disease clinics come from a higher-risk population than do private physician patients, then this hypothesis might also be helpful in explaining why gonorrhea is diagnosed more frequently than NGU in venereal disease clinics than in private physician offices.

In conclusion, the epidemiology of NGU is enigmatic and quite variable from environment to environment. Etiological studies should make it possible to discuss the epidemiology of specific forms of NGU in more specific terms in the future.

REFERENCES

1. Melton LJ: Comparative incidence of gonorrhea and nongonococcal urethritis. Am J Epidemiol 104:535–542, 1976
2. Volk J, Kraus SJ: Nongonococcal urethritis. Arch Intern Med 134:511–514, 1974
3. Dans PE, Judson F: The establishment of a venereal disease clinic. II. An appraisal of current diagnostic methods in uncomplicated urogenital and rectal gonorrhea. J Am Vener Dis Assoc 1:107–112, 1975
4. Holmes KK, Handsfield HH, Wang SP, et al: Etiology of nongonococcal urethritis. N Engl J Med 292:1199–1205, 1975
5. Smith TF, Weed LA, Segura JW, et al: Isolation of Chlamydia from patients with urethritis. Mayo Clin Proc 50:105–110, 1975
6. Errard JR: Sexually transmitted diseases on the college campus. J Am Coll Health Assoc 22:427–430, 1974
7. McChesnay JA, Zedd A, King H, et al: Acute urethritis in male college students. J Am Med Assoc 226:37–39, 1973

Diagnosis of Nongonococcal Urethritis

STEPHEN L. SWARTZ

The diagnosis of nongonococcal urethritis (NGU) is established by correlating the patients' signs and symptoms with results of the appropriate laboratory tests. Classically, when patients with NGU seek medical attention, they have had dysuria or discharge, or both, for several days. The dysuria may occur by itself (15%) or in association with a discharge (38%) (1). The discharge, which occurs without dysuria in 47% of patients with NGU, is typically clear or white in color, thin and mucoid in consistency, and scant; it often is demonstrable only after penile stripping (i.e., it is not a spontaneous discharge). As many as 29% of patients with NGU are conscious of the discharge only in the early morning (2). Less commonly observed symptoms include frequency of urination, urgency, and genital itching. No males in our study have complained of hematuria, impotence, fever, or flank pain.

In the differential diagnosis of NGU, gonococcal urethritis must be ruled out. The clinical characteristics of gonococcal urethritis differ from those of NGU in the following ways: (i) gonococcal urethritis is more often associated with both dysuria and discharge (71%), is less frequently accompanied by only a urethral discharge (27%), and only rarely occurs with dysuria alone (2%) (1); (ii) the discharge of gonococcal urethritis is more often purulent, profuse, and spontaneous (i.e., it does not require penile stripping for its demonstration); and (iii) the symptomatology of gonococcal urethritis is more acute and severe, as reflected in the observation that 76% of these patients are seen in the clinic within 4 days after the onset of their symptoms in comparison to 46% of patients with NGU. Previous episodes of documented urethritis are common in both groups of patients. However, patients with NGU more frequently report an episode of urethritis within the past 3 months, possibly an indication of the relapsing nature of this disease.

The patient's history, symptoms, and signs may suggest the diagnosis of either gonococcal urethritis or NGU, but the clinical impression must be confirmed by laboratory evaluation of the urethral secretions. Such secretions are obtained by compressing the urethra from the bulb distally. A calcium alginate swab is then inserted 2 to 3 cm into the anterior urethra, withdrawn, rolled over a glass slide, and inoculated on Thayer-Martin selective media (3). The slides are then stained by Gram's method and examined for the presence of polymorphonuclear cells (PMN) and gram-negative diplococci (GND); the morphology of these bacteria, their location, and possible association with other gram-negative bacteria are noted. *Neisseria gonorrhoeae* is a coccus which usually occurs in pairs or tetrads with flattening of the side of the bacterium adjacent to its pair. These organisms can be located either within or outside the PMN. They may be the only bacteria on the slide or they may be accompanied by pleomorphic GND. The latter are essentially always extracellular.

The identification of typical GND within PMN correlates with the diagnosis of gonorrhea in 98% of the cases (gonorrhea being defined as urethritis from which *N. gonorrhoeae* can be cultured). Likewise, failure to find typical GND either within or outside PMN correlates with the diagnosis of NGU in 98% of the cases (NGU being defined as urethritis from which *N. gonorrhoeae* cannot be cultured). These two Gram stain patterns of urethral exudate account for 84% of urethral smears obtained from patients with urethritis seen at the DeKalb County (Georgia) Venereal Disease Clinic. Thus, the simple and inexpensive Gram stain of a smear of urethral exudate allows for a rapid and accurate differentiation between gonococcal urethritis and NGU in the majority of men with urethritis.

However, urethral smears from 15% of men with urethritis are not diagnostic of either gonorrhea or NGU. The bacteria in these smears are all extracellular and appear as a mixture of pleomorphic gram-negative bacteria, some of which are diplococci with flattening of adjacent surfaces. Only 21% of patients with such smears have gonorrhea; the remaining 79% have NGU. These patients therefore require a culture of the urethral exudate to differentiate gonococcal urethritis from NGU. It is important to note that the presence of these extracellular, gram-negative bacteria does not alter the association of typical intracellular GND with gonococcal urethritis.

An infrequent pattern observed with smears of urethral exudate (approximately 1%) consists of the finding of only typical GND (i.e., no pleomorphic gram-negative diplococci), all located outside the PMN. The low incidence of this phenomenon has not allowed its diagnostic significance to be determined, and therefore such samples must be cultured to differentiate gonorrhea from NGU.

When a patient diagnosed as having NGU does not respond to an adequate course of tetracycline therapy, a more thorough evaluation may be indicated (4). A wet mount and culture of a fresh morning urethral specimen should be obtained and examined for trichomonads. It is especially important that this sample be obtained before first micturition, because as many as 30% of patients with trichomonal urethritis have had no evidence of urethral disease when examined during the day, but an early-morning examination with smear and culture established the diagnosis (5). A second sample of exudate should be mixed with 10% potassium hydroxide and examined for spores and mycelia of *Candida*. The next step is urinalysis and urine culture to rule out a urinary tract infection. In very resistant cases, urethroscopy is performed to rule out structural abnormalities.

Evaluation of patients with dysuria and discharge involves more than a differentiation between gonococcal urethritis and NGU; it also requires the differentiation between urethritis and prostatitis. This is attempted in the laboratory by performing the two-glass test on a urine sample (6). For this test, the first 10 ml of voided urine is collected in one container and the remainder of the urine in another. After centrifugation of equal samples from the two glasses, 0.5 ml of each sediment is examined microscopically. This is done with the expectation that an anterior urethritis will result in more pyuria and urinary threads in the first glass, whereas a prostatitis will result in relatively equal amounts of pyuria and urinary threads in both glasses (7). If both glasses are clear, the patients are instructed to return the following morning after having held their urine overnight. The two-glass test is then repeated, and in addition a prostatic massage is performed and a smear of the expressed fluid is made. Unfortunately, all investigators do not agree on the criteria required to distinguish normal from abnormal results on the prostatic smear. Suggestions have been made to disregard total PMN counts and instead to examine five preparations from one specimen of freshly expressed prostatic fluid. Then a diagnosis of prostatitis is made if either PMN clumping or more than 10 PMN per microscopic field are present in any one preparation (7, 8). Much of the enthusiasm for this approach was dampened, however, when it was found that, in 50% of patients clinically diagnosed as having prostatitis, PMN counts were normal, and that changes in the number of white cells in the prostatic fluid during therapy and during remission or exacerbation of symptoms could not be demonstrated.

Another recently suggested laboratory procedure to aid in the diagnosis of prostatitis was to irrigate the anterior urethra and then perform the two-glass test; the presence of urinary threads in the second glass was considered evidence of prostatitis (9). However, when the results from this technique were correlated with evidence of prostatitis (i.e., clumping of PMN) seen on the prostatic smears, the presence of urinary threads correlated with the diagnosis of prostatitis only 28% of the time, and false-negative results with clear urine were obtained in 22% of the cases.

Possibly the most difficult aspect in diagnosis involves that group of patients who present with dysuria or urethral discharge and in whom a thorough laboratory evaluation, as previously outlined, reveals only a borderline PMN response on their urethral smears or urinary sediments. In the scientific literature, 10 PMN per high-power field on the urethral smear (10) or 20 PMN per high dry field in the urinary sediment (6) have been accepted as the criteria indicating inflammation of the urethra. The clinical status of this group of symptomatic patients with less than the required number of PMN needed to denote inflammation now remains undefined. Likewise, the advisability of prescribing therapy for these patients and their sexual partners is uncertain. We clearly do not want to over-treat patients and thereby reinforce neuroses, nor do we want to under-treat and allow spread of disease. Long term follow-up evaluation of these patients is needed to define their natural history.

In summary, the diagnosis of gonococcal urethritis or NGU is often suggested by history and physical examination, and, in the majority of cases, is confirmed by a simple Gram stain of a smear of urethral exudate. With those patients in whom the diagnosis cannot be determined by examination of the urethral smear, culture of the exudate will establish the diagnosis. The laboratory differentiation between urethritis and prostatitis is not clearly defined at present. The clinical status of symptomatic patients with borderline urethral smears is also unclear and awaits further investigation.

REFERENCES

1. Jacobs NF, Kraus, SJ: Gonococcal and nongonococcal urethritis in men: clinical and laboratory differentiation. Ann Intern Med 82:7–12, 1975
2. Boyd JT, Csonka GW, Oates JK: Epidemiology of non-specific urethritis. Br J Vener Dis 34:40–43, 1958
3. Thayer MD, Martin JE: Improved medium selective for cultivation of N. gonorrhoeae and N. meningitidis. Public Health Rep 81:559–562, 1966
4. Kaufman RE, Wiesman PJ: Current concepts: non-specific urethritis. N Engl J Med 291:1175–1177, 1974
5. Catterall RD: Diagnosis and treatment of trichomonal urethritis in men. Br Med J 3:113–115, 1960
6. Holmes KK, Johnson DW, Floyd TM: Studies of venereal disease: II. Observations on the incidence, etiology, and treatment of the post-gonococcal urethritis syndrome. J Am Med Assoc 202:131–137, 1967

7. King A: Nonspecific urethritis. Med Clin North Am 56:1193–1202, 1972
8. Csonka GW: Non-gonococcal urethritis. Br J Vener Dis 41:1–8, 1965
9. Messent JJ: Association of threads in the urine with non-gonococcal urethritis and prostatitis. Br J Vener Dis 46:1, 1970
10. Oriel JD, Reeve P, Wright JT, et al: Chlamydial infection of the male urethra. Br J Vener Dis 52:46–51, 1976

Etiology of Nongonococcal Urethritis

WILLIAM R. BOWIE, SAN-PIN WANG,
E. RUSSELL ALEXANDER, AND KING K. HOLMES

In this manuscript, we shall address three questions:

1. Does *Chlamydia trachomatis* cause nongonococcal urethritis (NGU), or is its presence in the urethra coincidental?
2. Does *C. trachomatis* also cause those cases of NGU in which chlamydiae cannot be isolated?
3. If *C. trachomatis* does not cause chlamydia-negative NGU, what does?

We shall review previously published studies and also present data from recent studies in Seattle, Wash., which bear on these questions.

WHEN *C. TRACHOMATIS* IS ISOLATED FROM THE URETHRA OF A MAN WITH NGU, IS IT THE CAUSE OF THE NGU?

Richmond et al. (1) have postulated that some agent other than *C. trachomatis* may initiate NGU and reactivate a latent chlamydia infection. If this were true, chlamydia would be present coincidentally or as a secondary pathogen.

There are at least five types of evidence that the association of *C. trachomatis* with NGU in humans is a causal one. The evidence includes micro-immunofluorescent antibody (micro-IFA) data (2–4), the results of examination of sex partners (2), differences in the response of chlamydia-positive and chlamydia-negative cases of NGU to each of three antimicrobial drugs (5–7), prospective cohort studies of men before and after acquisition of NGU (Wang, Alexander, and Holmes, unpublished data), and experimental urethral inoculations conducted in attempting to fulfill Koch's postulates (8). We shall review the first three types of data.

Micro-IFA Data

To study the serological response to *C. trachomatis* in a population with the least likelihood of having prior antibody to chlamydiae, we studied a group of men experiencing their first known episode of urethritis (6). Men with prior urethritis, either gonococcal or nongonococcal, were excluded. All had urethral discharge noted on examination, and all had pyuria. All were Caucasian, under 36 years of age, had no more than three sex partners in the past 3 months, had not been on antimicrobials for the past 3 months, and had symptoms for less than 1 month. A comparison group without symptoms, discharge, or pyuria who fulfilled the other criteria were also studied. All were initially examined by Gram stain and culture of the urethral secretions, to exclude gonorrhea, and were asked to return for chlamydial cultures the following morning before urinating. Endourethral specimens were obtained for isolation of *C. trachomatis* with calcium alginate swabs which were placed in transport medium and frozen at −70°C until tested on iododeoxyuridine-treated McCoy cells (6).

Chlamydia serology was performed by a simplified micro-IFA technique, with the use of a set of eight antigens grown in egg yolk sacs. Initial screening for antibody was performed with commercial fluorescein conjugates of anti-human immunoglobulin. Subsequently, sera with positive antibody were tested for immunoglobulin M (IgM)-IFA by use of a commercial fluorescein conjugate specific for IgM (6).

The study included 69 men with NGU and 39 men without urethritis; 23 (33%) of the men with NGU and 1 (3%) of the comparison group had positive urethral cultures for *Chlamydia*. Table 1 summarizes the micro-IFA titers in the 20 culture-positive and the 39 culture-negative men who had paired sera available. Each group is subdivided according to duration of symptoms, ≤10 or >10 days. All of 20 men who were chlamydia culture positive developed detectable antibody, whereas only 8 (21%) of 39 chlamydia culture-negative men had detectable antibody. Looking specifically at the chlamydia culture-positive group who were examined early, within 10 days after onset of symptoms, 9 of 10 seroconverted. This suggests that the acquisition of chlamydiae was recent.

IgM-IFA was detected in 16 of 20 men from whom *C. trachomatis* was isolated, and from 3 chlamydia culture-negative men. Two of these three also showed a fourfold rise in antibody titer or seroconversion. Preliminary studies on this patient population showed that the IgM-IFA was usually detectable only transiently. Previous studies by Wang and Grayston with experimental ocular infection with *C. trachomatis* in Taiwan monkeys (*Macaca cyclopis*) showed that IgM-IFA was transiently detectable with primary inoculation, but was not demonstrable with repeat challenge with the same immunotype (9). Therefore, the presence of IgM-IFA supports the concept that the exposure to chlamydiae was primary. Combining these serological results, 19 culture-positive and 3 culture-negative men had serological evidence for recent chlamydia infection.

Examination of Sex Partners

In a group of sex partners of 45 men with NGU examined in Seattle, *C. trachomatis* was recovered from the cervix of the sex partners of 15 (63%) of 24 culture-positive or seropositive men and only 1 (5%) of the 21 culture-negative, seronegative men (2). Oriel et al. found that *C. trachomatis* was recovered more often from women who were suspected source contacts than from exposed

TABLE 1. *Acute and convalescent micro-IFA to C. trachomatis in men according to the duration of symptoms of NGU prior to onset of treatment*

Parameter	*C. trachomatis* isolated		*C. trachomatis* not isolated	
	≤10 days[a] (*n* = 10)[b]	>10 days (*n* = 10)	≤10 days (*n* = 21)	>10 days (*n* = 18)
Seroconversion	9 (7)[c]	1 (0)	1 (1)	
Fourfold increase		1 (1)		1 (1)
Titer ≥ 1:8, stable	1 (0)	8 (8)	2 (1)	4 (0)
Titer ≤ 1:8, stable			18 (0)	13 (0)

[a] Duration of symptoms.
[b] Number of men.
[c] Figure in parentheses equals number with IgM-IFA present at a titer of ≥1:8.

secondary contacts (10). Thus, if one postulates that the presence of chlamydiae in men with NGU represents reactivation of latent urethral infection, one would have to postulate simultaneous reactivation of endocervical infection in the female sex partner. The simpler alternative explanation is that the female source contacts transmitted the infection to the men.

Response to Antimicrobial Therapy

The third type of evidence that chlamydiae cause NGU concerns the differential response of chlamydia-positive and chlamydia-negative NGU to antimicrobial therapy. Handsfield et al. (5) followed 64 men for 2 to 6 weeks after treatment of NGU with tetracycline HCl (500 mg four times per day for 1 week). Persistence or recurrence of signs of NGU developed within 2 to 6 weeks in 17% of chlamydia-positive men and 47% of the chlamydia-negative men ($P < 0.025$).

To further explore the etiological implications of antimicrobial efficacy in NGU, we decided to evaluate sulfonamides. Since sulfonamides are active against *C. trachomatis* (11), yet overall have a poor record for the treatment of NGU (6), we postulated that sulfonamide therapy might also differentiate NGU into a more responsive chlamydia-positive group and a less responsive chlamydia-negative group.

The group of 69 men with first-episode urethritis, described above, were treated with a 10-day course of sulfisoxazole, 2.0 g per day. They were evaluated for response according to objective criteria 7 to 12 days after treatment was started. Total disappearance of discharge and pyuria was considered a complete response. No response indicated no improvement. Partial response included individuals whose discharge decreased in amount and became clear and nonpurulent, but did not completely disappear. Of those who returned for follow-up, there were 24 men who had positive urethral chlamydia cultures and/or a significant antibody response (seroconversion, fourfold change, or IgM-IFA positive) and 37 men who were culture negative and had no antibody response. Among 22 culture-positive men recultured 3 to 8 days after initiation of sulfisoxazole therapy, cultures for *C. trachomatis* became negative in all 22. Clinical response was seen in 88% of the chlamydia-positive men and in 49% of the chlamydia-negative men ($P < 0.005$). These results are also consistent with the evidence discussed below that most chlamydia culture-negative NGU is not caused by chlamydiae.

DOES *C. TRACHOMATIS* CAUSE NGU EVEN IN THOSE CASES WHERE IT CANNOT BE RECOVERED?

First, as mentioned earlier in the study of men with first-episode urethritis, antibody to *C. trachomatis* was demonstrated in 8 (21%) of 39 NGU patients who had urethral cultures negative for *C. trachomatis*, but only 3 (8%) of 39 had seroconversion, fourfold titer rise, or IgM-IFA. These results suggest that no less than 8% and probably no more than 21% of culture-negative cases actually had *C. trachomatis* infection. Second, cervical cultures for *C. trachomatis* were negative in the female sex partners of 95% of NGU patients who had urethral cultures negative for *C. trachomatis* (2). Third, Handsfield et al. obtained urethral cultures for *C. trachomatis* from 34 NGU patients before and after 1 week of therapy with placebo (5). All 19 men who were initially chlamydia culture negative remained negative at the follow-up visit. Finally, the differential response of NGU to therapy with antimicrobials suggests that chlamydia-negative NGU cases have a separate etiology. In particular, the results of sulfonamide therapy suggest that most cases of chlamydia-negative NGU are caused by a sulfonamide-resistant organism. The results of aminocyclitol therapy are also of interest and will be discussed subsequently.

WHAT CAUSES CHLAMYDIA-NEGATIVE NGU?

Herpes simplex virus and *Trichomonas vaginalis* are uncommon causes of NGU (2). To study the relationship of NGU to bacteria other than chlamydiae, the group of 69 men with first-episode NGU and the comparison group of 39 men without urethritis underwent aerobic and anaerobic bacteriological cultures of endourethral swabs, of first-voided urine specimens, and of urine obtained immediately after prostatic massage (6; W. R. Bowie et al., in preparation).

Lactobacilli, *Streptococcus viridans*, and *Haemophilus vaginalis* were isolated significantly more often from the group without urethritis than from either the chlamydia-positive or the chlamydia-negative NGU groups. The *H. vaginalis* results are of particular interest. This organism has been mentioned as a possible cause of NGU (12, 13), and it *is* sulfonamide resistant, but these data show that *H. vaginalis* is part of the normal flora of healthy, sexually active males. Because of reports that novobiocin-resistant, coagulase-negative staphylococci cause lower urinary tract infections in women (14) and thus may be pathogenic in the genitourinary tract, 62 coagulase-negative staphylococci isolated from men with NGU were tested for novobiocin resistance. None was resistant.

Diphtheroids were not significantly more common among NGU patients than among controls. Because diphtheroids are a heterogeneous and poorly characterized group, it remains possible that some subgroup could be associated with NGU. Parenthetically, available evidence does not suggest that the so-called "NSU corynebacteria" described by Furness et al. are a likely cause of NGU, since they have been recovered more often from the urethras of normal men than from men with NGU (15).

One or more anaerobic species were isolated from 91% of the comparison group without urethritis. Significantly fewer cultures from men with NGU yielded anaerobes. The distribution of organisms isolated was similar in all three groups, except that *Bacteroides* species were isolated significantly more often from the "no urethritis" group ($P < 0.01$) (Bowie et al., in preparation).

Although Hafiz et al. (16) reported a striking correlation of NGU with urethral infection by an anaerobe named *Clostridium difficile*, no gram-positive rods with the morphology of *Clostridium* sp. were seen on Gram stain or were isolated on anaerobic culture in this study.

Evidence for an Etiological Role of *Ureaplasma urealyticum* in NGU

The results of the mycoplasma cultures are shown in Table 2. *Mycoplasma hominis* was isolated from 19 to 22% of all groups and cannot be considered a cause of NGU. Previous studies of highly sexually active venereal disease clinic populations, including populations studied previously in Seattle, have not shown any difference in the rate of isolation of *Ureaplasma* from patients with

TABLE 2. *Percent isolation of mycoplasma from the first-voided urine of men with NGU and from a comparison group without urethritis*

Organism	Chlamydia-positive NGU[a] ($n = 26$)[b]		Chlamydia-negative NGU[c] ($n = 43$)		No urethritis ($n = 39$)
M. hominis	19		19		22
U. urealyticum	42	$P < 0.005$	81	$P < 0.005$	59

[a] Culture positive or serological evidence of recent infection (seroconversion, fourfold rise in IFA titer, or IgM-IFA positive).
[b] Number of men.
[c] Culture negative and no serological evidence of recent infection.

chlamydia-positive or chlamydia-negative NGU (2). We obtained very different results from our study of 69 men with first-episode urethritis. *U. urealyticum* was isolated significantly more frequently from men with chlamydia-negative NGU than from those with chlamydia-positive NGU or from those without urethritis.

Shepard has previously suggested that quantitative cultures for ureaplasma might help to elucidate the role of ureaplasma in NGU (17). Our quantitative results (Table 3) support this view and amplify the difference between the groups. The number of ureaplasmas was quantitated by serially diluting first-voided urine in phenol red broth, and results were expressed as color-changing units in phenol red broth per milliliter of first-voided urine (6). At all concentrations of ureaplasma, the number of men who were ureaplasma positive was significantly greater for the chlamydia-negative NGU group than for the comparison group with no urethritis or for the chlamydia-positive group. There was no difference between the chlamydia-positive NGU group and the comparison group. Thus, in this study, there appeared to be a strong association between chlamydia-negative NGU and the presence of ureaplasma.

Although all men studied had three or fewer sex partners in the preceding 3 months, the comparison group of 39 men without urethritis turned out to have a significantly higher total number of previous sex partners than did the 69 men with NGU (6). Because acquisition of ureaplasma is strongly correlated with the total number of different sex partners (18), we controlled for sexual experience by consecutively matching chlamydia-negative NGU patients with patients from the comparison group on the basis of equal number of total sex partners and age ± 2

TABLE 3. *Quantitative recovery of U. urealyticum expressed in color-changing units (CCU) per milliliter of first-voided urine in men with nongonococcal urethritis subdivided by chlamydia culture and serology, and in a comparison group*

U. urealyticum (CCU/ml)	Nongonococcal urethritis		Comparison group, no urethritis
	Chlamydia positive[a]	Chlamydia negative[b]	
10^1	11/26[c]	35/43	23/39
10^2	7/26	28/42	15/39
$\geq 10^3$	3/26	21/42	8/39

[a,b] See footnotes *a* and *c* of Table 2.
[c] Number of men/total tested.

years. Ureaplasma was isolated from 17 of 20 chlamydia-negative men with NGU and 9 of 20 matched controls without urethritis ($P < 0.02$). It should be noted that the difference in ureaplasma isolation between men with chlamydia-negative NGU and men without urethritis could be explained if first-voided urine specimens are less sensitive in detecting urethral ureaplasma infection in men without urethritis than in men with urethritis, as Tarr et al. have suggested (19). However, this would not account for the differences observed between the chlamydia-positive and chlamydia-negative NGU groups. We attempted to correlate isolation of ureaplasma with complement-dependent ureaplasmacidal antibody as tested by Joey-Shin Lin. Elevated titers were infrequently seen. Fourfold changes in titer with added complement were seen in four men without and two men with culture-proven ureaplasma infection. Five ureaplasma-positive men demonstrated fourfold changes without complement. These serological results do not provide support for an etiological role of ureaplasma in NGU.

Differential Therapeutic Response of Chlamydia-Positive, Ureaplasma-Negative Versus Chlamydia-Negative, Ureaplasma-Positive NGU to Sulfonamide

Nonetheless, the association of ureaplasma with chlamydia-negative NGU is of particular interest because ureaplasmas are sulfonamide resistant, both in vitro (20) and in vivo. The results of the sulfisoxazole treatment study involving 69 men with first-episode urethritis (6) were therefore analyzed with respect to initial culture results for chlamydia and ureaplasma (Table 4). Ureaplasmas were still present in the urethra 7 to 12 days after onset of sulfisoxazole therapy in 22 (92%) of 24 men who were initially ureaplasma culture negative. To assess the response to sulfonamides, the clearest groups to analyze are the chlamydia-positive, ureaplasma-negative and the chlamydia-negative, ureaplasma-positive men. A response was seen in all 13 chlamydia-positive, ureaplasma-negative men and in 14 of 30 chlamydia-negative, ureaplasma-positive men ($P < 0.002$). Overall, disregarding the chlamydia studies, a complete or partial response to sulfisoxazole was seen in 22 (54%) of 41 ureaplasma-positive men and in 17 (85%) of 20 ureaplasma-negative men ($P < 0.05$). These results are consistent with ureaplasma being a cause of some cases of chlamydia-negative NGU.

Differential Response of Chlamydia-Positive, Ureaplasma-Negative NGU Versus Chlamydia-Negative, Ureaplasma-Positive NGU to Randomly Assigned Sulfonamide or Aminocyclitol

As a means of evaluating the etiological significance of ureaplasma in NGU, we next decided to study the differential therapeutic response of NGU to treatment, by using an antibiotic that would selectively eradicate ureaplasmas but not chlamydiae. The antibiotics selected were the aminocyclitols: streptomycin and, later, spectinomycin. In this study (7), men with a history of prior urethritis were not excluded. Ninety men with NGU were randomly assigned to treatment with a 10-day course of sulfisoxazole (the "anti-chlamydial regimen") or to an aminocyclitol (the "anti-ureaplasma regimen"). Those who did not respond clinically were crossed over to the other drug.

This population of men was typical of the usual venereal disease clinic population

TABLE 4. *Objective response of men with NGU divided according to the pretreatment chlamydia (C) status, by culture and serology,[a] and the pretreatment ureaplasma (U) status, by culture, 7 to 12 days after beginning sulfisoxazole treatment*

Objective response	C + U+	C + U−	C − U+	C − U−
Complete	4	4	3	1
Partial	4	9	11	3
None	3	0 $P < 0.002$	16	3

[a] C+, Culture positive or serological evidence of recent infection (seroconversion, fourfold rise in IFA titer, or IgM-IFA positive). C−, Culture negative and no serological evidence of recent infection.

when consecutive men with NGU are studied. *C. trachomatis* was isolated from 36 (40%), and ureaplasma was isolated from 64 (70%). Among Caucasian men in this second study who were experiencing their first episode of urethritis, the rate of isolation of ureaplasma was again higher for those who were chlamydia culture negative than for those who were chlamydia culture positive, but the difference was not significant at the 0.05 level, perhaps because the number of such men was small (7).

In this study we also compared the isolation rate of ureaplasma when specimens were obtained by endourethral swab and first-voided urine. Of 89 men from whom satisfactory ureaplasma cultures were obtained at both the initial visit (swab) and the day of treatment (urine), *U. urealyticum* was recovered from both swab and urine in 55 cases whereas both cultures were negative in 29 cases. The swab alone was positive in two, and the urine alone was positive in three.

The initial aminocyclitol used was streptomycin, which was given as a single 2-g intramuscular dose. This caused dizziness and was discontinued. We next tried spectinomycin, since Lee et al. had reported eradication of ureaplasma from the urethras of men treated for gonorrhea with spectinomycin (21). Although it has been reported that spectinomycin has activity against *C. trachomatis* in vitro (E.R. Alexander et al., Abstr. Intersci. Conf. Antimicrob. Agents Chemother., 14th, San Francisco, 1974), the activity was marginal. We employed intramuscular doses of 2 g given either as a single dose or repeated once or twice at 12-h intervals.

Ureaplasma cultures became negative in 3 of 28 men receiving sulfisoxazole and 20 of 32 receiving aminocyclitols ($P < 0.002$). Overall, one or two doses of aminocyclitol produced a negative culture in 7 of 14 initially ureaplasma-positive men, and three doses of spectinomycin produced negative cultures in 13 of 19 (not significant).

In all 13 chlamydia culture-positive men, sulfisoxazole rendered subsequent cultures negative. In four of the 13 chlamydia culture-positive men given aminocyclitols, follow-up cultures at 3 to 7 days were negative. In two of these four, treatment with sulfisoxazole was begun because of failure to respond. Treatment was withheld in the other two, and a subsequent culture at 12 days was positive. Another four had negative cultures 1 or 2 days after starting spectinomycin, but all four again became culture positive by 6 to 12 days. Thus, aminocyclitols may transiently render chlamydia cultures negative, but, in the doses we used, probably do not eradicate the organism.

A clinical response was defined as complete disappearance of discharge, or

TABLE 5. *Objective response of men with chlamydia-negative, ureaplasma-positive NGU*[a]

U. urealyticum	No. responding/no. treated	
	Random allocation	Random and crossover
Eradicated	9/11	17/19
Persisted	0/8	1/11
	$P < 0.002$	$P < 0.002$

[a] Response assessed 7 to 14 days after treatment with aminocyclitol, according to whether or not ureaplasma was eradicated.

change from a purulent to a clear discharge, with < 20 white blood cells per high-power field in each of five ×400 microscopic fields in the first-voided urine sediment 7 to 14 days after initiation of therapy. Response to aminocyclitols apparently depended upon the effectiveness of aminocyclitol treatment in eradicating the ureaplasmas. Table 5 shows the response to aminocyclitol of chlamydia culture-negative patients in relation to whether ureaplasma was or was not eradicated. There was a significantly better response with eradication of ureaplasma, whether only the first course of therapy assigned according to the random allocation was included or all courses of therapy, including both the random and crossover treatment episodes, were included.

Table 6 compares the response in the chlamydia-positive, ureaplasma-negative men and the chlamydia-negative, ureaplasma-positive men, according to which treatment they were randomly assigned. Patients who were given aminocyclitol, but had persistent ureaplasma infection, were excluded from this analysis since the purpose of the study was to determine whether treatment of ureaplasma-positive men with a drug which eradicated ureaplasma was more effective than treatment with a drug that did not. Again in this study of consecutive venereal disease clinic patients, as in the previous study of 69 men with first-episode urethritis, sulfisoxazole was effective for chlamydia-positive NGU but not for ureaplasma-positive NGU. The reverse was true for aminocyclitols.

DISCUSSION

Sulfonamides and aminocyclitols were used in these studies to gain insight into the etiology of NGU. Neither drug alone is indicated as primary treatment of NGU. Sulfonamides produced a clinical response in 37% of patients in this study, but the

TABLE 6. *Objective response at 7 to 14 days after beginning random treatment with sulfisoxazole or aminocyclitols of men with NGU divided according to the results of pretreatment cultures for chlamydia (C) and ureaplasma (U)*

Therapy	No. responding/no. treated	
	C + U−	C − U+
Sulfonamide	7/7	5/19 $P < 0.01$
Aminocyclitol	0/6	9/11[a] $P < 0.01$
	$P < 0.002$	$P < 0.01$

[a] Analysis limited to cases in which ureaplasma was eradicated.

reported range of response varied from 22 to 90% (22–27). Streptomycin (22–24, 28, 29), kanamycin (30), and spectinomycin (31) have been used in various doses with reported cure rates of 41 to 67%. Aminocyclitols have often been used by venereologists in combination with sulfonamide for the treatment of NGU, and the present study gives some explanation of why the combination may be more effective than either drug alone. The combination has been reported to be effective in 44 to 85% of men with NGU (22, 24, 29, 31–33) and is probably not as effective as tetracycline for treatment of NGU (25).

The results of sulfonamide therapy repeated in two separate studies and the results of aminocyclitol therapy support the etiological importance of both *C. trachomatis* and *U. urealyticum* in NGU. It is possible that aminocyclitols eradicate some other organism besides ureaplasma, but the response to aminocyclitols was seen only if ureaplasma was eradicated.

Serological studies offer strong additional evidence that primary chlamydial infection causes first-episode NGU (6). However, attempts to show an immune response to ureaplasma in NGU have usually been unsuccessful (34–36).

In summary, the evidence that *C. trachomatis* causes NGU has become rather compelling. The evidence concerning the etiological role of ureaplasma is circumstantial. Although ureaplasmas are clearly still in the running as causal agents in NGU, some doubts remain. The major observation to reconcile is the evidence that most individuals from whom ureaplasma is isolated harbor the organism without any disease. Additional evidence for pathogenicity of ureaplasma could be sought by: (i) further typing of ureaplasma to look for particular strains associated with NGU, (ii) experimental inoculation of ureaplasma into animals to reproduce urethritis, (iii) observation of a cohort of men to correlate acquisition of ureaplasma with the possible development of urethritis, and (iv) the development or application of more sensitive and specific serological tests.

ACKNOWLEDGMENTS

This research was supported by a Fellowship from the Medical Research Council of Canada (W.R.B.), by Bureau of Medical Services grant SEA 75-6-72, by Public Health Service program project AI-12192 from the National Institute of Allergy and Infectious Diseases, and by Public Health Service research grant EY00219 from the National Eye Institute.

REFERENCES

1. Richmond SJ, Hilton AL, Clarke SKR: Chlamydial infection, role of subgroup A in nongonococcal and postgonococcal urethritis. Br J Vener Dis 48:437–44, 1972
2. Holmes KK, Handsfield, HH, Wang SP, et al: Etiology of nongonococcal urethritis. N Engl J Med 292:1199–1205, 1975
3. Reeve P, Gerloff RK, Casper E, et al: Serological studies on the role of chlamydia in the etiology of nonspecific urethritis. Br J Vener Dis 50:136–139, 1974
4. Philip RN, Casper EA, Gordon FB, et al: Fluorescent antibody responses to chlamydia infection in patients with lymphogranuloma venereum and urethritis. J Immunol 112:2126–2134, 1974
5. Handsfield HH, Alexander ER, Wang S-P, et al: Differences in the therapeutic response of chlamydia-positive and chlamydia-negative forms of nongonococcal urethritis. J Am Vener Dis Assoc 2:5–9, 1976

6. Bowie WR, Alexander ER, Wang S-P, et al: Nongonococcal urethritis: roles of chlamydia and ureaplasma. J Clin Invest, in press
7. Bowie WR, Alexander ER, Floyd JF, et al: Differential response of chlamydial and ureaplasma associated urethritis to sulphafurazole (sulfisoxazole) and aminocyclitols. Lancet 2:1276–1278, 1976
8. DiGiacomo RF, Gale JL, Wang S-P, et al: Chlamydial infection of the male baboon urethra. Br J Vener Dis 51:310–313, 1975
9. Wang S-P, Grayston JT: Local and systemic antibody response to trachoma eye infection in monkeys, p 217–232. *In* RL Nichols (ed), Trachoma and related disorders caused by chlamydial agents. Excerpta Medica, Amsterdam, 1971
10. Oriel JD, Powis PA, Reeve P, et al: Chlamydial infections of the cervix. Br J Vener Dis 50:11–16, 1974
11. Barker LF: Determination of antibiotic susceptibility of rickettsiae and chlamydiae in BS-C-1 cell cultures. Antimicrob Agents Chemother 1968, p 425–428, 1969
12. Leopold S: Heretofore undescribed organism isolated from the genitourinary system. US Armed Forces Med J 4:263–266, 1953
13. Dunkelberg WE, Woolvin SC: *Hemophilus vaginalis* relative to gonorrhea and male urethritis. Mil Med 128:1098–1101, 1963
14. Sellin M, Cooke DI, Gillespie WA, et al: Micrococcal urinary tract infections in young women. Lancet 2:570–572, 1975
15. Furness G, Kamat MH, Kaminski Z, et al: An investigation of the relationship of nonspecific urethritis corynebacteria to the other microorganisms found in the urogenital tract by means of a modified chocolate agar medium. Invest Urol 10: 387–391, 1973
16. Hafiz S, McEntergart MG, Morton RS, et al: *Clostridium difficile* in the urogenital tract of males and females. Lancet 1:420–421, 1975
17. Shepard MC: Quantitative relationship of *Ureaplasma urealyticum* to the clinical course of nongonococcal urethritis in the human male. INSERM 33:375–380, 1974
18. McCormack WM, Lee Y-H, Zinner SH: Sexual experience and urethral colonization with genital mycoplasmas. A study in normal men. Ann Intern Med 78:696–698, 1973
19. Tarr P, Lee Y-H, Alpert S, et al: Comparison of methods for the isolation of genital mycoplasmas from men. J Infect Dis 133:419–423, 1976
20. Shepard MC, Lunceford CD, Ford DK, et al: *Ureaplasma urealyticum* gen. nov., sp. nov: proposed nomenclature for the human T (T-strain) mycoplasma. Int J Syst Bacteriol 24:160–171, 1974
21. Lee Y-H, Alpert S, Bailey A, et al: *In vitro* and *in vivo* activity of spectinomycin against the genital mycoplasmas. J Am Vener Dis Assoc 1:37–39, 1974
22. Willcox RR: "Triple tetracycline" in the treatment of non-gonococcal urethritis in males. Br J Vener Dis 48:137–140, 1972
23. Doyle JO, Gill AJ, Laird SM: Treatment of non-gonococcal urethritis. Br J Vener Dis 33:100–104, 1957
24. Jelinek G: Treatment of nonspecific urethritis from the point of view of cost and efficiency. Br J Vener Dis 33:156, 1957
25. Morton RS, Read L: Non-gonococcal urethritis. An investigation to determine factors in the host influencing response to treatment and recurrence of symptoms. Br J Vener Dis 33:223–227, 1957
26. Fowler W: Evaluation of therapy in non-gonococcal urethritis. Br J Vener Dis 34:107–112, 1958
27. Babione RW, Graham RS: Nongonoccal urethritis in the navy. Am J Syph 36:480–482, 1952
28. Gartman E, Leibovitz A: Study of nongonococcal urethritis, presumably venereal in origin based upon 588 infections on 528 patients. Br J Vener Dis 31:92–97, 1955
29. Prebble EE: Treatment of non-gonococcal urethritis. Br J Vener Dis 33:43–46, 1957

30. Csonka GW: Kanamycin in non-gonococcal urethritis. Postgrad Med J, p 63–64, 1967
31. Willcox RR: Trobicin (actinospectacin) in nongonococcal urethritis. Br J Vener Dis 39:21–23, 1963
32. Fowler W: Studies in non-gonococcal urethritis therapy. The long term value of tetracycline. Br J Vener Dis 46:464–468, 1970
33. Lyall T: Comprehensive treatment for gonococcal and non-gonococcal urethritis. Br J Vener Dis 29:151–161, 1953
34. Purcell RH, Taylor-Robinson D, Wong D, Chanock RM: Color test for the measurement of antibody to T-strain mycoplasmas. J Bacteriol 92:6–12, 1966
35. Hill DA, Philip RN, Greaves AB, et al: Mycoplasmas recovered from patients with urethritis. Br J Vener Dis 49:524–530, 1973
36. Jansson E, Lassus A, Stubbs S, et al: Studies on T-mycoplasmas in nongonococcal urethritis. Br J Vener Dis 47:122–125, 1971

Possible Role of Ureaplasmas in Nongonococcal Urethritis

DAVID TAYLOR-ROBINSON

T-strain mycoplasmas were first detected by Shepard (1) in the urethral discharge of patients suffering from nongonococcal urethritis (NGU). These organisms were later often termed T-mycoplasmas (2). Because of their unique ability to metabolize urea, they are now referred to as ureaplasmas, and those of human origin have recently been ascribed to a new genus and species, *Ureaplasma urealyticum* (3). Since the original isolation of ureaplasmas, there have been numerous attempts to associate them etiologically with NGU and other diseases. The present aim is to discuss the significance of the results of studies which have been used towards this end in the past and most recently, and to consider the approaches which might be most profitable in leading to an understanding of the part, if any, played by ureaplasmas in NGU.

The lines of investigation which, in general, prove helpful in defining the relationship of an infectious agent to disease comprise the following: (i) a comparison of microorganisms isolated from persons with and without disease; (ii) an assessment of whether there is a relationship between clinical improvement following antibiotic treatment and the disappearance of microorganisms from the urogenital tract; (iii) immunological studies, including the use of vaccines; (iv) organ culture experiments; (v) use of animal models; and (vi) studies in human volunteers. Each of these aspects will be considered in relation to the role of ureaplasmas in NGU.

ISOLATION OF UREAPLASMAS FROM PERSONS WITH AND WITHOUT DISEASE

Since ureaplasmas were first isolated, there have been at least 20 separate studies of this kind (4, 5). In just over half the investigations, ureaplasmas have been isolated significantly more frequently from patients suffering from NGU than from control subjects apparently free from disease, whereas in the other studies the rate of isolation in the two groups of persons has been about the same. The factor which has probably contributed most to the difference between the results of one study and another is the selection of inappropriate controls. Subjects in the control group should be free from NGU and have sexual habits similar to those of patients suffering from the disease. This seems to be important because it has been shown (6) that the mycoplasma isolation rate increases with increased sexual promiscuity. It is difficult to obtain groups of persons who have been equally promiscuous, and the overall equivocal results of the various studies certainly do not provide any incentive to continue this comparative approach. However, if further investigations are to be carried out along comparative lines, they should certainly be of a quantitative nature. One of the failures of previous studies is that they have been qualitative rather than quantitative, the numbers of organisms isolated from individual patients not having been documented. If ureaplasmas are involved in the pathogenic process, it would be reasonable to expect them to be present in larger numbers than if they had a commensal role only. In saying this, however, it is necessary to be aware of the possibility that a larger number of organisms might be isolated from a patient who has a discharge than from one who has not, as a mere consequence of culturing a larger volume of material.

EFFECT OF ANTIBIOTIC TREATMENT ON THE COURSE OF DISEASE AND ON THE OCCURRENCE OF MICROORGANISMS

Suboptimal Therapy

Shepard (7) used suboptimal doses of doxycycline to treat patients with NGU. There was a temporary disappearance of both symptoms and ureaplasmas, and reappearance of the organisms in the urine in numbers similar to those found before treatment was associated with a return of symptoms. A similar phenomenon was also noted (8) in a man who was given a short course of tetracycline therapy for NGU which he had developed while en route to an Antarctic base. Because of the seclusion, the recurrence of symptoms and ureaplasmas about 3 weeks after therapy was not due to sexual reexposure. Although it is tempting to consider the concomitant return of symptoms and ureaplasmas as being indicative of a causative role for these organisms in NGU, this cannot be inferred because other tetracycline-sensitive organisms, such as chlamydiae, were not studied.

Placebo-Controlled Trial of Tetracycline

Although there have been many trials of tetracycline in the treatment of NGU (9), few have been placebo controlled and, until recently, none has been carried out in association with a comprehensive microbiological investigation. However, in a recent placebo-controlled trial of minocycline in NGU (10, 11), we obtained clear evidence that chlamydiae were etiologically involved in some cases. In addition, we noted that there was a significant association between minocycline therapy and the resolution of symptoms and signs in patients from whom only ureaplasmas were isolated, providing some evidence for the pathogenicity of these organisms in the male urethra.

Use of Differential Antibiotic Therapy

Another valuable approach to understanding the etiology of NGU is to treat patients with antibiotics which affect one but not the other potentially pathogenic microorganism. Lincomycin, which inhibits the growth of *Mycoplasma hominis* but not that of ureaplasmas or chlamydiae, was used by Csonka and Spitzer (12). Since neither of the latter microorganisms is affected and *M. hominis* was isolated from only 12% of the patients, it is perhaps not surprising that the cure rate of 25.5% was no better than is to be expected from giving a placebo. We are currently comparing the effect of rifampin with that of minocycline on the course of NGU. In vitro, rifampin does not inhibit the growth of ureaplasmas (3; Prentice and Taylor-Robinson, unpublished data), but it is cidal for chlamydiae. If ureaplasmas are in any way involved in causing the disease, one might expect treatment with rifampin to be less effective than tetracycline, which inhibits both microorganisms. Clearly, much might also be learned from using an antibiotic which has the reverse effect, that is, inhibition of the growth of ureaplasmas but not chlamydiae.

Tetracycline-Resistant Ureaplasmas

The occurrence of tetracycline-resistant ureaplasmas probably provides a differential antibiotic situation. Ford and Smith (13) were the first to isolate such an organism from a patient whose urethritis did not respond to this antibiotic. However, there was a clinical response to erythromycin, to which the ureaplasma was sensitive, which suggested that it was the cause of the disease. Unfortunately, chlamydiae were not sought. The opportunity to repeat this observation in conjunction with more extensive microbiological investigations is likely to arise since tetracycline-resistant ureaplasmas have been isolated from cases of NGU by others (10).

IMMUNOLOGICAL STUDIES

Humoral Antibody Responses

Several groups of workers have attempted to detect rises in the titer of antibody in sera from subjects with NGU. Nearly all attempts by means of the metabolism-inhibition technique have been unsuccessful (14, 15; Sang and Taylor-Robinson, unpublished data). However, Ford (16) and Black and Krogsgaard-Jensen (17) detected a few responses by this method, and the latter authors also found responses by using an immunofluorescence technique. In addition, Jansson et al. (18) claimed to have detected some antibody responses by using indirect hemagglutination. Clearly, however, the detection of ureaplasma antibody responses in less than 10% of cases does not provide strong support for believing that ureaplasmas are responsible for disease. A point that should not be overlooked is that the inhibition of ureaplasma metabolism by sera in the metabolism-inhibition test sometimes may not be due to antibody, because the presence of antibiotic in sera following therapy causes the same effect. Of course, it must also be realized that even the fact that an antibody response is real does not in itself indicate that the microorganism which produced it is pathogenic.

Delayed Hypersensitivity

The development of delayed hypersensitivity, as indicated by skin tests and other tests carried out in vitro, has been demonstrated in several mycoplasmal respiratory diseases (19). Each of these respiratory diseases is known to be due to a mycoplasma for reasons other than a positive delayed hypersensitivity test. However, where there is doubt about the association between organism and disease, as in the case of ureaplasmas and NGU, it does not seem that the tests are likely to be of value in defining the etiology. A positive result would indicate that the subject had been infected with ureaplasmas previously and that sensitized T-cells were responding to the antigen, but not necessarily that the organisms were causing the disease.

Vaccine Administration

The production of antibody by vaccination and the demonstration that it prevents disease or reduces its incidence is evidence that the organisms constituting the vaccine are able to cause the disease. Although this has been demonstrated in infections due to *M. pneumoniae* and other mycoplasmas, it seems that attenuated live vaccines which stimulate local antibody may be more successful in preventing disease than parenteral inoculation of killed organisms. However, intraurethral inoculation of ureaplasmas without a well-defined means of testing their pathogenicity and attenuation, in addition to having to select from among numerous serotypes (3), would pose severe problems. This being so, vaccine administration would not seem to be a rewarding exercise in helping to define the relationship between ureaplasmas and NGU.

Serotyping of Ureaplasma Isolates

The available serological tests (17, 20) make it possible to serotype isolates and so determine whether a particular serotype or mixture of serotypes is associated with NGU. Those who have done this (20, 21) have not found any serotype to be predominant, although the total number of isolates that has been examined is small.

ORGAN CULTURE EXPERIMENTS

In an organ culture system, tissue can be maintained more or less in the same physiological condition as it occurs in vivo. Organ cultures of genital mucosa provide a valuable tool for looking at the relationship that exists between a microorganism and the epithelial cell surface. Human penile urethral mucosa is difficult to obtain, but this is not so in the case of prostatic urethral mucosa. The latter has been maintained for over 1 week in an apparently satisfactory condition as assessed by cytology (Taylor-Robinson, unpublished data), although the effect of ureaplasma infection has not been determined. In this system, assessment of cell damage is more difficult than in Fallopian tube organ cultures, where ciliary activity may be used as an index of cell viability. Ureaplasmas have been inoculated into Fallopian tube cultures without causing obvious damage (22), in contrast to the effect of gonococci (23). It is probably worthwhile to look again at the effect of ureaplasmas thought to be pathogenic on epidemiological grounds, by use of the quantitative technique of assessing ciliary activity described by McGee et al. (24). Failure to demonstrate loss of activity cannot necessarily be taken as meaning avirulence of the ureaplasmas since the organ culture system is divorced from the immunological mechanisms of the host which may play an integral part in pathogenesis.

ANIMAL MODELS

The notion that some ureaplasmas may be pathogenic in humans is in keeping with observations made in cattle, where it has been shown that some bovine ureaplasma strains, when inoculated into the bovine udder, produce mastitis whereas other strains do not (25). Ureaplasmas have been isolated from several animal species (26–28), and so there is an opportunity to determine whether they are pathogenic in the urethra of these animals. In this connection it should be noted that inoculation of a bovine genital ureaplasma into the urethra of a Caesarian-derived pathogen-free bull caused infection which failed to produce disease (29). On the other hand, preliminary results of collaborative investigations (Gourlay, Howard, and Taylor-Robinson, unpublished data), in which goats have been inoculated intraurethrally with caprine ureaplasmas, suggest that these organisms are capable of producing a urethritis. However, these initial observations need to be confirmed and extended before any firm conclusions are drawn. Although animal models such as these would seem to have a place in studying the role of ureaplasmas in producing the disease in humans, there is always the obvious question of whether the models mentioned can be related to the human situation. It may not be easy to overcome this objection, and it may be necessary to use subhuman primates such as chimpanzees. At the moment, however, it is encouraging that ureaplasmas seem to have pathogenic potential in animals, and this may provide an opportunity of looking not only for variations in pathogenicity between different strains but also at the various associated immunological aspects.

STUDIES IN HUMAN VOLUNTEERS

As indicated previously, animal models may have considerable value, but it seemed that the most certain way of determining whether ureaplasmas cause human urethritis would be a deliberate attempt to infect humans.

Volunteers have been used in many studies concerned with the etiology of infectious respiratory disease, but clearly there would be greater problems with any study in which infection of the genitourinary tract was proposed. It does not seem reasonable or ethical to ask volunteers to be inoculated. However, recently two of us, D.T.-R. and G.W.C., inoculated ourselves intraurethrally (Taylor-Robinson, Csonka, and Prentice, unpublished data). Each subject received a different ureaplasma strain, although both strains were identified subsequently as serotype 5. These strains had been isolated from patients from whom no other organisms could be recovered. Because it was necessary to make certain that the inocula contained ureaplasmas only, the strains to be inoculated were cloned three times during the course of nine passages on artificial media. Both subjects, free from ureaplasmas and other potential pathogenic microorganisms, received about 5×10^4 ureaplasma organisms. The first subject developed a urethritis, characterized by dysuria, frequency, and urethral discomfort; early-morning specimens of urine contained threads and had a hazy appearance, and there were polymorphonuclear leukocytes in the centrifuged deposit. Ureaplasmas were isolated consistently from the urine samples, but these and the associated symptoms and signs disappeared on institution of minocycline therapy and did not recur subsequently. The second subject also developed evidence of a mild urethritis; in this case, however, the predominant feature was the appearance of urinary threads in which polymorphonuclear leukocytes were regularly observed. The threads persisted for at least 6 months after minocycline therapy, which eliminated the organisms from meatal, urine, and semen samples. Analysis of fractionated semen samples indicated that the ureaplasmas had infected the prostate and, therefore, it is possible that the threads were of prostatic origin. However, their persistence following treatment which was apparently effective in eliminating viable organisms is mysterious. There was no evidence for the production of autoantibodies against prostatic tissue, but the results of immunofluorescence tests, performed with antiserum against the ureaplasmas that had originally been inoculated, suggested, but did not unequivocally prove, that the continued production of threads might be due to the persistence of ureaplasma antigen in a nonviable form.

These results clearly indicate that some ureaplasmas are likely to be pathogenic under natural conditions and also suggest that they may be capable of initiating chronic disease.

CONCLUSIONS

The results of (i) ureaplasma isolation studies on subjects with and without NGU, (ii) attempts to detect specific antibody responses, (iii) serotyping the ureaplasmas isolated, and (iv) organ culture experiments have not suggested a significant causal relationship between ureaplasma infection and NGU. On the other hand, the results of (i) certain microbiological investigations during antibiotic treatment of NGU, (ii) experiments in animals, and (iii) intraurethral inoculation of human subjects indicate that ureaplasmas of human origin are able to cause urethritis. If this suggestion is correct, how can it be reconciled with the results of the studies mentioned above which have failed to show a positive relationship between the organisms and disease? It is possible that an explanation for these results can be found in the persistence of ureaplasmas in the urogenital tract. This is known to occur because the organisms have been isolated repeatedly over a period of about 1 year from the urogenital tract of men not experiencing sexual contact during this time (8, 30). It is likely that only some ureaplasmas are able to produce disease and that some, pathogenic or otherwise, may remain in the urogenital tract longer than others. Because of the failure to distinguish between two or more different strains in the same subject, those that disappear may still seem to be present because of the persistence of the other strain(s). Since promiscuous persons without overt disease are likely to be carrying ureaplasmas as a result of previous infection, it is understandable that it may be impossible to distinguish between the prevalence of mycoplasma infection in those with and without disease. Indeed, the matching of groups for sexual promiscuity is likely to obscure the existence of pathogenic ureaplasmas in the group of persons suffering disease. On this basis can be explained the failure of Hare et al. (31) to demonstrate a difference between the prevalence of ureaplasmas in men with postgonococcal urethritis and in men without the disease, as well as the failure of those who have studied NGU and matched control groups.

The persistence of ureaplasmas, culminating in failure to select the strain that has infected recently, may be one of the factors concerned in the infrequent detection of antibody responses. In addition, a poor immunological response of the host to infection of the genital mucosa and relative insensitivity of presently available techniques for antibody measurement may be contributory factors.

There are several profitable lines of approach for studies aimed at firmly establishing ureaplasmas as a significant cause of NGU. One is the treatment of NGU with antibiotics which differentially affect one agent but not another. In this regard, it is essential that chlamydiae be sought. Serotyping of ureaplasmas is another approach. It is possible that pathogenic and nonpathogenic strains may fall into a single serotype and that cases of NGU are not caused by a few serotypes which predominate. However, until isolates are typed by the epiimmunofluorescence technique on a greater scale than hitherto, this point cannot be established. A third approach is the exploration of ureaplasma infection in animal models such as subhuman primates, goats, and possibly others (32), which might help to determine whether pathogenic and nonpathogenic strains exist, particularly within a single serotype, and if so whether diminished pathogenicity is a reflection of increased host immunity. Furthermore, information is required on whether in vitro passage of ureaplasmas leads to loss of virulence. Finally, there is a need to investigate the influence that chlamydiae and ureaplasmas might have upon each other in producing NGU rather than, as has been the tendency in the past, to consider the role of each separately.

REFERENCES

1. Shepard MC: The recovery of pleuropneumonia-like organisms from Negro men with and without nongonococcal urethritis. Am J Syph Gonorrea Vener Dis 38:113–124, 1954
2. Taylor-Robinson D, Addey JP, Hare MJ, et al: Mycoplasmas and 'nonspecific' genital infection. I. Previous studies and laboratory aspects. Br J Vener Dis 45:265–273, 1969
3. Shepard MC, Lunceford CD, Ford DK, et al: *Ureaplasma urealyticum* gen. nov., sp. nov.: proposed nomenclature for the human T (T-strain) mycoplasmas. Int J Syst Bacteriol 24:160–171, 1974
4. McCormack WM, Braun P, Lee Y-H, et al: The genital mycoplasmas. N Engl J Med 288:78–89, 1973
5. Taylor-Robinson D: Mycoplasmas in the male urogenital tract, p 258–262. *In* DI Williams and GD Chisholm (ed), Scientific foundations of urology. Heinemann, London, 1976
6. McCormack WM, Almeida PC, Bailey PE, et al: Sexual activity and vaginal colonization with genital mycoplasmas. J Am Med Assoc 221:1375–1377, 1972
7. Shepard MC: Quantitative relationship of *Ureaplasma urealyticum* to the clinical course of nongonococcal urethritis in the human male. Les Colloques INSERM 33:375–379, 1974
8. MacLeod AD, Furr PM, Taylor-Robinson D: Prolonged eradication of urogenital mycoplasmas after administration of tetracycline to men in the Antarctic. Br J Vener Dis. 52:337–340, 1976
9. Grimble, AS, Amarasuriya KL: Non-specific urethritis and the tetracyclines. Br J Vener Dis 51:198–205, 1975
10. Prentice MJ, Taylor-Robinson D, Csonka GW: Non-specific urethritis: a placebo-controlled trial of minocycline in conjunction with laboratory investigations. Br J Vener Dis 52:269–275, 1976
11. Prentice MJ, Taylor-Robinson D, Csonka GW: A double-blind placebo-controlled trial of minocycline in the treatment of non-specific urethritis, p 97–102. *In* JD Williams and AM Geddes (ed), Chemotherapy, vol 3. Plenum Press, New York, 1976
12. Csonka GW, Spitzer RJ: Lincomycin, non-gonococcal urethritis, and mycoplasmata. Br J Vener Dis 45:52–54, 1969
13. Ford DK, Smith JR: Non-specific urethritis associated with a tetracycline-resistant T-mycoplasma. Br J Vener Dis 50:373–374, 1974
14. Purcell RH, Taylor-Robinson D, Wong D, et al: Color test for the measurement of antibody to T-strain mycoplasmas. J Bacteriol 92:6–12, 1966
15. Hill DA, Philip RN, Greaves AB, et al: Mycoplasmas recovered from patients with urethritis. Br J Vener Dis 49:524–530, 1973
16. Ford DK: Relationship between mycoplasma and the etiology of nongonococcal urethritis and Reiter's syndrome. Ann NY Acad Sci 143:501–504, 1967
17. Black FT, Krogsgaard-Jensen A: Application of indirect immunofluorescence, indirect haemagglutination and polyacrylamide-gel electrophoresis to human T-mycoplasmas. Acta Pathol Microbiol Scand 82B:345–353, 1974
18. Jansson E, Lassus A, Stubb S, et al: Studies on T-strain mycoplasmas in nongonococcal urethritis. Br J Vener Dis 47:122–125, 1971
19. Taylor G, Taylor-Robinson D: The part played by cell-mediated immunity in mycoplasma respiratory infections. International symposium on immunity to infections of the respiratory system in man and animals. Dev Biol Stand 28:195–210, 1975
20. Lin JS, Kendrick HI, Kass EH: Serologic typing of human genital T-mycoplasmas by a complement-dependent mycoplasmacidal test. J Infect Dis 126:658–663, 1972

21. Black FT: Modifications of the growth inhibition test and its application to human T-mycoplasmas. Appl Microbiol 25:528–533, 1973
22. Taylor-Robinson D, Carney FE: Growth and effect of mycoplasmas in Fallopian tube organ cultures. Br J Vener Dis 50:212–216, 1974
23. Carney FE, Taylor-Robinson D: Growth and effect of *Neisseria gonorrhoeae* in organ cultures. Br J Vener Dis 49:435–440, 1973
24. McGee ZA, Johnson AP, Taylor-Robinson D: Human Fallopian tubes in organ culture: preparation, maintenance, and quantitation of damage by pathogenic microorganisms. Infect Immun 13:608–618, 1976
25. Howard CJ, Gourlay RN, Brownlie J: The virulence of T-mycoplasmas, isolated from various animal species, assayed by intramammary inoculation in cattle. J Hyg 71:163–170, 1973
26. Taylor-Robinson D, Furr PM: The distribution of T-mycoplasmas within and among various animal species. Ann NY Acad Sci 225:108–117, 1973
27. Kundsin RB, Rowell T, Shepard MC, et al: T-strain mycoplasmas and reproductive failure in monkeys. Lab Anim Sci 25:221–224, 1975
28. Brown WJ, Jacobs NF, Arum ES, et al: T-strain mycoplasma in the chimpanzee. Lab Anim Sci 26:81–83, 1976
29. Taylor-Robinson D: Mycoplasmas and the evidence for their pathogenicity in man. Proc R Soc Med 64:31–33, 1971
30. Holmes MJ, Furr PM, Taylor-Robinson D: The persistence of mycoplasmas in the urogenital tract of men in the Antarctic. J Hyg 72:355–363, 1974
31. Hare MJ, Dunlop EMC, Taylor-Robinson D: Mycoplasmas and "non-specific" genital infection. III. Post-gonococcal urethritis. A prospective study. Br J Vener Dis 45:282–286, 1969
32. Howard CJ, Anderson JC, Gourlay RN, et al: Production of mastitis in mice with human and bovine ureaplasmas (T-mycoplasmas). J Med Microbiol 8:523–529, 1975

Treatment of Nongonococcal Urethritis

J. DAVID ORIEL

Nongonococcal urethritis (NGU) has a bad reputation among both physicians and their patients for poor response to treatment and unpredictable relapse. During the past 25 years, much effort has been expended, particularly in Britain, on trials of many antimicrobials for the treatment of NGU. Since the causes were unknown, there was no alternative to empirical trials. In these, it is essential that the criteria for diagnosis and cure be clearly defined and that the studies be controlled. Unfortunately, in the majority neither of these conditions was fulfilled, and this literature must be read with caution. Nevertheless, this early work did give broad indications as to which drugs were likely to be successful in treatment and which were not, and, no less important, sufficient placebo trials were performed to indicate the course of NGU if untreated. Since it is unlikely that in the present climate of opinion many more placebo trials will be possible, these results are important (1). In 50 to 70% of men, the disease slowly regresses in 1 to 3 months. During this time, of course, the patient will have symptoms, there is the possibility of complications, and at least some female contacts are likely to become infected. Few would argue today, as was argued in the past, that antimicrobial therapy for NGU is unnecessary.

The majority of observers now believe that *Chlamydia trachomatis* causes approximately 50% of NGU (2, 3). It is interesting to note that the antimicrobials which empirical studies have shown to be the most effective for the treatment of NGU, tetracyclines and erythromycin (4), are those which have also been shown to be most active against *C. trachomatis* in vitro (5), whereas penicillin and ampicillin, ineffective clinically, show substantially less in vitro activity. Nevertheless, *C. trachomatis* is not the only cause of NGU, so, although cell culture for *Chlamydia* before and after therapy is essential in present-day treatment trials, empirical assessment is still necessary. It is most important that the criteria of diagnosis and cure be clearly stated. The diagnostic criteria have been described by Swartz (p. 15, this volume), and cure presumably means the permanent reversal of these. Specifically, the patient should be symptom-free and show no evidence of significant urethritis as defined either by microscopy of a Gram-stained urethral smear (3) or by the determination of pyuria (2), specimens being collected under standard conditions with respect to the time of last micturition. In the case of previously isolation-positive patients, cell culture for *C. trachomatis* will, of course, be negative.

Holmes et al. (6) were able to conduct an NGU treatment trial under ideal circumstances, in patients at sea in an aircraft carrier. This enabled the administration of therapy under supervision, the collection of specimens in the early morning before micturition, and avoidance of intercourse by the patients during the follow-up period. This classic double-blind study showed unequivocally the superiority of a 1-week course of tetracycline to a placebo in the short term. Under more usual conditions therapy cannot be directly supervised, but should be as simple as possible and clearly explained. Tests of cure are applied not less than 2 h after micturition, and the importance of abstention from intercourse until cure is established is impressed on the patient.

In these trials two important matters must be decided at the outset: first, how long should the patient be kept under observation, and second, how is the trial to be controlled? Ideally, a period of observation of perhaps 3 months might be postulated, to exclude the possibility of late relapse, and this may be practicable in special circumstances. It is suggested, however, that normally the follow-up period should not exceed 3 weeks after the completion of therapy (7). There are several reasons for this. (i) In other infections antimicrobial therapy, if effective, is seen to be effective early, unless there are particular problems in penetration, as occurs, for example, in tuberculosis. The possibility that the intracellular habitat of *C. trachomatis* may present such problems is admitted, although in vitro studies do not support it. (ii) Many patients cease to attend when they are feeling better, and those remaining may not be a representative sample of the original group. (iii) Sooner or later, men with NGU will resume intercourse. Reappearance of the disease after this is normally impossible to interpret in terms of recurrence or reinfection. Many patients can be persuaded to abstain from intercourse during a 3-week follow-up period, but it is unrealistic to expect indefinite abstinence. (iv) Spontaneous remission may not be a major consideration shortly after therapy, but becomes an increasingly important factor as time goes by. Prolonged follow-up will necessitate placebo control.

Control of NGU treatment trials is important but presents practical difficulties. Except in special circumstances, placebo control of a sexually transmissible disease raises major ethical difficulties and may be thought hard to justify. However, if a therapeutic regimen approaches 80% success in the short term, this regimen could be used to control other types of therapy. That tetracyclines can achieve this degree of success has been common knowledge for many years and has been shown again in a recent trial at University College Hospital, London, England, which included cell culture for *C. trachomatis* before and after therapy. In this study some of the long-term problems in following patients treated for NGU were also seen. A total of 75 men with NGU were treated with oxytetracycline, 250 mg every 6 h for 2 weeks. Among 34 initially *Chlamydia*-positive men observed 2 to 4 weeks after therapy was started, cell culture for *C. trachomatis* was negative in every case, and a favorable clinical response was seen in 33. The diminishing numbers of patients available for subsequent study restricted analysis, but among them were some who had NGU again after intercourse, whose assessment as "relapse" or "reinfection" was impossible. Others, however, exhibited a return of chlamydial infection without apparently any further exposure to risk of infection; this late recurrence of *C. trachomatis* is of interest and implies a true treatment failure which should be further investigated. It is impossible to assess the reasons for the return of *Chlamydia*-negative NGU, which occurred in a few patients, through lack of knowledge of etiology. However, it was concluded that oxytetracycline in the dosage used was effective against both *C. trachomatis* and clinical evidence of urethritis in the majority of these men, and that in the short term the cure rate approached 80%. In 41 men with initially *Chlamydia*-negative NGU, cure in the short term appeared satisfactory in 38, but again both relapse and possible reinfection after resumption of intercourse were seen in some patients.

It is suggested that a 2-week course of oxytetracycline therapy gives results sufficiently good to be used to control other treatment regimens. John (8) observed that a 3-week course of oxytetracycline apparently gave better results than a 2-week course. It may be unrealistic to expect many patients to persevere with medication every 6 h for 3 weeks, during much of which time they will be symptom-free. John's results have not been confirmed, and this experiment should be repeated, with a 2-week course of oxytetracycline as a control. Other tetracyclines such as minocycline could be similarly evaluated; this agent has the advantage of a twice-daily regimen, and in an uncontrolled study we found that it gave good results, although vestibular side effects are not uncommon (9). Doxycycline, with a once-daily dosage, has not yet been evaluated against genital chlamydial infections, and gastrointestinal side effects have been experienced by some patients. Erythromycin is active against *C. trachomatis* in vitro, and in a pilot study we have found it effective in the treatment of NGU.

There are adequate laboratory and clinical grounds for accepting certain antibacterials, particularly tetracyclines, for the treatment of men with NGU, and for considering others for future study. How then are female contacts of these men to be managed? There is a paradox that while men with NGU are traditionally advised against early resumption of intercourse, presumably because of the risk of infection, contact tracing in NGU is not energetically pursued in many centers, and indeed some physicians doubt the necessity of examining or treating female contacts at all. Rosedale (10) believed that "epidemiologic" treatment of contacts makes no difference to the rate of recurrence of NGU in men, but his study was uncontrolled and involved no fewer than six different treatment schedules in small groups of patients. The epidemiology of NGU has been largely unknown, but the recognition of *C. trachomatis* as one of the causes has made possible epidemiological studies of patients infected with this organism.

I suggest that our results leave no doubt of the value of investigating female contacts of men with NGU. Of 328 such contacts, 121 (37%) were found to be infected by *C. trachomatis*, *Neisseria gonorrhoeae*, or *Trichomonas vaginalis*, alone or in combination. In a further investigation we studied female contacts of men with both *Chlamydia*-positive and *Chlamydia*-negative NGU. It was seen that 59% of contacts of isolation-positive men and 11% of contacts of isolation-negative men were infected by *C. trachomatis*. Overall, approximately one-third of all female contacts were infected by this organism. It may be conjectured that at least some female contacts of men with isolation-negative NGU are probably infected with other agents. The case for epidemiological treatment of all female NGU contacts appears to be strong. It is sometimes suggested (11) that only women who show clinical evidence of cervicitis should be so treated. Our data indicate that, although chlamydial infection does show a statistical correlation with clinical cervicitis, about 20% of women with *C. trachomatis* as sole pathogen have a clinically normal cervix, and it seems wiser to treat all contacts irrespective of the clinical findings.

The results of treating 50 women with chlamydial infection of the cervix with a course of oxytetracycline, 250 mg every 6 h for 2 weeks, showed that only 1 of these women when reexamined 2 to 4 weeks after commencing therapy still yielded *C. trachomatis* on cell culture; the remaining 49 were isolation negative. It was possible to reexamine only 20 of these women 4 to 8 weeks after therapy; none yielded *C. trachomatis*.

In conclusion, recommendations for the routine clinical management of men with NGU are given. For the most part these recommendations can be justified by the evidence already given, but they are also based in part on clinical experience. Once the diagnosis has been made, the patient starts a course of oxytetracycline, 250 mg every 6 h, and is told that at least 2 weeks of therapy is essential for cure. He is advised to abstain from intercourse for at least 2 and preferably 3 weeks, and is asked to persuade all his female contacts to attend for examination. Patients with NGU are traditionally advised to abstain from alcohol, although there is not a single controlled trial to support this. Nevertheless, alcohol does appear to aggravate urethritis in some patients and is best avoided or reduced during the first 2 weeks of therapy.

Female sexual contacts are examined fully as soon as possible. Those found to have gonorrhea or trichomoniasis will, of course, require specific treatment for these infections, and their male partners will need to be reassessed. The remainder are treated with a 2-week course of oxytetracycline unless they are pregnant, in which case erythromycin is substituted. They, like their male partners, are advised to avoid intercourse during therapy.

A man being treated for NGU should be reexamined at approximately weekly intervals. If the measures outlined are applied with careful attention to detail, about 80% of men will be symptom-free, with no evidence of significant urethritis, at the end of a 3-week follow-up period. Some, however, will fail to respond. They may have failed to take their medication correctly, or they may have had intercourse again; in either case, a further course of oxytetracycline is indicated. Other patients fail despite correct medication and avoidance of intercourse. Further diagnostic tests to exclude gonorrhea and, some would say, trichomoniasis should be performed, together with cell culture for *C. trachomatis* if available. Empirically, it is then worth retreating these patients with erythromycin for 2 weeks. With this drug clinical response is often obtained in these circumstances, although the reasons are impossible to define at present.

The problem of patients whose NGU persistently recurs despite apparently adequate antimicrobial therapy is most intractable. In some, particularly the more promiscuous, reinfection is a continuing possibility, and contact tracing, which is a dynamic rather than a static procedure, must continue. In others, however, continued recurrence cannot be ascribed to reinfection: there may have been no further intercourse, or current contacts may have been adequately investigated and treated. After 1 or 2 months, particularly in those who give a past history of gonorrhea or NGU, a full-scale urological investigation is warranted. The patient may then be found to have prostatitis, urethral stricture, or other urogenital disease requiring appropriate therapy.

In contrast to the treatment of gonorrhea, the treatment of NGU is not at present an exercise in applied microbiology. The clinician must be prepared to spend much time on investigation—on the search for unacknowledged contacts, the identification of unusual pathogens such as *T. vaginalis* or secondary bacterial invaders, and the detection of structural complications. Even then the causes of recurrent NGU are in some cases quite obscure, and their elucidation will depend on further research. Despite this apparently rather depressing conclusion, the majority of men with NGU *can* be cured. Further research by microbiologists and immunologists will no doubt in time enable the cure rate to be raised to an even higher level.

REFERENCES

1. Csonka GW: Nongonococcal urethritis. Br J. Vener Dis 41:1–8, 1965
2. Holmes KK, Handsfield HH, Wang SP, et al: Etiology of nongonococcal urethritis. N Engl J Med 292:1199–2005, 1975
3. Oriel JD, Reeve P, Wright JT, et al: Chlamydial infection of the male urethra. Br J Vener Dis 52:46–51, 1976
4. Willcox RR: 'Triple tetracycline' in the treatment of nongonococcal urethritis. Br J Vener Dis 48:137–140, 1972
5. Ridgway GL, Owen JM, Oriel JD: A method for testing the antibiotic susceptibility of *Chlamydia trachomatis* in a cell culture system. J Antimicrob Chemother 2:71–76, 1976
6. Holmes KK, Johnson DW, Floyd TM: Studies in venereal disease. III. Double-blind comparison of tetracycline hydrochloride and placebo in treatment of nongonococcal urethritis. J Am Med Assoc 202:138–140, 1967
7. Lassus A, Perko R-L, Stubb S, et al: Doxycycline treatment of nongonococcal urethritis with special reference to T-strain mycoplasmas. Br J Vener Dis 47:126–130, 1971
8. John J: Efficacy of prolonged regimens of oxytetracycline in the treatment of nongonococcal urethritis. Br J Vener Dis 47:266–268, 1971
9. Oriel JD, Reeve P, Nicol CS: Minocycline in the treatment of non-gonococcal urethritis: its effect on *Chlamydia trachomatis*. J Am Vener Dis Assoc 2:17–22, 1975
10. Rosedale N: Female consorts of men with nongonococcal urethritis. Br J Vener Dis 35:245–248, 1959
11. Willcox RR: 'Epidemiological' treatment in venereal diseases other than syphilis. Br J Vener Dis 49:116–122, 1973

Problems in Assigning a Causative Role to Chlamydiae Isolated in Nongonococcal Urethritis

SHIRLEY J. RICHMOND AND SUZANNE K. R. CLARKE

There are three possible roles genital chlamydiae may have in nongonococcal urethritis (NGU):

1. They may merely be commensals in the male urethra, not causally related to NGU at all.
2. They may cause NGU when they infect an otherwise normal urethra for the first time.
3. They may be acting as opportunistic pathogens in the urethra, only causing significant symptoms and signs in the presence of some other factor(s).

There is now ample published evidence supporting the pathogenic role of chlamydiae in urethritis, both NGU and postgonococcal urethritis (1). The question is not, therefore, whether genital chlamydiae are pathogens, but rather what sort of pathogens they are. Chlamydiae can undoubtedly act as primary pathogens (role 2 above) in selected patients with first-episode NGU (Bowie et al., this volume, p. 19), but the primary etiological role of chlamydiae in the majority of culture-positive men with NGU attending venereal disease clinics is not yet proven. To do this, one must be able to demonstrate a significant difference in the chlamydial isolation rates, not only between men with NGU and similarly exposed men without urethritis, but also between men with NGU and men with gonorrhea. Furthermore, since the chlamydial isolation rate in men with urethritis increases significantly as the duration of the untreated urethral discharge increases (2–4), and since men with gonorrhea present for treatment earlier on average than men with NGU (5), patients must be matched for the duration of their symptoms before comparing the chlamydial isolation rates in these two conditions. When this was done, on the rather crude basis of whether men had had symptoms for more or less than 7 days before attending for treatment, there was no real difference at all between the isolation rates in these two conditions (Table 1). The difference which emerged in the overall study was due to the fact that more men with NGU presented for treatment late than did men with gonorrhea, and in the group with long-standing untreated symptoms the isolation rate was significantly higher than in men who presented for treatment early. We therefore believe that the alternative possibility (role 3 above), that genital chlamydiae may be acting as opportunistic pathogens under certain circumstances, should also be considered.

Table 2 illustrates, with the use of entirely hypothetical data, the isolation rates which might be expected (i) with a primary etiological agent of NGU, designated organism *X* (role 2 above), and (ii) with an opportunistic pathogen which complicates urethritis, designated organism *Y* (role 3 above). Four possible groups of men may be studied, if one assumes mixed infections of NGU and gonorrhea occur, but, since the diagnosis of NGU is made by excluding gonorrhea, in

TABLE 1. *Effect of duration of urethral discharge on chlamydial isolation rates in men with NGU and gonorrhea (2)*

Men studied	Duration of symptoms (culture-positive/total patients)	
	0–7 days	>7 days
NGU	22/70 (31%)	14/23 (61%)
Gonorrhea	22/84 (26%)	8/12 (67%)

practice in any study the mixed NGU and gonorrhea group would fall into the gonorrhea category. Isolations of *X* would be made both in men with NGU and in men with gonorrhea, but the isolation rate would be significantly lower in the total gonorrhea group, since no isolations would be expected from men with gonorrhea only. Agent *Y*, in contrast, is a very common organism in sexually exposed populations, and it is capable of forming persistent infections. It multiplies actively, so that it can be isolated from the urethra only when the urethral mucosa is inflamed by gonorrhea or as yet undefined nonspecific inflammatory factors. Under these conditions, the infection with *Y* is pathogenic. In similarly exposed men without urethritis, it is postulated that the infection is often present, but it is clinically silent and the organism cannot usually be isolated.

We are proposing that genital chlamydiae frequently act as agent *Y*, and only occasionally as agent *X*, when NGU occurs in men attending venereal disease clinics. We suggest that chlamydiae form persistent infections in the urethra and that whether or not the infection produces overt symptoms and signs of urethritis depends on the state of the urethral mucosa. An inflamed urethra provides the right conditions for an active (culture-positive) chlamydial infection, which may be a primary infection, an exogenous reinfection, or a reactivated endogenous infection. This active infection will then itself prolong and exacerbate the inflammation.

This hypothesis would explain the following observations:

1. The increase in chlamydial isolation rates as the duration of symptomatic urethritis increases (2–4).

2. The similar isolation rates found in men with NGU and gonorrhea when patients are matched for the duration of their untreated urethral discharge (Table 1).

TABLE 2. *Expected isolation rates of primary etiological agent (X) and opportunistic pathogen (Y) in NGU*[a]

Men studied	Isolation rate (%)	
	X	*Y*
Gonorrhea only	nil	40
Gonorrhea and NGU	85	40
NGU only	85	40
No urethritis[b]	nil	(40)[c]

[a] Isolation rates are hypothetical.
[b] Matched for sexual exposure.
[c] Quiescent, not isolated.

3. The absence of a clearly defined incubation period in cases of NGU from which *C. trachomatis* has been isolated (6), and the isolation of chlamydiae from some men with NGU who deny any recent exposure at all (2, 4, 6).

4. The absence of positive correlation between chlamydial isolation rates and the number of recent (3 months) sexual partners, both in men with NGU (S. J. Richmond, M.D. thesis, Cambridge University, Cambridge, England, 1975) and in women attending venereal disease clinics (7), despite the fact that these organisms are sexually transmitted.

5. The recent failure to produce signs or symptoms of urethritis in male baboons experimentally infected with genital chlamydiae (8).

This hypothesis also fits in with the epidemiology of many other chlamydial infections. The *Chlamydia* genus is notorious for its ability to form latent infections (9, 10), and it is often only when the host-parasite relationship is disturbed in some way that severe manifestations of disease occur in the host. This, for instance, occurs in trachoma; blinding hyperendemic trachoma only occurs under conditions of what Jones has called "ocular promiscuity" (11), that is, conditions where the conjunctival epithelium is repeatedly damaged by agents other than the chlamydial organisms themselves.

If true, what are the practical implications of this hypothesis? It is likely that in sexually exposed populations an extremely complicated epidemiological situation exists, with primary chlamydial infections, exogenous reinfections, and endogenous reactivations all occurring. In all men in whom NGU is associated with an active (culture-positive) chlamydial infection, adequate treatment of the chlamydial infection is required, irrespective of whether the infection is a primary one, a reinfection, or a reactivation.

If this latency hypothesis is correct, however, it does have certain important implications.

1. The proportion of cases of NGU associated with genital chlamydiae will depend on the extent the particular population studied has been previously sexually exposed. A preliminary report on chlamydial isolation rates in young American airmen with NGU suggests that this is the case (12).

2. Chlamydial infections in the community will not be controlled simply by identifying and treating the culture-positive patients.

3. Measures undertaken to control NGU should include consideration of factors which may alter the relation between host cell and chlamydial parasite.

REFERENCES

1. Richmond SJ, Sparling PF: Genital chlamydial infections. Am J Epidemiol 103: 428–435, 1976
2. Richmond SJ, Hilton AL, Clarke SKR: Role of *Chlamydia* subgroup A in non-gonococcal and post-gonococcal urethritis. Br J Vener Dis 48:437–444, 1972
3. Holmes KK, Handsfield HH, Wang SP, et al: Etiology of nongonococcal urethritis. N Engl J Med 292:1199–1205, 1975
4. Oriel JD, Reeve P, Wright JT, et al: Chlamydial infections of the male urethra. Br J Vener Dis 42:46–51, 1976
5. Boyd JT, Csonka GW, Oates JK: Epidemiology of non-specific urethritis. Br J Vener Dis 34:40–43, 1958
6. Oriel JD, Reeve P, Povis P, et al: Isolation of *Chlamydia* from patients with non-specific genital infection. Br J Vener Dis 48:429–436, 1972

7. Hilton AL, Richmond SJ, Milne JD, et al: *Chlamydia* A in the female genital tract. Br J Vener Dis 50:1–10, 1974
8. Digiacomo RF, Gale JL, Wang SP, et al: Chlamydial infection of the male baboon urethra. Br J Vener Dis 51:310–313, 1975
9. Meyer KF: The host spectrum of psittacosis-lymphogranuloma verereum (PL) agents. Am J Ophthalmol 63:1225–1246, 1967
10. Storz J: *Chlamydia* and *Chlamydia*-induced diseases. Charles C Thomas, Publisher, Springfield, Ill., 1971, p 259–261
11. Jones BR: The prevention of blindness from trachoma. Trans Ophthalmol Soc UK 95:16–33, 1975
12. Grayston JT, Wang SP: New knowledge of chlamydiae and the diseases they cause. J Infect Dis 132:87–105, 1975

Chairman's Report and Discussion

E. RUSSELL ALEXANDER

P. J. Wiesner reviewed selected aspects of the epidemiology of nongonococcal urethritis (NGU). Private physicians in the United States saw an estimated 397,000 cases of NGU in 1974. In the United Kingdom, NGU increased 133% from 1966 to 1974 and has exceeded gonorrhea in incidence since 1965. A prospective study of a Navy cohort showed that several risk factors, including race, age, marital status, past history of urethritis, and number of previous sex partners, were correlated with risk of acquiring NGU. J. Gale said that he and W. Hinds had recently carried out a retrospective postal survey of 90% of physicians and a prospective inquiry of a 20% random sample of physicians who serve the 1.2 million people of King County, Washington. They found that the ratio of NGU to gonococcal urethritis was approximately 2:1. The ratio was 3:1 for patients of private physicians and 2:1 for Public Health Clinic patients. He went on to point out that private physicians saw two-thirds of the cases of urethritis diagnosed in the county versus one-third diagnosed by public health facilities. In agreement with Dr. Wiesner, they found that internists and family practitioners saw 90% of all urethritis cases seen in private practice. Finally, some 40 NGU cases that were diagnosed by private physicians were cultured, and *Chlamydia trachomatis* was isolated from 20% of these patients.

D. Oriel agreed with K. Holmes that the phrase "nonspecific genital infection," particularly in women, should be interred, because it is fairly meaningless. Regarding United Kingdom statistics, Dr. Oriel noted that there are 120 clinics in Britain, and their diagnostic criteria for NGU are not uniform. It was his opinion that, although United Kingdom statistics could reflect an over-diagnosing of NGU, they are comparable from year to year and do reflect a real increase. Dr. Wiesner replied that the apparent recent decrease of NGU in the United Kingdom occurred simultaneously with a change in the system of classification of genital infection.

S. L. Swartz discussed the diagnosis of NGU. The identification of typical gram-negative diplococci (GND) within polymorphonuclear leukocytes (PMN) was confirmed by isolation of *Neisseria gonorrhoeae* in 98% of cases. Likewise, failure to find GND either within or outside PMN was associated with failure to isolate *N. gonorrhoeae* in 98% of cases. Only atypical gram-negative bacteria were found in 15% of patients with urethritis, and about one in five of these equivocal smears was associated with isolation of *N. gonorrhoeae*. More difficult aspects of diagnosis include differentiation of NGU from prostatitis and establishing a diagnosis of urethritis in the patient with clinical symptoms but borderline urethral smears. E. R. Alexander noted that in a study of NGU he had examined Giemsa-stained urethral smears of approximately 40 cases positive for *C. trachomatis* and found none with inclusions. C. Mordhorst reported that he had attempted this hundreds of times, often 5 to 10 times per patient, and on only one occasion had he had positive results—in a man who also had an eye infection. S. Darougar reported that he had found only 2 cases with inclusions among 300 cases when Giemsa staining was used. When the same smears were examined by the fluorescent-antibody technique, the yield was 10 of 300, or 3%. J. Schachter suggested that it depended on the type of specimen. He preferred urethral scrapings rather than smears. Scrapings yield inclusions by Giemsa stain in 15% of positive cases, and inclusions are found roughly three times as often by fluorescent-antibody techniques. Dr. Wiesner commented that he did not think urethral scrapings were acceptable any more in the United States. W. McCormack commented that the specificity of the Gram-stained urethral smear is determined mainly by the skill of the person who reads the smear. In his hospital, in the venereal disease clinic, the data are exactly like those reported by Dr. Swartz, but for smears read by house officers in a walk-in clinic there is only about an 80% concordance between positive smears and positive cultures. Dr. Swartz noted that they had found that the prevalence of *Trichomonas vaginalis* in NGU was between 1 and 2%. Therefore, he normally looked for *Trichomonas* only among patients who failed therapy. D. Ford quoted S. Toma of Ontario, who believes that in the male urethra *Trichomonas* is nonmotile and that an intensified Giemsa stain is needed to see *Trichomonas* with any degree of reliability. M. Shepard reported that 8% of NGU patients at Camp Lejeune had *T. vaginalis* infection, as determined by an agar culture method that has not yet been reported. Regarding urethral smears, he commented that ureaplasmas were difficult to identify in Giemsa-stained smears. D. Taylor-Robinson noted that Z. McGee had worked on a modification of the Papanicolaou stain which differentiated cell nuclei from gonococci better than did the Gram stain (1). Dr. Oriel asked whether the diagnosis of NGU should be regarded as invalid unless the patient had urethral discharge. If so, what about gonorrhea, a substantial proportion of which is asymptomatic? W. Bowie and E. Rees each had observed a few patients with no discharge and few leukocytes in the smear or spun sediment who had chlamydiae recovered from the urethra.

Dr. Bowie described the Seattle studies of the etiology of NGU. Chlamydiae were isolated from 118 of 305 (39%) men with NGU and 1 of 60 (2%) men with no urethritis. The serum antibody response to *C. trachomatis* in men experiencing their first attack of NGU was consistent with a primary chlamydial infection. In men with first-episode NGU, first-voided urine from *Chlamydia*-negative men contained *Ureaplasma* significantly more often than the urine from *Chlamydia*-positive men, suggesting a possible etiological role for *Ureaplasma* in *Chlamydia*-negative NGU. The response to therapy with sulfonamides (more active against *Chlamydia* than against *Ureaplasma*) and with aminocyclitols (more active against *Ureaplasma* than against *Chlamydia*) was consistent with separate etiological roles of *Chlamydia* and *Ureaplasma* in NGU. J. Tully emphasized the need for serotyping of ureaplasmas. Dr. Shepard reported that at Camp Lejeune serotype 4 was most frequently recovered from NGU patients (51%), whereas in controls serotype 8 (35%) and other serotypes were found. There was more heterogeneity in the controls. Dr. Ford commented that in his experience there seems to be a random distribution of types over any disease, including NGU or Reiter's syndrome.

A. Braude asked if gonococcal L-forms had been looked for. He stated that a good case can be made for L-forms. They would be hard to isolate by usual cultural techniques, they would fail to respond to penicillin, they might not give rise to antipilus antibody, and they might be responsible for the low-grade inflammatory response seen in NGU. Dr. Holmes commented that in this study gonococcal L-forms had been looked for without success, using methods previously capable of inducing and propagating gonococcal L-forms in vitro. One of the media tried was one that was used to isolate from joints L-colonies that subsequently had reverted to the classical form of *N. gonorrhoeae*. Drs. Bowie and Holmes agreed that gonococcal L-forms had not been ruled out as a cause of NGU. W. McCormack stated that he found urine culture to be less productive than urethal cultures for *Ureaplasma*, especially among patients who did not have urethritis. The patient with urethritis would be expected to have more of these cell-associated organisms in the urine.

Dr. Bowie replied that urine cultures for *Ureaplasma* were used in both of his studies. He had not looked at a comparison of swabs versus first-voided urine among men with no urethritis, but he suggested that stripping the urethra might increase the yield in the first-voided urine. He then presented data comparing urethral swabs versus first-voided urine in men who did have urethritis, and the data showed no difference in the yield of *Ureaplasma*. He concluded that use of urine cultures was valid for comparison of *Chlamydia*-positive versus *Chlamydia*-negative cases of NGU, since both groups had urethritis.

Dr. Taylor-Robinson reviewed the previously published evidence concerning the role of ureaplasmas in NGU and suggested profitable avenues of research for further examination of their etiological role. He also described the occurrence of urethritis after intraurethral inoculation of two human volunteers with an inoculum of 5×10^4 ureaplasma cells of serotype 5. One subject developed dysuria, frequency, and a mucoid discharge. Both had urinary threads containing PMN. S. Razin discussed the use of in vitro organ cultures. He stated that infection of bovine oviduct by ureaplasmas had been reported, yielding titers as high as 10^8 or 10^9, which is unusual. Dr. Taylor-Robinson replied that 10^6 was the most that they could recover from organ cultures of the fallopian tube. He did not think that one would expect to recover concentrations of 10^8 or 10^9 from the human urethra. Dr. Holmes asked whether the threads observed in the two human volunteers might have arisen from periurethral glands rather than the prostate. He reported that in a previous study of about 25 infertile men and 25 fertile men (M. Ulstein et al., Andrologia, in press) ureaplasmas were often recovered and were consistently present in the ejaculate from both groups of men in concentrations one or two logs higher than in the first-voided urines, perhaps because they were being expressed from the periurethral glands during masturbation. Dr. Taylor-Robinson acknowledged that he was not certain where the threads originated in the volunteers. W. Fair agreed that the data on split ejaculates from Dr. Taylor-Robinson's volunteers were consistent with the site of ureaplasma infection being in the prostate. Dr. Fair also stated he had seen only one man whom he thought had true prostatitis caused by T-mycoplasma, on the basis of localization cultures, which showed more than 10^4 *Ureaplasma* cells per ml of prostatic fluid but fewer than 10^3 in the first-voided urine. Dr. Shepard suggested that his A7 differential agar medium could be used to determine whether the threads that were associated with leukocytes also contained ureaplasmas. The threads could be "fished out" and planted on the surface of the agar, and could be examined subsequently for growth of *U. urealyticum*.

Dr. Oriel described the results of treating 75 men with NGU (34 *Chlamydia* positive and 41 *Chlamydia* negative) with a 2-week course of oxytetracycline, 250 mg every 6 h. The treatment gave good short-term results in 80% of cases, with eradication of *Chlamydia* in the men who were initially infected with this organism. Epidemiological data indicate that approximately one-third of female contacts of men with NGU are infected by *C. trachomatis*, and the results of treating 50 of these infected women with oxytetracycline showed short-term cure in 49 of them. Dr. Oriel outlined an approach to the routine management of men with NGU and their female contacts. Dr. Shepard commented that at Camp Lejeune they had in the past cautioned patients about alcohol during and after treatment. Beer drinking is an old established test-of-cure among treated NGU patients at Camp Lejeune. It was his observation if patients survived a weekend of beer drinking with no symptoms they were cured. Symptoms may return after beer drinking, but not after whiskey. Dr. Oriel agreed with the concept of beer as a test-of-cure. He believes that it is used to this day in French clinics. He also agreed that the type of drink was significant, suggesting that it was not alcohol but some other factor that produced symptoms. Dr. Holmes commented that, when alcohol causes exacerbations of NGU, it may not be a direct effect of the alcohol on the urethra but rather on the sexual behavior of the individual. In reply, Dr. Oriel recalled that it used to be said that alcohol, in fact, would inflame people's sexual urges. In reply to a question concerning the differentiation of NGU and prostatitis, Dr. Oriel stated that he believed that it was a distinction without a difference because most patients with gonococcal urethritis or NGU before treatment have prostatitis as well. It is very difficult to perform diagnostic tests. He accepts prostatitis as a common reason for treatment failure in NGU.

Dr. McCormack questioned whether there was any good evidence that it really makes any difference whether partners of patients with NGU are studied and treated.

S. Richmond then discussed three possible roles that genital chlamydiae may have in NGU: (i) that of commensal nonpathogens, (ii) that of opportunistic pathogens which form persistent latent infections that reactivate to produce urethritis in the presence of some other factor, and (iii) that of primary pathogens. She advanced the hypothesis that chlamydiae frequently act as opportunistic pathogens and only occasionally as primary pathogens in the urethra, and she reviewed evidence which supports this hypothesis. In response to a question regarding the compatibility of her hypothesis with experimental primate chlamydial urethral infection, Dr. Richmond stated that in the recent report on infection in the baboon urethra there were no signs of urethral inflammation, although infection was clearly established. Thus, one of the postulates has not yet been fulfilled. Dr. Holmes stated that, like Dr. Richmond, he had noted an increasing rate of isolation of *Chlamydia* with increasing duration of symptoms of gonorrhea or NGU. An alternative explanation to Dr. Richmond's hypothesis of reactivation might be that *N. gonorrhoeae* and *C. trachomatis* are acquired simultaneously but have different incubation periods. The individual may present within 3 days with gonorrhea, but with no detectable chlamydiae as yet. The phenomenon of recovering chlamydiae frequently from patients with gonorrhea could mean that, because these agents are both sexually transmitted, they tend to become prevalent in the same population of sexually active people. Dr. Holmes said he believed that there is much evidence that the rate of recovery of chlamydiae from men with gonorrhea is significantly lower than from men with NGU. Finally, concerning the argument about whether chlamydiae are primary or opportunistic pathogens, the data of Drs. Wang and Bowie on the immunofluorescent-antibody seroconversion, and the presence of immunoglobulin M antibody, suggest that, at least among men experiencing first-episode NGU, *C. trachomatis* is a primary pathogen. Dr. Richmond replied that either primary infection or a reactivated chlamydial infection can cause symptomatic urethritis, but whether or not it does may depend on the state of the mucosa. She has not demonstrated a difference in isolation rate of *Chlamydia* from men with gonorrhea versus men with NGU. She felt that it would be difficult to explain recovery of *C. trachomatis* as early as 1 day after onset of symptoms of gonorrhea if *C. trachomatis* and *N. gonorrhoeae* had been acquired simultaneously.

B. Jones commented that he thought it was interesting to think about genital infections with *C. trachomatis* in the light of experience of these infections in the eye. There is no doubt that *C. trachomatis* is a pathogen in the human conjunctival mucosa. It becomes easier to isolate the organism after inoculation during the initial period. It reaches a plateau, continues for a variable period, and then it goes into a phase in which it becomes progressively more difficult to isolate. Disease may continue for a very long time, so that, if one studies a community that is saturated with *C. trachomatis* infection in the eye, one finds a lot of individuals suffering with disease, but one can isolate the organism from a diminishing proportion of them as age increases. There is evidence from L. Hanna and others that traumatic inflammatory insults to the conjunctiva can lead to the shedding of more chlamydial agent. This does not in any way indicate that the chlamydial agent is not a primary pathogen in the conjunctiva. Dr. Jones is sure the same consideration applies to the genital tract. It would seem not unlikely that other infections might have a similar effect in making it easier to isolate *Chlamydia* in a person with a chronic chlamydial infection in the genital tract. It is important to bear in mind that chlamydial agent affects the conjunctiva, first in an acute phase and then as a chronic disease. Surely the same thing should happen in the urethral mucosa? Dr. Jones suggested that there must be millions of people in the world who have a chronic, more or less subclinical disease of the urethral mucosa which may become symptomatic when something else happens. He did not see that in any way these phenomena, if they are confirmed, militate against *C. trachomatis* being a primary pathogen in the genital area.

REFERENCE

1. Sowter C, McGee ZA: Evaluation of a new technique for the demonstration of gonococci and other organisms in host cells. J Clin Pathol 29:433–437, 1976

Session II

RELATED GENITAL INFECTIONS

Epididymitis and Prostatitis

WILLIAM R. FAIR

In this, an overview of the problem seen by the clinician faced with making the definitive diagnosis in a given patient presenting with symptoms suggestive of epididymitis or prostatitis, I shall begin with a brief review of the human anatomy involved, in the hope that this might serve as a useful review to those not dealing directly with patients.

EPIDIDYMITIS

The epididymis enfolds the posterior aspect of the testicle. The inferior, or proximal, portion of the epididymis is referred to as the globus minor and the larger superior pole, as the globus major. The vasa deferentia (or, in the singular, vas deferens) are muscular tubes which carry testicular and epididymal contributions to semen and enter the urethra by coursing obliquely downward through the prostate. The seminal vesicles are near the termination of the vas and are applied to the posterior aspect of the prostate gland. The common channel formed by the junction of the vas deferens with the terminal portion of the seminal vesicle is referred to as the ejaculatory duct. The paired ejaculatory ducts enter the urethra at the level of the verumontanum. In the normal anatomic situation, the oblique passage of the ejaculatory ducts through the body of the prostate is thought to prevent reflux of urine from the urethra into the vas deferens.

A variety of etiological agents are thought to be responsible for epididymitis. Congenital defects and retrograde urine flow have been implicated as well as the diseases gonorrhea and tuberculosis and the microbial agents meningococcus, pneumococcus, *Ureaplasma urealyticum*, nonspecific urethritis (NSU) corynebacteria, viruses, *Trichomonas*, and *Chlamydia*.

It appears beyond reasonable doubt that mechanical factors can at times be responsible for the development of epididymitis, especially in young children with congenital urethral obstruction (1, 2). As a clinical point, the finding of epididymitis in a young child warrants complete X-ray and cystoscopic evaluation to exclude urethral obstruction.

The prophylactic value of performing bilateral vasectomies prior to prostatectomy to reduce the incidence of post-prostatectomy epididymitis seems clearly established. In a study of 1,029 men who had unilateral vasectomy done before prostatectomy, Rinker and colleagues (3) noted a 4.2% rate of inflammation of the epididymis on the side with the intact vas, but less than a 0.2% rate of epididymitis on the side that had the vasectomy. Active controversy exists as to whether or not reflux of urine into the vas deferens, particularly during straining, can produce epididymitis. O'Connor (4) confirmed that bladder urine can reflux into the vas by finding silver nitrate solution in the vasa of two patients within 24 h after bladder instillation. Similarly, Graves and Engel (5) performed some elegant studies in dogs in which they injected sterile urine taken from the same dog retrograde down the vas into the epididymis. As a control, physiological saline solution was introduced in the same manner into the opposite vas. Gross and microscopic evidence of inflammation of the epididymis was evident on the side of the urine injection, but no similar changes were observed in the control epididymis. Finally, the frequently observed clinical situation in which the patient relates his first symptoms of the problem as pain noticed while coughing, sneezing, or during rigorous exercise, followed within a few hours to days with a swollen, painful epididymis, is strong circumstantial evidence that retrograde flow of sterile urine may, in some instances, be responsible for the development of epididymitis.

Gonorrhea and tuberculosis are two diseases well recognized as etiological agents. Before the advent of modern chemotherapy, gonorrheal epididymitis, as a complication of gonorrheal urethritis, was a common occurrence in urological practice. Although still seen occasionally in modern-day practice, the gonococcus probably accounts for fewer than 5% of all cases of epididymitis (6). When tuberculosis is found involving the urogenital tract, there is epididymal involvement in 65 to 75% of patients.

A wide variety of microorganisms have been implicated as causing epididymitis. Virtually every bacterium known to be pathogenic for humans has, at one time or another, been responsible for epididymitis. Meningococcal or pneumococcal epididymitis probably represents hematogenous spread of organisms from other sites of infection in the body.

I need not dwell on this list for this audience except to point out that the *Chlamydia* group has been implicated in at least one report (7), although direct culture evidence of the organism's presence was lacking. All the above etiologies probably account for not more than 50% of cases of epididymitis. In at least half the instances, no definite etiological cause can be found and the disease is labeled idiopathic. However, a complete microbiological assessment of patients with epididymitis is lacking in virtually all reports published thus far. Perhaps such a study utilizing adequate sample collection and culture techniques, not only for the usual urinary tract bacterial pathogens but to include *Ureaplasma*, NSU corynebacteria, and *Chlamydia* cultures, may show that a larger percentage of cases can actually be related to a specific infectious agent.

PROSTATITIS

As in epididymitis, only a small fraction of patients presenting with a symptom complex consisting of various degrees of perineal aching, low back pain, and irritating voiding symptoms will be shown to have a bacterial cause for their condition. Abacterial prostatitis, or "prostatosis," causes symptoms that simulate bacterial prostatitis, but no bacterial pathogen can be recovered from urine or prostatic fluid cultures. In one of the most complete microbiological studies of this condition published to date, Meares (8) obtained historical and bacteriological data from 21 patients with prostatosis and compared it with similar data from patients with chronic bacterial prostatitis. Urethral, midstream, and post-prostatic massage urine specimens, and, when obtainable, expressed prostatic secretions (EPS), were cultured for *Neisseria*, T-strain mycoplasmas, and aerobic and anaerobic bacteria. Despite careful specimen collecting and culture technique, no infectious agent was isolated from the patients with prostatosis. Nielson et al. (9), using similar techniques but including viral and fungal cultures, likewise failed to isolate an infectious agent in 18 patients with a history of prostatosis. One-half of the patients were found to have chronic inflammatory changes noted on transperineal prostatic biopsy. However, 50% of patients with chronic urethritis and no symptoms referable to the prostate had similar changes noted on biopsy.

Although less common than prostatosis, chronic bacterial prostatitis causes significantly greater morbidity. In the absence of renal calculi or obstruction, chronic bacterial prostatitis is the chief cause of recurrent bacteriuria in males. The important point in differentiating the two conditions is the presence of urinary tract infection—the *sine qua non*, without which true bacterial prostatitis cannot be entertained as a diagnosis. The techniques of localizing lower tract infections are easily performed (10). The diagnosis of prostatitis is only made when the post-prostatic massage specimens or the EPS have greater than a 10-fold increase in bacteria as compared with the urethral or midstream urine specimens. Although the diagnosis can usually be made without difficulty, the treatment of chronic bacterial prostatitis poses a different problem. Because of the difference in pH between plasma and prostatic fluid, most currently available antibiotics will not accumulate in prostatic fluid in concentrations to render it bactericidal. Our studies into the etiology and treatment of prostatitis have led us to some interesting observations on the antibacterial activity of normal prostatic fluid which I would like to share with you at this time.

Our initial investigation into the antibacterial activity of human prostatic fluid was stimulated by the observation that urine from a normal, healthy male, obtained immediately after a vigorous prostatic massage, was bactericidal to *Escherichia coli*, whereas the same patient's urine before manipulation of the prostate easily supported bacterial growth and multiplication. This observation supported our earlier studies in which canine prostatic fluid was demonstrated to have a pronounced antibacterial activity against a wide variety of urinary tract pathogens (11). Subsequent studies in our laboratory confirmed a similar antibacterial action in the fluid of the dorsolateral lobe of the rat prostate (12).

The following is a review of the current status of our research in the area.

ISOLATION AND PURIFICATION OF THE ANTIBACTERIAL FACTOR

The prostatic antibacterial factor (PAF) was isolated from prostatic fluid or semen by cation-exchange chromatography using a Dowex 50WX2 (H^+ form) column (13). The PAF eluted from the Dowex column was resistant to heating at 100°C for 15 min at pH 7.0. However, when the proteins in the whole prostatic fluid were coagulated by heating, no antibacterial activity was demonstrated in the supernatant fluid of heated whole prostatic fluid. Antibacterial activity could be recovered in the PAF fraction when the flocculant precipitate obtained by heating intact prostatic fluid or seminal plasma was treated with perchloric acid and centrifuged, after which the supernatant was placed on the Dowex column and eluted with HCl. This indicated that the loss of activity was the result of binding of the PAF, in neutral solution, to the heat-precipitated protein.

The PAF appeared to be freely soluble in water. Attempts to extract the antibacterial activity into a variety of organic solvents, e.g., $CHCl_3$, ether, toluene, 1-butanol, 2-propanol, tertiary butyl alcohol, cyclohexane, ethyl acetate, and methylisobutylketone, were uniformly successful over a wide pH range.

ANTIBACTERIAL ACTIVITY OF HUMAN SEMEN AND EPS

The PAF fraction obtained via ion-exchange chromatography (7) was used to quantitate the antibacterial activity of human semen or EPS in normal males and patients with documented bacterial prostatitis. Early in the study, observations were made on secretions following prostatic massage. In 10 of 13 samples (77%) obtained from five patients documented as having chronic bacterial prostatitis, no PAF was demonstrable; when bactericidal activity was present, the diameter of the zone of inhibition averaged only 12.4 mm. Six samples of EPS from normal males all had evidence of antibacterial activity, with zones of inhibition averaging of 21.2 mm in diameter. Because of the small volume of fluid obtained on prostatic massage, the quantitative difference in PAF activities between patients and controls was more easily and more accurately assessed by using seminal plasma. In Fig. 1, the diameter of the zone of inhibition shown by the PAF fractions from 63 seminal plasma specimens obtained from 49 controls is compared with the PAF activity of 33 seminal plasmas obtained from 15 men with chronic bacterial prostatitis. The diameter of the zone of inhibition obtained with the undiluted PAF fraction isolated from the seminal plasma of the controls averaged 30.7 ± 4.0 mm. Eight of 33 "patient" seminal plasma specimens had no demonstrable antibacterial activity. Even in samples where antibacterial activity was seen, the diameter of the zone of inhibition was much less (mean = 18.9 ± 6.8 mm) than that found in the control specimens. A comparison on the dilution curve (Fig. 1) enabled us to quantitate PAF activity in the seminal plasma of patients. Hence, a 1:4 dilution of a "normal" seminal plasma gave a larger zone size than the undiluted sample from the patient group, indicating that the mean PAF level in the patients' semen was less than 25% of that expected to be present in normal subjects. However, serial semen samples from the same normal individual assayed over a period of several months varied by as much as 5 mm from sample to sample, indicating a moderate degree of fluctuation in the amount of PAF normally present. Successive assays on the same sample over a 2-month period did not vary by more than 4%.

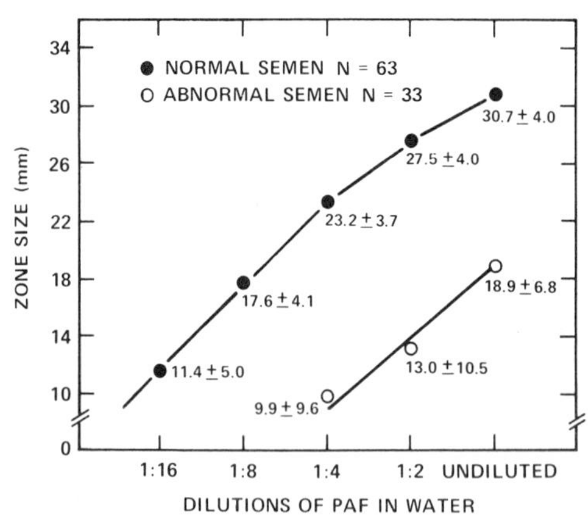

FIG. 1. *Comparison of the diameters of inhibition zones shown by prostatic antibacterial factor (PAF) fractions from 63 seminal plasma specimens obtained from 49 controls and 33 plasma specimens obtained from 15 men with chronic bacterial prostatitis. (From Urology 7:169, 1976).*

CHEMICAL IDENTIFICATION OF THE PAF

The PAF fraction, purified at least 5,000-fold on a weight basis (13), was readily crystallized by increasing the pH to above 9 by adding 1 N NaOH. After separation, no antibacterial activity was observed in the supernatant fraction. The resulting crystals were submitted for ion probe and mass spectrographic analysis. Both analytical methods indicated that the crystals were predominantly zinc chloride. The PAF fraction was not composed entirely of zinc salts, as was shown by the results of amino acid analysis and gas chromatography on the acid-hydrolyzed fraction. Under these conditions the following amino acids were also identified: alanine, glutamine, glycine, isoleucine, lysine, phenylalanine, proline, serine, threonine, tyrosine, and valine.

RELATIONSHIP OF ANTIBACTERIAL ACTIVITY TO ZINC CONCENTRATIONS

The experiments done after identification of zinc salts in the PAF fraction proved that the fraction of canine and human prostatic fluid possessing antibacterial activity was eluted from ion-exchange and gel-filtration chromatography columns in an identical value of eluate as was a zinc chloride standard solution. Those substances previously demonstrated as inhibitors of PAF activity, i.e., agar, casein, tartrate, or other anionic agents (11), also inhibited the normal bactericidal action of zinc salts in solution. The zones of inhibition measured on the PAF assay of both canine and human prostatic fluid were directly proportional to the zinc concentration of these fluids. The addition of a zinc chloride solution in various concentrations to the prostatic fluid gave an increase in the bactericidal action which was proportional to the amount of zinc added (Fig. 2). Additional indirect evidence that the antibacterial action of prostatic fluid was directly related to its zinc content was the abolishment of the antibacterial effect when equimolar amounts of EDTA (ethylenediaminetetraacetic acid), a cation chelating agent, were added to the prostatic fluid (Fig. 2).

ZINC LEVELS IN HUMAN PROSTATIC FLUID

EPS collected from control males and patients with chronic bacterial prostatitis were analyzed for zinc. The zinc concentration of the prostatic fluid showed a wide variation among individuals. The range of zinc levels in 65 EPS specimens obtained from 49 men free from bacterial prostatic infection (normal individuals plus benign prostatic hypoplasia [BPH] group) was 150 to 1,000 μg per ml, with a mean value of 448 μg per ml. In sharp contrast, the zinc concentration in 61 specimens of EPS obtained from 15 patients with documented chronic bacterial prostatitis averaged only 50 μg per ml. Several of the specimens had no detectable zinc present by this sensitive atomic absorption spectrophotometric method, and in no instance was a zinc level higher than 139 μg per ml obtained. In the absence of such data in the literature, we chose to establish a value of 150 μg per ml as the "lower limit of normal" for zinc in expressed prostatic secretions. Table 1 compares the zinc levels in prostatic fluid of noninfected men with the same number of specimens obtained from patients with documented bacterial prostatitis with respect to this value.

The zinc level in the prostatic fluid of a given individual appears to be relatively

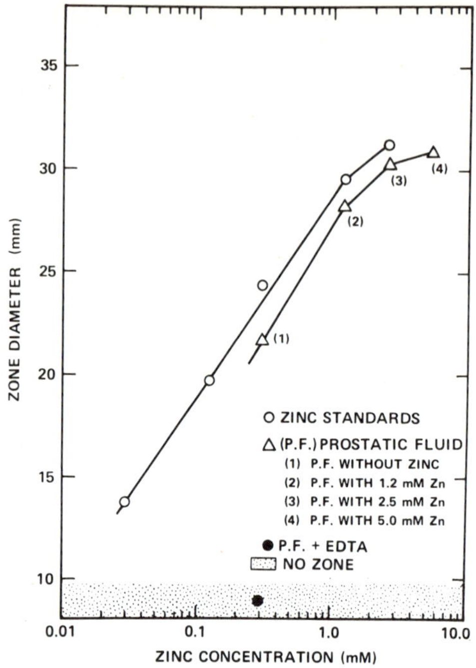

FIG. 2. *Effect of addition of zinc chloride solution on the bactericidal action of prostatic fluid. (From Urology 7:169, 1976).*

constant, particularly in the patient with prostatitis. Figure 3 illustrates typical serial prostatic fluid zinc levels found in two men (one patient and one normal) from whom we obtained multiple EPS collections, each of which was preceded by at least 3 days of sexual abstinence. Considering the value of 150 μg per ml suggested above as the lower limit of normal for zinc concentration in the EPS of normal men, it is apparent that the zinc level of the EPS in the normal man remained well above this value in multiple EPS specimens obtained during a 12-month period. Similarly, the patient with a documented chronic bacterial prostatitis continued to have a low EPS zinc level over the course of the follow-up. In the example shown in Fig. 3, the patient was asymptomatic and had sterile urine while on prolonged suppressive therapy with 100 mg of minocycline HCl twice daily, although direct cultures of his EPS confirmed the presence of small numbers of *Pseudomonas* cells and the

TABLE 1. *Zinc concentration of human prostatic secretion*[a]

Study group	No. of samples	No. with zinc concn:	
		>150 μg/ml	<150 μg/ml
Control group			
Normal men	36	36	0
BPH patients	29	29	0
Prostatitis patients	65	0	65

[a] $P < 0.001$.

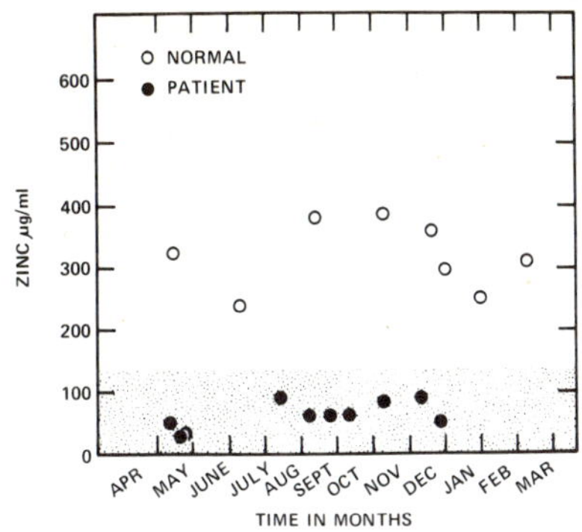

FIG. 3. *Zinc levels in the prostatic fluid of one normal man and one man with chronic bacterial prostatis over a 12-month period. (From Urology 7:169, 1976).*

persistence of chronic bacterial prostatitis which was unaffected by the antibacterial therapy.

SERUM ZINC LEVELS IN NORMAL SUBJECTS AND PROSTATITIS PATIENTS

Table 2 lists the serum zinc levels found in 61 samples from control men and 42 samples from patients with bacterial prostatitis. As with the EPS levels, there was no appreciable difference between the ''normal'' males and those with BPH with respect to serum zinc levels; both are included in the control group. The mean value of 0.89 μg per ml found in the controls was not significantly different from the mean value of 0.80 μg per ml found in the patient population. Hence, we cannot classify a given individual as being in the control or patient groups solely on the basis of the serum zinc level.

EFFECTS OF EXOGENOUS ZINC SALTS ON SERUM AND PROSTATIC FLUID LEVELS

We studied the effect of exogenously administered zinc salt on serum and prostatic fluid zinc levels in seven patients with bacterial prostatitis. The patients

TABLE 2. *Serum zinc levels*

Group	No. of samples	Age (years)	Concn (μg/ml)
Control	61	41–81 (62)[c]	0.89 ± 0.17[a] (0.60–1.28)[b]
Prostatitis patients	42	27–70 (52)	0.80 ± 0.14 (0.58–1.16)

[a] Standard deviation.
[b] Range.
[c] Mean.

received between 50 and 100 mg of elemental zinc orally per day for periods of 3 to 6 months. Serum zinc levels increased from a mean value of 0.80 μg per ml to 1.08 μg per ml, but prostatic fluid zinc levels did not undergo any appreciable change from the pretreatment levels while the patients were on this supplementation. More importantly, the additional oral zinc had no effect on the course of the infection, and the prostatic fluid cultures were still diagnostic of chronic bacterial prostatitis.

IN VITRO ANTIBACTERIAL ACTIVITY OF ZINC SALTS

The antibacterial activity of various concentrations of a zinc sulfate solution was tested against a variety of bacteria isolated from the urine of patients with urinary tract infections. The organisms were deemed "sensitive" to a particular level of zinc only if *all* the strains tested at that level had a 50% or greater decrease in the number of bacteria within 6 h after the addition of the zinc salt to a culture of the organism in MOPS medium.

As shown in Table 3, the sensitivity of the organisms varied widely, but, with the exception of *Proteus mirabilis* and *Streptococcus faecalis*, all of the organisms were sensitive to levels of zinc that are easily attained in the prostatic fluid of most human males, even those with chronic bacterial prostatitis.

The 66 strains tested varied in sensitivity to zinc from 5.1 μg per ml for *Klebsiella* to >650 μg per ml for *S. faecalis*. Despite the relative resistance of *S. faecalis, Staphylococcus albus*, the other gram-positive organism tested, was very sensitive at the level of 20.4 μg per ml. No attempt was made to study the effect of protein binding or the possibility of an in vivo zinc inhibitor in this study.

To determine whether a differential response to zinc could be of any significance in the observation that relatively few strains of *Escherichia coli* are responsible for the majority of human urinary tract infections, we also quantified the response of the strains commonly found as urinary tract pathogens, i.e., "common *E. coli*" (O1, O4, O6, O7, O75), versus the sensitivity of strains rarely found to be uropathogens (O34, O43, O53, O63, O103). The "uncommon" *E. coli* strains were slightly more sensitive to the effect of the zinc salt, but the differences between the two groups were not significant.

To sum up our experience to date, we have identified the PAF, responsible for the antibacterial activity of normal prostatic fluid, as a zinc compound. It appears as if the bactericidal activity of the EPS is related to the amount of zinc present in the fluid and that this may play a role in the natural resistance of the male urinary tract to infection. In addition, the determination of the zinc content of the EPS may be a

TABLE 3. *In vitro antibacterial activity of zinc sulfate*

Organism	No. of strains tested	Sensitivity (μg/ml)
Escherichia coli, "common"	19	20.4
E. coli, "uncommon"	21	10.2
Pseudomonas sp.	5	40.8
Proteus mirabilis	6	81.6
Klebsiella sp.	5	5.1
Streptococcus faecalis	5	>650.0
Staphylococcus albus	5	20.4

useful test in diagnosing patients with chronic bacterial prostatitis or those who are likely to be susceptible to prostatitis. The factors responsible for the marked decrease of zinc in the prostatic fluid of patients with bacterial prostatitis and methods of altering the zinc level in the fluid as a possible means of eradicating chronic bacterial prostatitis or increasing the resistance of the patient to the disease are important questions requiring further study.

REFERENCES

1. Kiviat MD, Shurtleff D, Ansell JS: Urinary reflux via the vas deferens: unusual cause of epididymitis in infancy. J Pediatr 80:476–479, 1972
2. Megalli M, Gursel E, Lattimer JK: Reflux of urine into ejaculatory ducts as a cause of recurring epididymitis in children. J Urol 108:978–979, 1972
3. Rinker JR, Hancock CV, Henderson WD: A statistical study of unilateral prophylactic vasectomy in the prevention of epididymitis: 1029 cases. J Urol 104:303, 1970
4. O'Connor VJ: Silver solution in the lumen of the vas after bladder instillation. J Urol 33:422–425, 1935
5. Graves RS, Engel WJ: Experimental production of epididymitis with sterile urine: clinical implications. J Urol 64:601–613, 1950
6. Mittemeyer BT, Lennox, KW, Borski AA: Epididymitis: a review of 610 cases. J Urol 95:390–392, 1966
7. Heap G: Acute epididymitis attributable to chlamydial infection—preliminary report. Med J Aust 1:718–719, 1975
8. Meares EM: Bacterial prostatitis vs "prostatosis." A clinical and bacteriological study. J Am Med Assoc 224:1372–1379, 1973
9. Nielson ML, Asnaes S, Hattel T: Inflammatory changes in the noninfected prostate gland. A clinical, microbiological and histological investigation. J Urol 110:423–426, 1973
10. Meares EM, Stamey TA: Bacteriologic localization patterns in bacterial prostatitis and urethritis. Invest Urol 5:492–518, 1968
11. Fair WR, Stamey TA: Bactericidal properties of prostatic fluid, p 199–211. *In* Bacterial Infections of Male Genital System (Workshop), National Research Council, National Academy of Sciences, Warrenton, Va., October 1967
12. Levy BJ, Fair WR: The location of antibacterial activity in the rat prostatic secretions. Invest Urol 11:173–177, 1973
13. Fair WR, Couch J, Wehner N: The purification and assay of the prostatic antibacterial factor (PAF). Biochem Med 8:329–339, 1973

Reiter's Syndrome: Current Concepts of Etiology and Pathogenesis

DENYS K. FORD

The term "Reiter's syndrome" can be defined in a variety of ways, but this discussion will use the following definition: a clinical syndrome characterized by the close temporal association of genitourinary signs of "infection" with arthritis and either conjunctivitis or the characteristic mucocutaneous lesions of circinate balanitis or keratodermia blenorrhagica. It should be stated at the outset that there is no satisfactory understanding of the cause of the condition and that none of the many possible pathogenetic hypotheses has direct support. This discussion will outline the main facts around which hypotheses can be constructed.

The recent finding that about 80% of patients with Reiter's syndrome have the histocompatibility antigen HLA-B27 (1, 2) clearly demonstrates a genetic predisposition to the syndrome, but the fact that perhaps 20% of patients do not have the B27 antigen cannot be ignored in any hypothesis concerning the pathogenesis of the disease. Moreover, genetic predisposition would seem to include a familial association with psoriasis (3), although no data have yet been reported to indicate that patients with Reiter's syndrome have an increased frequency of the HLA antigens 13, 16, and 17, which have been linked with psoriasis, and results of HLA typing of 28 patients with the syndrome in Vancouver, B.C., Canada, have shown no such relationships (Table 1). The mechanisms whereby specific HLA characteristics could predispose to diseases are unknown, although they are under intense study at this time. The current hypotheses involve closely associated immune response genes, common antigenicity between organisms and HLA antigens, or cell-surface HLA-determined receptors for pathogenic microorganisms. Unfortunately, these hypotheses seem not to have become adequately defined at this time to assist in the understanding of Reiter's syndrome. As a consequence, the present discussion will focus on the inciting infections that precipitate the syndrome.

Reiter's syndrome follows *Shigella* dysentery, *Yersinia enterocolitica* infections, and venereally acquired nongonococcal urethritis (NGU). It may also rarely follow gonorrhea, and I believe I have seen a very few patients with "gonococcal Reiter's syndrome," in whom a rapid and complete response of the urethritis to penicillin G would indicate that a simple gonococcal urethritis, unassociated with any NGU, was the precipitating cause.

There is general agreement that the manifestations of the disease are "aseptic" and that the organisms causing the precipitating infection cannot be cultured from the joints, eyes, or mucocutaneous lesions. Consequently, indirect pathogenetic mechanisms such as toxins or aberrant immunological reactions have been implicated as means whereby the organisms could produce aseptic lesions. In the absence of any supportive evidence for hypothetical toxins and in the presence of great current enthusiasm for immunopathogenesis as a feature of many diseases of uncertain etiology, the latter has more appeal.

TABLE 1. *HLA typing of 28 patients with Reiter's syndrome*

HLA antigen	No. of patients
B27	22
B13	1
B16	0
B17	0

The largest series of postdysenteric Reiter's syndrome is that of Paronen (4), involving 344 cases from about 50,000 patients with Flexner dysentery. An important statistic from this experience was that about 80% of the patients had genitourinary symptoms in spite of the dysenteric origin of their disease. Would tetracycline have any demonstrable effect on the urethritis of postdysenteric Reiter's syndrome? An answer to this would help to settle the hypothesis that the dysenteric infection causes the syndrome by activating a latent infection by an organism similar to that responsible for the venereally acquired infection.

The best epidemiological study of the postdysenteric syndrome is that of Noer (5), who carefully recorded nine cases arising among 602 individuals who contracted an untyped *Shigella* dysentery on board a U.S. naval ship. A recent follow-up study of five of these nine patients showed that four were HLA-B27 positive (6), and it was calculated that the risk of developing Reiter's syndrome in B27-positive individuals with dysentery was about 20%.

Whereas the most frequent form of post-*Yersinia* arthritis is that associated with erythema nodosum, which is clinically distinct from Reiter's syndrome, the latter has been reported from Finland (7), Norway (8), and Sweden (9), secondary to infection with *Y. enterocolitica* serotypes O3 and O9, the commonest types in Scandinavia. *Yersinia* arthritis is clearly related to the HLA-B27 histocompatibility antigen (10), but there are presently no clues as to why one patient might develop Reiter's syndrome and another erythema nodosum with arthritis.

Shigella organisms have some common O antigens with *Escherichia coli* (11), and *Y. enterocolitica* serotype O9 has antigenic cross-reactions with *Brucella abortus* (12); there are, however, at this time no defined antigenic cross-relationships between *Shigella* and *Yersinia* organisms to support the immunopathogenetic hypothesis for the induction of the clinically identical postdysenteric and postyersinial Reiter's syndromes.

In turning to venereally acquired Reiter's syndrome, the following assumptions probably should be made: first, that the urethritis of Reiter's syndrome responds to tetracycline, and second, that both chlamydial trachoma-inclusion conjunctivitis (TRIC) agents and human ureaplasmas (T-mycoplasmas) are fairly frequent causes of NGU and, together, account for the great majority of such cases seen in a venereal disease control clinic. Unfortunately, at this time there is no satisfactory study of fresh cases of Reiter's syndrome in which the presence of both TRIC and human *Ureaplasma* organisms in the urethra and the response of urethritis to tetracycline are unequivocally defined. Such a study would be required before it could be determined whether an additional tetracycline-sensitive agent, or a tetracycline-resistant agent, was possibly involved in causing the syndrome.

There appears to be no direct substantiated evidence to support the view that either chlamydial TRIC agents or human ureaplasmas are causally related to the extragenital manifestations of the syndrome. Both agents are found in the urethras of

TABLE 2. *Serum antibody titers to human ureaplasma, serotype 5*[a]

Group	Total no.	Percent male	Avg age (years)	Percentage of sera with titer of:		
				0	1:4	1:8
Red Cross donors	108	68	37	66	21	14
Venereal disease clinic controls	137	100	31	59	30	15
Reiter's syndrome patients	40	100	36	25	45	25

[a] Study of antibody levels to other serotypes gave similar results, as a result of broad cross-reactivity of human antisera.

patients with the syndrome about as frequently as in the urethras of patients with uncomplicated NGU. There are no satisfactory immunological studies of either B- or T-cell responses to either organism in patients with the syndrome. In my laboratory, low-level metabolic-inhibiting antibody against human ureaplasmas has been demonstrated in a somewhat higher proportion of patients with Reiter's syndrome than in patients attending a venereal disease clinic for a variety of reasons, but the differences are not striking (Table 2). Nevertheless, because it is my clinical impression that the abacterial urethritis of venereally acquired Reiter's syndrome is tetracycline sensitive, and because both chlamydiae and mycoplasmas are tetracycline-sensitive organisms, it seems most probable that the extragenital manifestations are secondary to a primary chlamydial or mycoplasmal genital infection.

REFERENCES

1. Brewerton DA, Nichols A, Oates JK, et al: Reiter's disease and HL-A 27. Lancet 2:996–998, 1973
2. Morris R, Metzger AL, Bluestone R, et al: HL-A W27-A clue to the diagnosis and pathogenesis of Reiter's syndrome. N Engl J Med 290:554–556, 1974
3. Lawrence JS: Family survey of Reiter's syndrome. Br J Vener Dis 50:140–145, 1974
4. Paronen I: Reiter's disease: study of 344 cases observed in Finland. Acta Med Scand Suppl 212, 1948
5. Noer HR: An "experimental" epidemic of Reiter's syndrome. J Am Med Assoc 197:693–698, 1966
6. Calin A, Fries JF: Epidemic Reiter's syndrome: genetics and environment. Arthritis Rheum 18:390–391, 1975
7. Ahvonen P, Sievers K, Aho K: Arthritis associated with Yersinia enterocolitica infection. Acta Rheumatol Scand 15:232–253, 1969
8. Solem JH, Lassen J: Reiter's disease following Yersinia enterocolitica infection. Scand J Infect Dis 3:83–85, 1971
9. Winblad S: Arthritis associated with Yersinia enterocolitica infection. Scand J Infect Dis 7:191–195, 1975
10. Aho K, Ahvonen P, Lassus A, et al: HL-A antigen 27 and reactive arthritis. Lancet 2:157, 1973
11. Wilson GS, Miles AA: Shigella, p 907–909. *In* GS Wilson and AA Miles (ed), Principles of bacteriology, virology and immunity, 6th ed. Edward Arnold Publishers, London, 1975
12. Ahvonen P, Jansson E, Aho K: Marked cross-agglutination between Brucellae and a subtype of Yersinia enterocolitica. Acta Pathol Microbiol Scand 75:291–295, 1969

Chlamydia in Relation to Cervical Infection and Pelvic Inflammatory Disease

ELISABETH REES, I. ANNE TAIT, DEREK HOBSON, AND F. WILLIAM A. JOHNSON

Assessment of the pathogenic role of *Chlamydia trachomatis* in the cervix is important for three reasons: (i) because of the possible carcinogenic effect of a deoxyribonucleic acid vector which has an intracellular developmental sequence and is associated with coitus; (ii) because of the effect on the fetus—prenatal or intrapartum; and (iii) because of the effect on fertility if ascending infection to the fallopian tubes occurs.

CERVICAL INFECTION

Novak and Woodruff (1), reviewing the pathogenesis of carcinoma of the cervix, stated that at present the evidence is very strong in favor of the theory that infection, possibly viral, plays a major role in the production of this disease, which accounts for 50% of all primary malignancy in the pelvis. Evidence that an inflammatory process can occur in the cervix in association with the presence of *C. trachomatis* was provided by Dunlop et al. in 1966 (2). They described follicles at the margins of the cervix seen with a colposcope, and later Ashton et al. (3) and Swanson and his colleagues (4) reported inflammatory changes in biopsy specimens from these areas and the identification of chlamydiae on electron microscopic examination.

The association of "cervicitis," variously defined, but including simple erosion, has been reported in several studies. Hilton et al. (5) found the cervix eroded and/or appearing "infected" to the naked eye in 82% of 34 *Chlamydia*-positive women, but 56% of their control group of 54 women attending a family planning clinic showed similar changes in the absence of *Chlamydia* or other organisms, as did 56% of 115 women at a clinic for sexually transmitted diseases. Similar findings were reported by Oriel and his colleagues (6). The high incidence in the control group is probably due to the inclusion of simple erosion in the category "abnormal," although this is not considered to be a pathological condition in its common nonulcerative form, but simply ectopic columnar epithelium (7, 8). The present study was undertaken to determine the possible association of clinical signs in the cervix with the presence of *C. trachomatis*, to determine the effect of treatment on the signs when present and on the subsequent isolation of the organism, and to observe whether or not *C. trachomatis* persists in the cervix in the absence of treatment and of other infection.

Materials and Methods

Patients. We have isolated *C. trachomatis* from 228 of 729 (31.2%) selected patients examined in a clinic for sexually transmitted diseases and from 52 of 145 (35.9%) consecutive contacts of patients with nongonococcal urethritis (NGU) (primary and secondary contacts of first and recurrent attacks). For the purpose of observing clinical signs, we selected the whole group of 127 *Chlamydia*-positive NGU contacts because we wished to exclude known cases or contacts of gonococcal infection, which also causes endocervicitis. They have been paired with 127 consecutive *Chlamydia*-negative NGU contacts, none of whom had received antibiotics for 4 weeks prior to their attendance. All patients were investigated for gonorrhea, trichomoniasis, and candidiasis by conventional methods.

Specimens for culture of *Neisseria gonorrhoeae* were taken on at least two occasions and were inoculated directly onto modified Thayer-Martin medium. Specimens for *C. trachomatis* culture were placed in transport medium and inoculated onto cover-slip cultures of replicating McCoy cells, i.e., not pretreated by irradiation or other antimitotic agents (9). The total number of chlamydial inclusions per cover slip (IFU/WC) was counted on every specimen after 48 h of incubation.

Endocervical smears were not examined for a pus cell count, as pus cells are a normal constitutent of cervical secretion and their presence is not a reliable guide to an inflammatory process. Their numbers exhibit physiological variation with the stage of the menstrual cycle and the presence of degenerating spermatazoa.

The clinical examination of the vaginal cervix (ectocervix) was recorded as no erosion, simple erosion (ectopic columnar epithelium), hypertrophic erosion (edematous and congested erosion), and chronic cervicitis (lacerated with ectropion or nabothian follicles with metaplasia). The endocervical contents were recorded as clear mucus, cloudy mucus, mucoid discharge (white), or mucopus (yellow). The findings were recorded at the time of taking the specimen for *Chlamydia* on both the laboratory request form and the case sheet. Colposcopic examination was not carried out.

As a basis of comparison with a known pathogen of the cervix, the cervical findings in 100 consecutive cases of gonorrhea diagnosed during the period of this study were analyzed. Routine tests for *C. trachomatis* were not carried out in this group, and it will, therefore, contain some cases of concurrent infection.

The presenting symptoms in the 127 *Chlamydia*-positive NGU contacts are compared with those of 326 consecutive gonorrhea patients attending the same clinic. In some cases symptoms may be due to concurrent trichomoniasis or candidiasis rather than gonococcal or chlamydial infection.

Treatment. Routine treatment consisted of the oral administration of oxytetracycline hydrochloride, 250 mg every 6 h for 21 days. Pregnant women were given a similar course of erythromycin.

Follow-up. All *Chlamydia*-positive patients were requested to attend for observation for 12 months unless it was judged that such a prolonged period of attendance would be bad for psychological reasons. Finnerty et al. (10) have shown the importance of clear communication with the patient in reducing the drop-out rate during observation, and care was taken to explain to patients that we were trying to assess the effect of treatment in a new condition which we were now able to recognize and that their cooperation would be of great assistance. They were asked to refrain from sexual intercourse for the 21 days of treatment, if possible, and then to resume when their husband or consort had permission. Alcohol was not forbidden. Patients were asked to attend for post-treatment tests at 2, 5, 8, 14, 26, 38, and 52 weeks during the latter part of the study, but were seen more frequently during the early part. Some patients volunteered to continue for longer periods and others re-attended with fresh symptoms or following new contact. Swabs for

TABLE 1. *Clinical signs in the cervix in 254 NGU contacts (127 C. trachomatis positive and 127 C. trachomatis negative)*

Clinical sign	No. of contacts		Percent positive
	Chlamydia positive	*Chlamydia* negative	
Ectocervix			
No erosion	28	61	31.5
Erosion			
Simple	53	51	50.9
Hypertrophic	40	6	86.9
"Chronic cervicitis"	5	9	
Not recorded	1	0	
Endocervical contents			
Clear mucus	31	76	28.9
Cloudy mucus	28	36	43.7
Mucoid	5	0	
Mucopus	59	11	84.0
Blood	4	3	
Not recorded	0	1	

culture of *C. trachomatis*, *N. gonorrhoeae, Trichomonas vaginalis*, and *Candida* species were taken on each occasion.

Results

Clinical findings. *C. trachomatis* was isolated from cervices exhibiting no clinical signs of infection (Table 1) and in this is closely paralleled by gonococcal infection (Table 2). The high isolation rate associated with hypertrophic erosion (86.9%) and endocervical mucopus (84%) suggested that these signs might be attributable to the presence of the organism and could be observed for a response to treatment.

The number of IFU/WC varied between 1 and 112,000. The number was not directly related to the clinical findings. The highest count was recorded in an uneroded cervix containing sticky mucopus.

Isolation of other organisms. In the group of 127 *Chlamydia*-positive NGU contacts, *N. gonorrhoeae* was isolated from 4 (3 on the second time of testing and

TABLE 2. *Clinical signs in the cervix in 100 consecutive patients with gonorrhea*

Clinical sign	No. of patients
Ectocervix	
No erosion	57
Erosion	
Simple	34
Hypertrophic	7
"Chronic cervicitis"	2
Endocervical contents	
Clear mucus	42
Cloudy mucus	19
Mucoid	5
Mucopus	34

from the urethra as the only positive site in 2). *T. vaginalis* was present in 17 patients and *Candida* species, in 45.

Presenting symptoms. As shown in Table 3, 49% of *Chlamydia*-positive patients and 53% of patients with gonorrhea denied any symptoms (Table 3). When there was more than one symptom, the first or more severe was recorded as the presenting symptom. A lower incidence of lower abdominal pain, which indicates possible pelvic infection, was found in the *Chlamydia*-positive group. Postcoital bleeding, found in 3.2% of the latter group, was associated with hypertrophic erosion or ectopy of endocervical mucosa.

The gonorrhea group was not screened for *C. trachomatis*.

Effect of treatment on clinical signs. Thirty-one cases exhibiting hypertrophic erosion have so far been followed for longer than 4 weeks. In 27 cases the hypertrophic erosion became and remained simple (Fig. 1 and 2). In four cases (one of whom was initially pregnant), no erosion was present at the end of the period of follow-up, metaplasia having taken place. Forty-five cases in which mucoid or mucopurulent discharge was found in the endocervix have completed at least 4 weeks of follow-up. The cervical secretion became clear in 25 cases (Fig. 3), became cloudy in 19, and remained mucoid in one patient from whom *Chlamydia* was reisolated. The change in the type of cervical secretion took place during the course of treatment.

Effect of treatment on the isolation of *C. trachomatis*. The method of culture for *C. trachomatis* used in this study has been shown to be highly reproducible and sensitive (Johnson et al., p. 309, this volume). A total of 161 *Chlamydia*-positive sexually transmitted disease patients have been followed for periods up to 20 months. Patients who attended for fewer than two post-treatment tests have been excluded. Two to four tests were carried out on those attending for less than 3 months, and 3 to 12 tests were made on those attending for a longer period. *C. trachomatis* was reisolated in only 16 (Table 4). In this group of 16 patients, three re-attended as contacts of men treated for gonorrhea and four as contacts of men with NGU (one *Chlamydia* negative, three not tested). When the asymptomatic consorts of three were examined, pus cells were present in shreds in the urine and *C. trachomatis* was isolated. One consort was said to have attended a clinic with symptoms but was not traced. Three patients had a new consort who did not attend for examination. Both of the remaining two patients were having intercourse with their original consort, neither of whom attended for examination.

TABLE 3. *Presenting symptoms in 127 Chlamydia-positive NGU contacts and 326 consecutive cases of gonorrhea*

Symptom	NGU contacts (%)	Gonorrhea patients (%)
None	48.9	54
Discharge	32.2	31
Abdominal pain	3.9	9
Urinary	3.9	5
Irritation/soreness	6.4	
Postcoital bleeding	3.2	2
Warts	1.5	
Abscess	0	

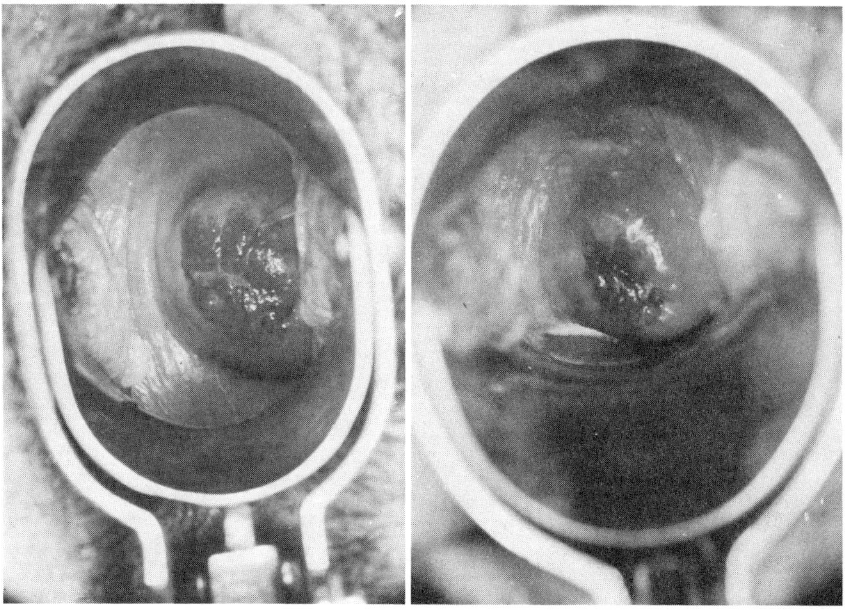

FIG. 1. *(A) Hypertrophic erosion with endocervical mucopus before treatment. C. trachomatis isolated. (B) After treatment. Simple erosion with clear mucus in the endocervix. C. trachomatis not isolated.*

Effect of physiological stress. *C. trachomatis* was not reisolated from one patient who became pregnant and three who were pregnant at the time of treatment. All were observed until after parturition, and no isolate was obtained from the mothers or from the babies on conjunctival swabbing. No isolate was obtained from one patient who acquired gonorrhea.

Persistence of *C. trachomatis* in the absence of treatment. Having early in the study observed the resolution of clinical signs after treatment, on ethical grounds we did not feel that treatment could be withheld once the diagnosis was made. However, in 19 cases either the patient failed to re-attend in 1 week for the results of her investigation, or treatment was delayed because she was under investigation for another condition, and *Chlamydia* was reisolated on one or two subsequent occasions (Table 5).

Discussion

The clinical findings suggest that edema, congestion, and mucopurulent discharge are associated with chlamydial infection of the cervix in a proportion of cases and that these signs regress after treatment. In the majority of cases, hypertrophic erosion became and remained simple during the period of observation. In 4 of the 31 cases metaplasia took place. If there should be an association between carcinoma of the cervix and *C. trachomatis* infection, it is possible that these four cases would carry the greatest risk if the hypothesis of Coppleson and Reid (11) is accepted.

The lack of correlation between the number of IFU/WC and the tissue response

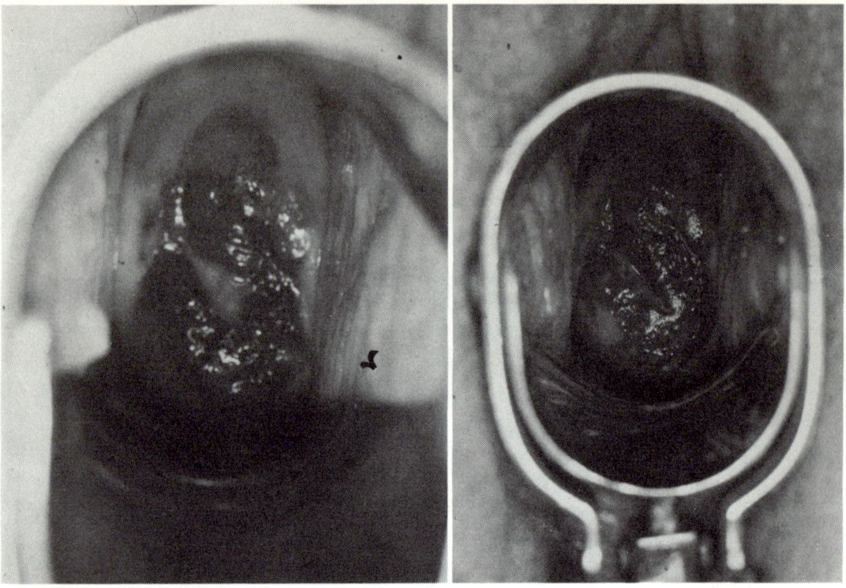

FIG. 2. *(A) Hypertrophic erosion with marginal follicles and endocervical mucopus before treatment. C. trachomatis isolated. (B) After treatment. Simple erosion with areas of metaplasia. Cloudy mucus in the endocervix. C. trachomatis not isolated.*

suggests that the latter may be due to a host factor or to variations in response to different immunotypes of *C. trachomatis*. It is unlikely to be due to sampling, as repeat specimens, usually taken 1 week later, showed little variation in count. The higher isolation rate from NGU contacts with simple erosion compared with no erosion may be due to the fact that columnar epithelium is more receptive to *C. trachomatis* and infection is more easily achieved. This would explain the higher isolation in women on oral contraceptives reported by Hilton et al. (5), in that the development of ectopic columnar epithelium is influenced by hormonal factors (12).

Our findings do not support the view of Hilton et al. (5) that *C. trachomatis* infection is latent but reactivated by other infection. We found signs of severe inflammation in the cervices of a group of women free from other known infection, and these signs responded to therapy. Reisolation of *C. trachomatis* occurred in only 16 of 161 women followed for periods up to 20 months and was associated with a high probability of reinfection rather than relapse. The physiological stress of pregnancy and parturition in four cases and gonococcal infection in one did not result in reisolation. Conversely, the persistence of *C. trachomatis* with isolation on two or three occasions over periods up to 19 months in 18 patients and 12 months in one suggests its persistence in an infective form in this small group of patients available for observation. The isolation of *C. trachomatis* from three asymptomatic male consorts examined because of reisolation of the organism in women on long follow-up suggests that the asymptomatic male may be a source of infection to the female.

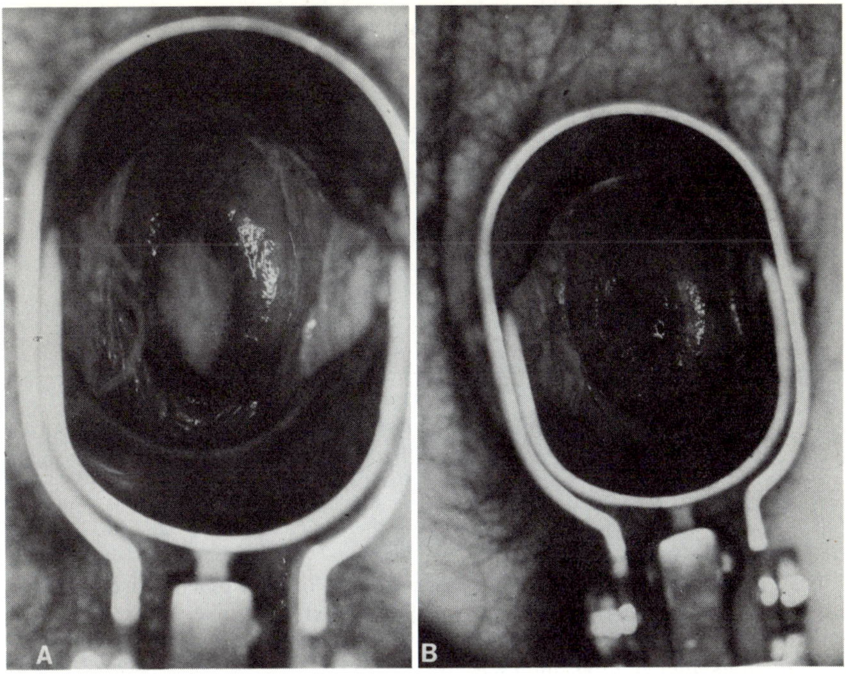

FIG. 3. *(A) Simple erosion with endocervical mucopus before treatment. C. trachomatis isolated. (B) After treatment. Simple erosion, clear mucus in the endocervix. C. trachomatis not isolated.*

PELVIC INFLAMMATORY DISEASE

There are two major difficulties in considering a possible relationship between the presence of *C. trachomatis* in the cervix and pelvic inflammatory disease: (i) the difficulties associated with the differential diagnosis of pelvic pain of a subacute character, which are appreciated by physicians in all the disciplines involved—gynecologists, urologists, venereologists, surgeons, and general physicians; and (ii) the low isolation rate of organisms from fallopian tubes which show objective signs of infection at laparoscopy or laparotomy, combined with the fact that these procedures are not usually carried out prior to antibiotic therapy when the diagnosis is clear. Curtis (13), in the pre-chemotherapeutic era, investigated the bacteriology and pathology of fallopian tubes removed at operation and concluded that gonococci live only a short time in the tube. Mårdh and Weström (14) confirmed a clinical diagnosis of salpingitis on laparoscopic examination and took swabs for bacteriological investigation from 50 patients in whom there was marked reddening and swelling of the tubes. No growth was obtained in 39 (78%). These findings suggest the presence of potent antibacterial factors in luminal fluid and the difficulties likely to be encountered in isolating *C. trachomatis* from the fallopian tube of a women with salpingitis. There is historical evidence of an association between chlamydial cervicitis and pelvic inflammatory disease, particularly in relation to maternal infection and neonatal chlamydial conjunctivitis (2, 15). We

TABLE 4. *Follow-up of 161 sexually transmitted disease clinic female patients after treatment for C. trachomatis infection of the cervix*

Time after treatment (months)	No. of cases	No. with *C. trachomatis* reisolated
<3	68	9
3–6	37	3
6–9	13	2
9–12	14	2
12–15	17	0
15–18	8	0
18–20	4	0
Total	161	16

have investigated the prevalence of pelvic infection in 127 *Chlamydia*-positive NGU contacts and 24 mothers of babies with neonatal conjunctivitis from whom *C. trachomatis* was isolated.

Results

NGU contacts. Pelvic infection was diagnosed in 5 of 127 women with chlamydial cervicitis. Two were wives of men who admitted extramarital contact. Both were found to have a hypertrophic erosion of the cervix and endocervical mucopus. The erythrocyte sedimentation rate (Westergren) (ESR) was 13 and 57, respectively. Tests for *N. gonorrhoeae* were negative. There was no previous history of pelvic infection. The symptoms settled rapidly with tetracycline therapy. Two women were the regular consorts of men with NGU. They admitted no other recent sexual contact. On examination neither cervix was eroded; one contained blood-stained mucus and the other, cloudy mucus. The ESR was 13 and 25, respectively. Tests for *N. gonorrhoeae* were negative. There was no previous history of pelvic infection, but one had a history of gonorrhea. The fifth case gave a history of discharge and of cramplike pain in the right and left iliac fossae for 3 weeks, severe at onset but now low grade. On examination of the lower genital tract, there was a simple erosion of the cervix. The endocervix contained clear mucus but bled easily. Mucopus was present on the rectal mucosa. The ESR was 19. Tests for *N. gonorrhoeae* were negative. *C. trachomatis* was isolated from the cervical and rectal swabs. The

TABLE 5. *Persistence of C. trachomatis in the absence of treatment*

Time after initial isolation (weeks)	No. of cases	Isolations	
		×2	×3
52	1	1	
19	2	1	1
14	1		1
12	3	2	1
7–8	4	4	
5–6	3	2	1
3–4	5	2	3

consort had NGU, and *C. trachomatis* was isolated. The psittacosis/lymphogranuloma venereum complement fixation test was carried out at the start of treatment, and 1 and 4 months later. The titer was 320, 320, and 80, respectively.

Mothers of babies with neonatal chlamydial conjunctivitis. Among 24 mothers of babies with neonatal chlamydial conjunctivitis examined, 16 developed postpartum infection. Seven were already receiving antibiotics for infected lochia or pyrexia. Nine of the remaining 17 mothers developed pelvic infection. They had already been discharged from hospital when the diagnosis of chlamydial conjunctivitis was made in the babies who had remained in hospital because of prematurity. The onset of pain varied between 13 and 38 days postpartum and was of such severity in two cases that they were admitted to hospital for curettage to exclude retained products of conception. None was found on histological examination of the curettings. *N. gonorrhoeae* was isolated from two mothers whose babies had concurrent gonococcal and chlamydial conjunctivitis. *C. trachomatis* was isolated from the cervix of seven of the nine mothers. The two negative mothers were already taking antibiotics prescribed by the family doctor for pelvic infection. In four cases blood was sent for psittacosis/lymphogranuloma venereum group antibody estimation before treatment and again after treatment in two cases (Table 6).

Discussion

The prevalence of pelvic infection in the 127 NGU contacts with chlamydial cervicitis was 3.9%. This is lower than is found with gonococcal cervicitis (16, 17). The frequency of pelvic infection in mothers of infected infants causes concern. The group antibody titers, high in two cases and showing conversion in one, suggest the association of the chlamydial infection with the pelvic inflammatory disease.

These findings lend support to the theory that chlamydiae may ascend to the upper genital tract. Further support is given by the development in a special-care unit of chlamydial conjunctivitis in a baby delivered by caesarean section 11 days previously (Rees et al., p. 140, this volume). If the organism can reach the fallopian tubes and if a similar pathological process can occur there as has been found in the cervix, then its etiological role in nongonococcal salpingitis would not be surprising.

TABLE 6. *Psittacosis/lymphogranuloma venereum group antibody titers in four women with postpartum pelvic infection*[a]

Patient	Day postpartum	Antibody titer	Contact
M.C. (−, AB)	39	640	Husband (+)
	90	320	
	135	80	
M.W. (+)	44	320	Husband (+)
M.L. (+)	9	<10	Consort (NE)
	26	20	
	46	10	
M.F. (−, AB)	28	40	Consort (+)

[a] For the patients and contacts: + = *C. trachomatis* isolated; − = *C. trachomatis* not isolated; AB = on antibiotics; NE = not examined.

CONCLUSION

The findings in this study suggest that *C. trachomatis* is a pathogen in the female genital tract and that it provokes, in some cases, a well-marked tissue response to infection in the cervix.

The treatment of all women from whom *C. trachomatis* is isolated is strongly indicated, and facilities for diagnosis should be generally available.

ACKNOWLEDGMENTS

We are indebted to Alan Percival and R. E. Rewell for bacteriological investigations, Bruce White for viral studies, and G. C. Turner for serological tests for syphilis.

This investigation was supported by a project grant from the Medical Research Council of Great Britain.

REFERENCES

1. Novak ER, Woodruff JD: Novak's gynecologic and obstetric pathology, 7th ed, p 97–98. W. B. Saunders Co., Philadelphia, 1974
2. Dunlop EMC, Al-Hussaini MK, Freedman A, et al: Infection by TRIC agent and other members of the Bedsonia group. Genital infection and disease of the eye. Trans Ophthalmol Soc UK 86:321–334, 1966
3. Ashton N, Dunlop EMC, Darougar S (in preparation, 1974). Quoted by Dunlop EMC: Non-specific genital infection, p 284. *In* RS Morton and JR Harris (ed), Recent advances in sexually transmitted diseases, 1st ed. Churchill Livingstone, Edinburgh, 1975
4. Swanson J, Eschenbach DA, Alexander ER, et al: Light and electron microscopic study of Chlamydia trachomatis infection of the cervix. J Infect Dis 131:678–687, 1975
5. Hilton AL, Richmond SJ, Milne JD, et al: Chlamydia A in the female genital tract. Br J Vener Dis 50:1–10, 1974
6. Oriel JD, Powis PA, Reeve P, et al: Chlamydial infections of the cervix. Br J Vener Dis 50:11–16, 1974
7. Jeffcoate TNA: Chronic epithelial dystrophies. Cervix. Principles of gynaecology, 4th ed, p 339–341. Butterworths, London, 1975
8. Coppleson M, Reid B: Physiological squamous metaplasia, p 133–134. *In* Preclinical carcinoma of the cervix uteri, 1st ed. Pergamon Press, Oxford, 1967
9. Hobson D, Johnson FWA, Rees E, et al: Simplified method of diagnosis of genital and ocular infections with chlamydia. Lancet 2:555–556, 1974
10. Finnerty FA Jr, Matt EC, Finnerty FA: Hypertension in the inner city. Analysis of clinic dropouts. Circulation 47:73–78, 1973
11. Coppleson M, Reid B: Cervical dysplasia. Atypical form of squamous metaplasia, p 148–149. *In* Preclinical carcinoma of the cervix uteri, 1st ed. Pergamon Press, Oxford, 1967
12. Coppleson M, Reid B: Physiological squamous metaplasia. Methods of exposure of columnar epithelium on the ectocervix, p 129. *In* Preclinical carcinoma of the cervix uteri, 1st ed. Pergamon Press, Oxford, 1967
13. Curtis AH: Bacteriology and pathology of fallopian tubes. Surg Gynecol Obstet 33:621–631, 1921
14. Mårdh PA, Weström L: Tubal and cervical cultures in acute salpingitis with special reference to *Mycoplasma hominis* and T-strain mycoplasmas. Br J Vener Dis 46:179–186, 1970
15. Mordhorst CH, Dawson C: Sequelae of neonatal inclusion conjunctivitis and associated disease in the parents. Am J Ophthalmol 71:861–867, 1971
16. Gisslén H, Hellgren L, Starck V: Incidence, age distribution and complications of gonorrhoea in Sweden. Bull WHO 24:367–371, 1961
17. Rees E, Annels EH: Gonococcal salpingitis. Br J Vener Dis 45:205–215, 1969

Chlamydia trachomatis as an Etiological Agent in Acute Salpingitis

PER-ANDERS MÅRDH, K. TORVALD RIPA, SAN-PIN WANG, AND LARS WESTRÖM

Chlamydia trachomatis is frequently recovered from urethral specimens of men with nongonococcal urethritis (NGU) (1, 2). It also occurs frequently in cervical specimens of women who are sexual partners of men with NGU (3).

A possible role for *C. trachomatis* in cervical infections has recently been considered (4, 5), but whether the organism is associated with infections in the upper female genital tract has not yet been explored. The cause of the tubal infection in acute salpingitis remains uncertain in a large proportion of cases (6).

The present study shows that *C. trachomatis* may be considered as a probable etiological agent of acute salpingitis.

MATERIALS AND METHODS

Patients. Twelve consecutive patients with acute salpingitis were studied after admission to the Department of Obstetrics and Gynecology, University Hospital of Lund, Lund, Sweden (Table 1). The patients were subjected to laparoscopy within 24 h of admission and before any treatment had been instituted. The laparoscopic technique and the criteria used for the classification of salpingitis as mild, moderate, or severe have been described elsewhere (7, 8).

Sampling technique. Before the gynecological examination and other diagnostic procedures were performed, the uterine portio was exposed by insertion of a sterile speculum into the vagina with the patient in lithotomy position. For the isolation of *Chlamydia*, a specimen was collected by rotating a sterile cotton-tipped swab in the cervical canal. For the isolation of *Neisseria gonorrhoeae*, another specimen was collected by the same technique except that a cotton-tipped swab treated with charcoal was used. From eight of the patients, the latter specimen was also used for the isolation of *Mycoplasma hominis*.

During laparoscopy, tubal material was aspirated via a baby-feeding catheter which was inserted through the abdominal ostium of one of the tubes. In four of the patients (L.K., E.L., A.H., and B.B.), purulent material was aspirated during laparoscopy by puncturing tubal abscesses with a long syringe which was inserted through the abdominal wall above the symphysis. The tubal specimens were cultured for the same organisms as those which were sought in the cervical specimens. In addition, the tubal specimens of five of the patients (I.N., E.L., A.H., A.C.O., and B.B.) were cultured for bacteria, including anaerobes. These specimens either were taken with a cotton-tipped swab treated with charcoal by the two test tube system as described in the *Anaerobe Laboratory Manual* (9) or were transferred through a rubber membrane to test tubes filled with CO_2.

The specimens for the isolation of chlamydiae were transferred to 2-SP (10) containing streptomycin (50 μg/ml), vancomycin (100 μg/ml), and amphotericin B (2.5 μg/ml). The specimens in 2-SP were frozen to $-80°C$ within 1 h of sampling.

TABLE 1. *Anamnestic data and some symptoms and signs in 12 consecutive salpingitis patients*

Patient	Age (years)	Ever preg-nant	Earlier salpin-gitis	Earlier treated for LGTI[a]	Intra-uterine device	Oral contra-ceptive	Day of laparoscopy in menstrual cycle				No. of days of ab-dominal pain before admission			Rectal temp 38°C	ESR 15 mm/h	Tubal inflammatory changes		
							1–10	11–16	17–28	28	3	4–14	14			Mild	Moderate	Severe
E.H.	16	–	–	–	–	+	+				+			+	–		+	
L.K.	20	–	–	+	+	–				+			+	+	–			+
E.R.	24	+	–	+	–	–	+						+	–	–	+		
I.N.	22	–	+	+	+	–			+			+		+	+	+		
B.M.E.	23	–	+	–	–	–	+					+		+	+		+	
E.L.	17	–	–	–	+	–		+				+		+	+			+
A.H.	19	+	–	+	–	+			+			+		–	+			+
A.C.O.	23	+	+	+	+	–			+			+		–	–	+		
A.B.	15	–	–	–	–	+			+			+		+	–	+		
G.J.	21	+	–	–	+	–		+			+			+	+	+		
E.K.O.	22	+	+	+	+	–			+				+	–	–	+		
B.B.	22	–	–	–	+	–			+			+		–	–	+		

[a] Lower genital tract infection.

Specimens for the isolation of gonococci, mycoplasmas, and ureaplasmas were transported in a modified Stuart medium (11).

Culture technique. For the isolation of *C. trachomatis*, cycloheximide-treated McCoy cells were used as described in detail elsewhere in this volume (Ripa and Mårdh, p. 323). The methods used for the isolation and identification of mycoplasmas, ureaplasmas, and *N. gonorrhoeae* have been reported earlier (7). The specimens for the isolation of facultatively and strictly anaerobic bacteria were plated within 60 min of sampling on three blood agar plates, which had been stored in GasPak jars (BBL) from within 1 h of production. One of the plates contained 100 μg of kanamycin per ml. The kanamycin-containing plate and one of the other blood agar plates were incubated at 37°C for 48 h in GasPak jars; the third plate was incubated in an atmosphere of 10% CO_2.

Chlamydia serology. Demonstration of serotype-specific immunoglobulin M and G (IgM and IgG) antibodies to *C. trachomatis* in sera from the two patients from whose Fallopian tubes the organism had been isolated was made by the micro-immunofluorescence (micro-IF) test (12) against a set of *C. trachomatis* antigens including immunotypes A, B, CJ, ED, GF, K, H, and I.

RESULTS

Culture studies. The results of the cultures for *C. trachomatis*, *N. gonorrhoeae*, and *M. hominis* from the cervix and the Fallopian tubes are shown in Table 2. The anaerobic bacterial cultures made from tubal specimens of five of the patients yielded no growth.

Serological studies. The results of the micro-IF test with the sera from the two patients in whom *C. trachomatis* was isolated from the Fallopian tubes are shown in Table 3. In both cases, a significant rise of antibody titer was demonstrated with IgM antibody pattern specific to immunotype ED. The presence of IgM antibodies in these sera indicates that both patients were in an early stage of ED infection (12, 13). The serological patterns of IgM and IgG antibodies from patient I.N. indicate

TABLE 2. *Cultures of Chlamydia trachomatis, Neisseria gonorrhoeae, and Mycoplasma hominis from the cervix and the Fallopian tubes of 12 consecutive patients with acute salpingitis*[a]

	Cervix			Fallopian tubes		
Patient	*C. trachomatis*	*N. gonorrhoeae*	*M. hominis*	*C. trachomatis*	*N. gonorrhoeae*	*M. hominis*
E.H.	ND	–	–	–	–	–
L.K.	S	–	+	–	–	–
E.R.	S	–	ND	–	–	ND
I.N.	S	–	ND	+	–	ND
B.M.E.	S	–	ND	S	–	ND
E.L.	S	–	+	S	–	–
A.H.	S	–	+	+	–	–
A.C.O.	+	–	–	–	–	–
A.B.	S	+	ND	–	–	ND
G.J.	S	+	+	–	+	–
E.K.O.	S	–	–	–	–	ND
B.B.	S	–	+	–	–	+

[a] ND = culture not done; S = cell culture spoiled by bacterial contamination.

TABLE 3. *Results of the micro-IF test with sera from two patients with acute salpingitis in whom C. trachomatis was isolated from Fallopian tubes*

Patient	Date of serum	IgM antibody vs. test antigens:								IgG antibody vs. test antigens:							
		CJ	A	H	I	K	B	ED	GF	CJ	A	H	I	K	B	ED	GF
I.N.	9 January 1975	0[a]	0	0	0	0	0	8	0	0	0	0	0	16	8	64	16
	14 January 1975	0	0	0	0	8	0	32	0	0	16	0	0	16	16	256	16
A.H.	13 August 1975	0	0	0	0	0	0	16	0	0	0	0	152	16	8	64	16
	21 August 1975	0	0	0	0	0	8	128	0	0	8	0	1,024	16	1,024	2,048	16

[a] Titer less than 8.

that her infection was due to type ED, whereas those from patient A.H. indicate that she had previously had type I infection, as shown by IgG antibody of her 13 August 1975 serum. The current ED infection brought an increased titer and broadly reacting pattern of antibody response in the latter patient.

CASE REPORTS

Case I.N. The patient was a 24-year-old woman who had never been pregnant. She had previously been in good health. In March 1974, she was treated for uncomplicated gonorrhea. An intrauterine device (IUD) (Cu-T 200) was inserted in September of the same year. She had regular menstrual periods. The last menstrual bleeding before admission was 22 December 1974.

She was admitted 8 January 1975, after having had abdominal pain for 5 days and a malodorous vaginal discharge for 1 week. On the day of admission, she had a copious vaginal bleeding. She presented abdominal tenderness at bimanual palpation of the adnexa but had no palpable inflammatory masses. Laparoscopy revealed mild salpingitis. The Fallopian tubes were reddened and swollen, and the abdominal ostia were open. The tubes were covered with seropurulent secretion. The rectal temperature was 37.4°C and erythrocyte sedimentation rate (ESR) was 35 mm/h.

Cultures for gonococci from the cervix, the urethra, the rectum, and the Fallopian tubes were all negative. *C. trachomatis* was isolated from the Fallopian tubes, but the cervical culture was spoiled by bacterial contamination. Meinicke's test was negative.

The patient was treated with bed rest for 14 days and was given a daily oral dose of 100 mg of doxycycline. Her body temperature never exceeded 37.8°C. Upon leaving the hospital on 22 January, she was free from symptoms and the gynecological examination revealed normal findings. ESR was 24 mm/h. At a control visit, on 6 February, the gynecological examination revealed nothing remarkable and she was free from symptoms. Her IUD was still in place. ESR was 7 mm/h.

Case A.H. The patient was a 19-year-old I-para with a normal delivery in July 1972. At the age of 3 years she had been treated conservatively for peritonitis. From March 1973 to December 1974, she had been treated three times for infections of the lower genital tract. Two of these infections were treated with doxycycline. On all three occasions, cultures for gonococci were negative, and *Trichomonas vaginalis* was diagnosed once. In January 1975, a Cu-T 200 IUD was inserted.

She was admitted because of abdominal pain that had lasted for 3 days. She was in the 10th day of her menstrual cycle. She had had an increased vaginal discharge for 2 months and dysuria for "some weeks" before the admission.

On admission, she had palpable tender inflammatory masses, the size of a golf ball, on both sides of a normal uterus. There was slight bleeding from the vagina. She had a body temperature of 37.6°C. ESR was 45 mm/h. Laparoscopy revealed pelvoperitonitis and pyosalpinx on the left side.

Cultures from the cervix, the urethra, the rectum, and the Fallopian tubes were all negative for gonococci. *C. trachomatis* was isolated from the pyosalpinx, whereas the cultures for bacteria, mycoplasmas, and ureaplasmas were all negative. The patient was treated in the clinic with bed rest for 16 days and was given a daily oral dose of 4 g of ampicillin. When leaving hospital, she was free from symptoms, and the gynecological examination revealed normal findings. ESR was 18 mm/h. The IUD was removed during her stay in the clinic.

Two months later the patient was readmitted because of abdominal pain. On this occasion, she was not subjected to laparoscopy because of influenza-like symptoms. *T. vaginalis* was demonstrated in a wet vaginal smear. She was treated on an outpatient basis with 100 mg of doxycycline daily for 10 days and with 600 mg of metronidazole orally daily for 1 week. She failed to return for the scheduled checkup.

DISCUSSION

In women 15 to 45 years of age, tubal infections are (except in a minority of cases) caused by an infection ascending from the lower genital tract. However, on the basis of the results of cultures from the cervix, it is not possible to establish the cause of the tubal infection. Use of laparoscopy in patients presenting symptoms suggestive of acute salpingitis offers not only a diagnostic means but also a possibility to obtain specimens directly from the Fallopian tubes (6).

Only in a minority of salpingitis cases have we been able to demonstrate bacteria, including gonococci and strictly anaerobic bacteria, in Fallopian tube specimens. In a series of 411 consecutive salpingitis patients, *N. gonorrhoeae* was isolated from the cervix in 134 (31.6%). However, gonococci could be demonstrated from the Fallopian tubes in only 17 of the 134 cases (12.7%). In the present series, 2 of the 12 patients had cervical gonorrhea, and in one of these gonococci were also recovered from the tubal specimen. In 1965, 50% of the salpingitis patients treated in our clinic had cervical gonorrhea, whereas the corresponding percentage in 1974 was 9% (Weström and Mårdh, p. 84, this volume), which certainly shows that *N. gonorrhoeae* is of only minor importance as a causative agent in acute salpingitis in our region at the present time.

Anaerobic bacteria have been associated with acute salpingitis (14). In five of the cases described here, we were unable to isolate strictly anaerobic bacteria from the Fallopian tube specimens. Results were the same in an earlier study of 20 other cases (6).

In an earlier study, we found evidence implicating *M. hominis* as one of the etiological agents of acute salpingitis (7). In that study, *M. hominis* was isolated from tubal specimens in 8% of the cases studied, whereas in the present study *M. hominis* was isolated by tubal culture from one of eight patients.

Ten of the 12 cervical cultures for *C. trachomatis* were spoiled by bacterial or mycotic contamination. During the period covered by the present study, we used streptomycin and vancomycin in the transport medium. This might explain the high contamination rate (15). It is now our experience that the use of gentamicin in the transport medium reduces the contamination rate significantly. When culturing the specimens, we used gentamicin in the culture medium, but this was apparently not sufficiently effective to keep the contamination rate at a reasonably low level. Whether the new isolation technique (Ripa and Mårdh, p. 323, this volume) for the culture of *C. trachomatis* used in this study constituted any advantage for tubal cultures could not be evaluated, as no comparative studies with earlier-described culture techniques were performed. It is our experience that this new technique is at least as sensitive for the isolation of chlamydia from the urethra of men as is the use of 5-iodo-2-deoxyuridine-treated McCoy cells (16).

Patient A.H. had, during the 6 months prior to the salpingitis, three episodes of cervicitis. On each occasion, she was culture negative for gonococci. As no cultures for *C. trachomatis* were made on these occasions, it is not known whether *Chlamydia* was associated with these infections. However, the antibody response during the salpingitis bout indicated an acute infection with *C. trachomatis*. As we were unaware of this *C. trachomatis* infection at the time of the salpingitis, the patient was treated with ampicillin, a therapy not optimal for the treatment of chlamydial infections. Whether a persisting chlamydial infection caused her readmission 2 months later is not known.

The knowledge that *C. trachomatis* may be associated with acute salpingitis further stresses the necessity of laboratory facilities for the diagnosis of chlamydial infections. The risk of spread of a cervical chlamydial infection to the tubes must be considered in the treatment of cervicitis.

In conclusion, the present study would seem to incriminate *C. trachomatis* as an etiological agent in acute salpingitis.

ACKNOWLEDGMENTS

This investigation was supported by grant B76-16X-04509 from the Swedish Medical Research Council and by Public Health Service research grant EY-00219 from the National Eye Institute.

REFERENCES

1. Holmes KK, Handsfield HH, Wang SP, et al: Etiology of nongonococcal urethritis. N Engl J Med 292:1199–1205, 1975
2. Oriel JD, Reeve P, Thomas BJ, et al: Infection with Chlamydia group A in men with urethritis due to *Neisseria gonorrhoeae*. J Infect Dis 131:376–382, 1975
3. Dunlop EMC, Vaughan-Jackson JD, Darougar S, et al: Chlamydial infection: incidence in "non-specific" urethritis. Br J Vener Dis 48:425–428, 1972
4. Hilton AL, Richmond SJ, Milne JD, et al: Chlamydia A in the female genital tract. Br J Vener Dis 50:1–10, 1974
5. Swanson J, Eschenbach DA, Alexander ER, et al: Light and electron microscopic study of *Chlamydia trachomatis* infection of the uterine cervix. J Infect Dis 131:678–687, 1975
6. Weström L, Mårdh PA: Acute salpingitis. Aspects on aetiology, diagnosis, and prognosis, p 157–167. *In* D Danielsson, L Juhlin, and PA Mårdh (ed), Genital

infections and their complications. Almqvist & Wiksell International, Stockholm, 1975

7. Mårdh PA, Weström L: Tubal and cervical cultures in acute salpingitis with special reference to *Mycoplasma hominis* and T-strain mycoplasmas. Br J Vener Dis 46:179–186, 1970
8. Weström L: Effect of acute pelvic inflammatory disease on fertility. Am J Obstet Gynecol 121:707–713, 1975
9. Holdeman LV, Moore WEC (ed): Anaerobe laboratory manual, 3rd ed. Virginia Polytechnic Institute and State University, Blacksburg, Va., 1975
10. Gordon FB, Dressler HR, Quan AL, et al: Effect of ionizing irradiation on susceptibility of McCoy cell cultures to *Chlamydia trachomatis*. Appl Microbiol 23:123–129, 1972
11. Gästrin B, Kallings LO, Marcetic A: The survival time for different bacteria in various transport media. Acta Pathol Microbiol Scand 74:371–380, 1968
12. Wang SP, Grayston JT: Human serology in *Chlamydia trachomatis* infection with microimmunofluorescence. J Infect Dis 130:388–397, 1974
13. Wang SP, Grayston JT: Local and systemic antibody response to trachoma eye infection in monkeys, p 217–232. *In* RL Nichols (ed), Trachoma and related disorders caused by chlamydial agents. Excerpta Medica, Amsterdam, 1971
14. Eschenbach DA, Buchanan TM, Pollock HM, et al: Polymicrobial etiology of acute pelvic inflammatory disease. N Engl J Med 293:166–171, 1975
15. Wentworth BB: Use of gentamicin in the isolation of subgroup A *Chlamydia*. Antimicrob Agents Chemother 3:698–702, 1973
16. Wentworth BB, Alexander ER: Isolation of *Chlamydia trachomatis* by use of 5-iodo-2-deoxyuridine-treated cells. Appl Microbiol 27:912–916, 1974

Epidemiology, Etiology, and Prognosis of Acute Salpingitis: a Study of 1,457 Laparoscopically Verified Cases

L. WESTRÖM AND P.-A. MÅRDH

Acute nontuberculous salpingitis (synonym: pelvic inflammatory disease) is a common disease and seems to be on the increase. It generally afflicts women in the first half of their reproductive period. The incidence of acute salpingitis is difficult to assess, not in the least because of the variety of criteria used to define the condition.

DEMOGRAPHIC DATA

The investigation presented here has concerned patients who have been treated for acute salpingitis at the Department of Obstetrics and Gynecology, University Hospital, Lund, Sweden, during 1960–1974. This department is the only gynecological clinic serving a mixed rural and urban region in Malmöhus County in southern Sweden. In this region, the registered population which consults this hospital totaled about 175,000 in 1960, increasing to about 205,000 in 1974. Of this population, 12% live in rural districts and 88% in urban areas (1970). The age group under investigation—women between 15 and 39 years of age—comprise 16.7 to 17.1% of the population. The area includes a town with a university which had 2,900 female students in 1960, increasing to 7,600 in 1970, and thereafter decreasing to 6,200 in 1974. An estimation of the demographic differences of the three 5-year periods 1960–1964, 1965–1969, and 1970–1974 in the women (students included) in the area, with the mean number of women in the various age groups in 1960–1964 indexed at 100, is given in Table 1.

DIAGNOSIS AND INCIDENCE

The conventional diagnosis of acute salpingitis is associated with great uncertainty, when based on the physician's interpretation of the anamnestic data, the physical examination of the patient, the body temperature, and the erythrocyte sedimentation rate (ESR) (1). The uncertainty surrounding the clinical criteria used for the diagnosis of the disease has caused a demand for reliable methods for diagnosing acute salpingitis. Direct inspection of the Fallopian tubes by laparoscopy is one such means (2).

Since 1960, laparoscopy has been routinely applied at this clinic to verify a presumptive clinical diagnosis of acute salpingitis in more than 2,000 cases. Laparoscopy was performed within 24 h of the patients' admission and before treatment was instituted.

Of 814 cases given a provisional diagnosis of acute salpingitis based on clincial findings, laparoscopy revealed other diseases in 12.0% and normal intrapelvic organs in 22.5%. The latter group—the visually normal group—presented

TABLE 1. *Demographic changes in the female population between 15 and 39 years of age in the region served by the Department of Obstetrics and Gynecology, Lund, and the corresponding changes in the number of patients hospitalized because of acute salpingitis*[a]

Age group (years)	Total female population			Salpingitis patients		
	1960–64	1965–69	1970–74	1960–64	1965–69	1970–74
15–19	100	101	99	100	109	108
20–24	100	122	140	100	155	197
25–29	100	126	145	100	163	167
30–39	100	99	105	100	89	107

[a] The mean number of women in the various age groups during 1960–1964 is indexed at 100.

symptoms and signs which differed significantly in only a few respects from verified cases of acute salpingitis (3). The "classic" clinical picture of acute salpingitis as described in most textbooks, i.e., pelvic pain, increased vaginal discharge, ESR > 15 mm/h, tender adnexal swellings, was present in only 20.7% of the visually verified cases of acute salpingitis.

The mean and median durations of abdominal pain prior to the laparoscopy, as reported by the patients, were 9.8 and 7 days, respectively. The duration of abdominal pain before admission is summarized in Table 2.

The annual number of patients with acute salpingitis treated at this clinic has increased during the past 15 years (Fig. 1). The mean age of the patients in the 5-year periods showed an insignificant increase from 21.8 years in 1960–1964 to 22.4 years in 1970–1974 (Table 3). A comparison of the demographic changes in the region with the number of salpingitis patients treated indicates an unchanged incidence of acute salpingitis in the age groups from 15 to 19 and 30 to 39 years, whereas a significant ($P < 0.001$) increase can be seen in women in the age group 20 to 29 years (Table 1).

ETIOLOGY

It is generally considered that the tubal infection in acute salpingitis in women 15 to 45 years of age, except in a minority of cases, is caused by an infection ascending from the lower genital tract, usually the cervix (4). In a minority (1%) of cases of salpingitis in women of fertile age, the tubal infection is secondary to an infection of

TABLE 2. *Reported period of abdominal pain before admission in patients with acute salpingitis*

Days of abdominal pain before admission	Percentage of cases studied
1–3	31.4
4–6	14.9
7–9	19.0
10–12	8.3
13–15	9.9
16–21	9.9
21	6.6

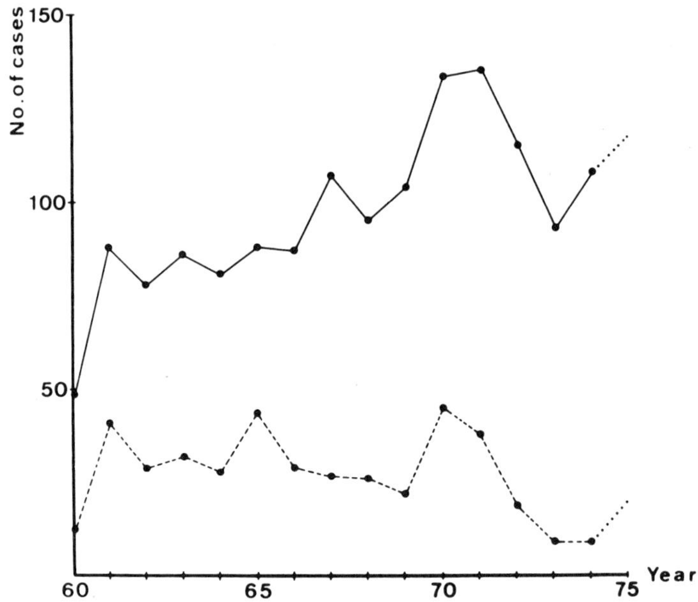

FIG. 1. *Yearly number of patients treated for acute salpingitis (solid line) and yearly number of salpingitis patients with gonorrhea (dotted line) at the Department of Obstetrics and Gynecology, Lund, Sweden, during 1960–1974.*

intrapelvic structures close to the uterine tubes, usually the appendix (5). Acute salpingitis is a rarity in women who are not sexually active (6).

It is common knowledge that cultures of cervical specimens in patients with acute salpingitis yield a variety of microorganisms. Thus, the results of cultures from the cervix cannot be used to establish the cause of an infection of the Fallopian tubes. Opening of the cervical canal or the introduction of foreign material into the uterine cavity facilitates an ascending spread of an infection from the cervix. Such procedures include abortion, exairesis, curettage, hystero-salpingography, tubal insufflation, and insertion of an intrauterine contraceptive device (IUD).

In the 623 salpingitis cases treated at our clinic during 1960–1967, such factors could have predisposed for the tubal infection in 14% of the patients. In these cases, any one of the above-mentioned procedures preceded by less than 4 weeks the laparoscopy that revealed the acute salpingitis. Exairesis in spontaneous abortions

TABLE 3. *Patients with laparoscopically verified acute salpingitis during 1960–1964, 1965–1969, and 1970–1974*

Period	Total no. of patients	Mean age (years)	Patients with cervical gonorrhea		
			No.	Percent	Mean age
1960–1964	382	21.8	142	37.2	19.5
1965–1969	481	21.8	148	30.8	20.2
1970–1974	584	22.4	120	20.5	21.7

TABLE 4. *Use of intrauterine contraceptive devices (IUDs) in salpingitis patients from whom gonococci were and were not isolated from the lower genital tract*

Gonococci	IUD users		Nonusers
	Copper IUD	Polythene IUD	
Isolated	13	4	65
Not isolated	84	24	325
Total	97	28	390

was most common (49 cases, i.e., 8%), whereas IUD insertion preceded the onset of the disease in 6 cases (1%) and legal abortions in 13 cases (2%). In the 525 salpingitis cases treated during 1971–1975, one of the above-mentioned factors preceded the onset of symptoms of acute salpingitis by less than 4 weeks in 11.5% of the cases. In this series, IUD insertions were most common (5.8%), followed by legal abortion (3.5%) and exairesis after spontaneous abortions (2%).

The frequency of women wearing an IUD in the above-mentioned 525 salpingitis patients was compared with the corresponding frequency in 741 control women, matched to the patients as to date of birth, domicile, and socioeconomic status. Using statistical methods for risk estimation, we estimated that IUD usage increased the risk of acquiring acute salpingitis threefold compared with nonusage (7). In women who had never been pregnant, the corresponding risk was sevenfold. In salpingitis patients wearing an IUD, the frequency of cervical gonorrhea was the same as in women not using an IUD, regardless of age. No difference in the frequency of gonorrhea was noted between salpingitis patients using a copper IUD and those using polythene devices (Table 4) (7).

The frequency of gonorrhea was the same in salpingitis patients reporting use of oral contraceptives as in those reporting no use of oral contraceptives at the onset of their salpingitis.

The frequency of salpingitis cases in which gonococci could be cultured from the lower genital tract varied between 50.0 and 8.3% per year. From 1965 to 1974, a discontinuous, but significant ($P < 0.001$), decrease in the percentage of salpingitis cases with gonorrhea was noted (Fig. 1). This was exclusively due to a decrease in the isolation rate of gonococci in the women < 20 years of age. During the same period, there was a numerical, but not significant, increase in the number of patients >19 years of age who had salpingitis and gonorrhea. It should be noted that the culture technique used for the isolation of gonococci was principally the same since 1960.

The changes in the frequency and age distribution of the salpingitis patients with gonorrhea were compared with the changes in the age distribution of the female population at large in the region served by this clinic during the three 5-year periods 1960–1964, 1965–1969, and 1970–1974. It was found that the incidence of salpingitis cases with gonorrhea seen at our department in the 15- to 19-year group during this period decreased, i.e., from 2.9 to 1.3 per 1,000 women per year ($P < 0.001$). In the 20- to 29-year group, the corresponding incidence has increased from 0.5 to 0.7 per 1,000 women per year (not significant). As a consequence of this, the mean age of the salpingitis patients with cervical gonorrhea increased from 19.5 years in 1960–1964 to 21.7 years in 1970–1974. As a comparison, the changes in the

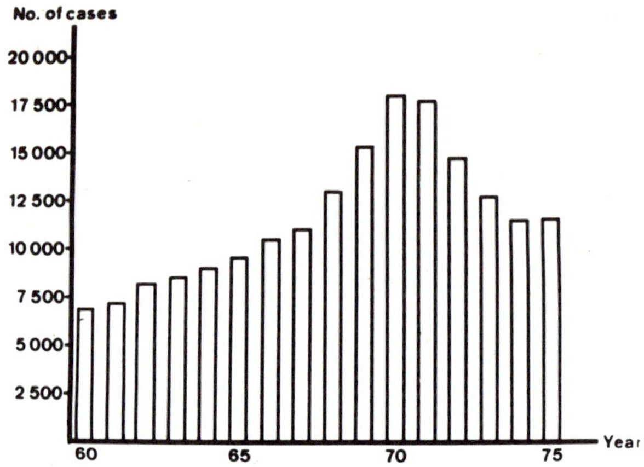

FIG. 2. *Reported number of cases of gonorrhea in women in Sweden during 1960–1975.*

annual number of reported cases of gonorrhea in females in Sweden during 1960–1975 are shown in Fig. 2.

Other bacteria, including strictly anaerobic bacteria, have also been associated with tubal infections (5, 8). It is our belief that bacteria, including strictly anaerobic bacteria, are responsible for a proportion of severe adnexal infections with abscess formation, but play a minor role in mild and moderately severe cases of the primary "venereal" type which constitutes the majority of salpingitis cases.

Other microorganisms, such as mycoplasmas (9) and chlamydiae (Mårdh et al., p. 77, this volume), may be isolated from the Fallopian tubes in women with acute salpingitis. These findings as well as serological studies suggest that these organisms are responsible for a proportion of salpingitis cases; this is further discussed by Mårdh et al. elsewhere in this volume (p. 77).

PROGNOSIS

The late sequelae of acute salpingitis are well known: involuntary childlessness, chronic abdominal pain, and an increased frequency of ectopic pregnancies.

In the earlier-mentioned 623 cases of salpingitis treated in 1960–1967, the disease afflicted 539 women. A follow-up study of these women was performed in 1974 (10).

TABLE 5. *Distribution of women who became pregnant and of those who were voluntarily or involuntarily childless after one or more infections, in salpingitis and control groups*

Group of women	Salpingitis		Controls		Difference
	No.	Percent	No.	Percent	P
Pregnant	263	63.4	84	84.0	<0.001
Not pregnant					
Voluntarily	64	15.4	13	13.0	>0.5
Involuntarily	88	21.2	3	3.0	<0.001
Total	415		100		

TABLE 6. *Number of women who became pregnant and those who were involuntarily childless because of tubal occlusion after one, two, and three or more tubal infections (follow-up period >9.5 years)*

Group	Acute salpingitis			
	Once	Twice	Three or more times	Total
Pregnant	217	40	6	263
Involuntarily childless	32 (12.9%)	22 (35.5%)	18 (75.5%)	72

Thirty-five women over 35 years of age at the onset of the disease and 19 women who had genital malformations or other disorders that rendered pregnancy improbable already at the time they had their salpingitis were excluded from the follow-up study. In the study of the remaining 485 women, the mean observation time was 9.5 years. In 70 cases, the information obtained was unreliable, or the women could not be traced. The remaining 415 women were analyzed and compared with 100 carefully selected age-matched controls who had never had acute salpingitis. The controls had been subjected to laparoscopy between 1960 and 1967 because of abdominal pain and infection of the lower genital tract. In all, laparoscopy had revealed normal conditions.

During the observation period, which ranged from 6 to 14 years after the salpingitis, 15.4% of the patients who had had salpingitis and 13% of the controls were voluntarily childless (not significant). Involuntary childlessness due to tubal occlusion, i.e., most probably a sequela of their salpingitis, was found in 72 (17.3%) of the 415 women with acute salpingitis. All the remaining involuntarily childless women with acute salpingitis had normal Fallopian tubes (Table 5). After exclusion of the voluntarily childless women, 72 (20.5%) of 351 women were involuntarily childless as a result of tubal occlusion. The frequency of tubal occlusion after acute salpingitis varied with the number of tubal infections that had afflicted the woman. Tubal damage, sufficient to render pregnancy impossible, was found in 12.8, 35.5, and 75% after one, two, or three or more tubal infections (Table 6).

The fertility prognosis was significantly better after mild than after moderate or severe tubal infections. The frequencies of tubal occlusions were 2.6, 13.1, and 28.6%, respectively.

In the patients with acute salpingitis and gonorrhea, the fertility prognosis was significantly ($P < 0.01$) better (6.1% tubal occlusion) than in patients in whom gonococci were not recovered (17.3% tubal occlusions) (Table 7).

TABLE 7. *Fertility after one episode of laparoscopically verified acute salpingitis with and without cervical gonorrhea (mean follow-up time, 9.5 years)*

Group	Acute salpingitis		
	With gonorrhea	Without gonorrhea	Total
Pregnant	93	124	217
Involuntarily not pregnant (tubal occlusion)	6	26	32
Total	99	150	249

Abdominal pain lasting for longer than 6 months, and causing the woman to seek medical advice, was reported by 18.1% of those women who had had acute salpingitis. The corresponding figure for the controls was 5%.

The ratio of ectopic to intrauterine pregnancies in the salpingitis patients during the observation period was 1:24. The corresponding ratio in the controls was 1:147.

To sum up, acute salpingitis is a disease on the increase, and every fourth woman afflicted by the disease suffers from sequelae.

ACKNOWLEDGMENT

This investigation was supported by grant B76-16X-04509 from the Swedish Medical Research Council.

REFERENCES

1. Jacobson L: Laparoscopy in the diagnosis of acute salpingitis. Acta Obstet Gynecol Scand 43:160–174, 1964
2. Sjövall A: Laparoskopi. K Fysiogr Saellsk Lund Foerh 32:131–135, 1962
3. Jacobson L, Weström L: Objectivized diagnosis of acute pelvic inflammatory disease. Am J Obstet Gynecol 105:1088–1098, 1969
4. Falk HC: Interpretation of the pathogenesis of pelvic infections as determined by cornual resection. Am J Obstet Gynecol 52:66–73, 1946
5. Weström L, Mårdh P-A: Acute salpingitis—aspects on aetiology, diagnosis, and prognosis, p 157–167. *In* D Danielsson, L Julin, and P-A Mårdh (ed), Genital infections and their complications. Almqvist & Wiksell International, Stockholm, 1975
6. Stemmer W: Ueber die Ursachen von Eileiterentzuendungen. Zentralbl Gynaekol 63:1062–1067, 1941
7. Weström L, Bengtsson LP, Mårdh P-A: IUD and PID. Estimations of the risk of acquiring pelvic inflammatory disease in women using intra-uterine contraceptive devices as compared to non-users. Lancet ii:221–224, 1976
8. Eschenbach DA, Buchanan TM, Pollock HM, et al: Polymicrobial etiology of acute pelvic inflammatory disease. N Engl J Med 293:166–171, 1975
9. Mårdh P-A, Weström L: Tubal and cervical cultures in acute salpingitis with special reference to *Mycoplasma hominis* and T-strain mycoplasmas. Br J Vener Dis 46:179–186, 1970
10. Weström L: Effect of acute pelvic inflammatory disease on fertility. Am J Obstet Gynecol 121:707–713, 1975

Lymphogranuloma Venereum and Other Nonocular *Chlamydia trachomatis* Infections

JULIUS SCHACHTER

Lymphogranuloma venereum (LGV) is a venereally transmitted systemic chlamydial disease. The agent causing LGV is almost always a *Chlamydia trachomatis* strain (1), although a *C. psittaci* isolate has been found in one instance to be associated with an LGV-like disease involving inguinal lymphadenopathy (2). The agents causing this disease are more invasive in vivo and in vitro than the related trachoma-inclusion conjunctivitis (TRIC) agents. Currently, three serotypes of LGV isolates are recognized, with L_2 apparently being the most common (3, 4).

LGV has been recognized as a clinical disease for well over 100 years. Early in this century, the disease was named lymphogranuloma by Durand, Nicolas, and Favre (1913), who noted its histological similarity to lymphadenomas. A major advance in the differential diagnosis of LGV from other sexually transmitted infections involving inguinal lymphadenopathy was the introduction of the Frei test, a specific skin test positive in patients with LGV.

LGV was once quite common, and its prevalence seems to have declined as a result of increasing modernization rather than as a reflection of specific control campaigns. The disease is worldwide in distribution, although apparently more prevalent in warmer climates. Infectivity is unknown but probably low, and it is possible that broad use of chemotherapeutic agents may have suppressed chlamydial levels sufficiently to reduce incidence.

Widespread serological screening tests have indicated the high prevalence of either asymptomatic LGV or previous exposure to LGV in sexually active populations. However, those tests were performed before our current state of knowledge concerning the common nature of the related TRIC agent genital tract infections. Thus, the many papers that refer to positive complement fixation (CF) reactor rates of 10, 20, or 30% in studies conducted in various venereal disease clinics or sexually active populations must be reinterpreted. The same holds true for results gained by using the Frei test. For example, King et al. (5) found that 18.4% (206 of 1,119) of their patients at a venereal disease clinic had positive skin tests and 2% (24) had high CF titers, but only 10 had clinical signs of LGV. Their positive reactions probably do not reflect LGV infection; many are most likely signs of exposure to the chlamydiae that cause nongonococcal urethritis and cervicitis.

CLINICAL MANIFESTATIONS OF LGV

The first consideration in differential diagnosis of LGV is the possibility that the patient may have another venereal disease with similar clinical symptomatology. But over and above the problem of a specific diagnosis is the consideration that many patients with LGV do have other venereal diseases; thus, proper clinical work-up will include at the very least tests for gonorrhea and syphilis. Gonorrhea has been diagnosed in a significant proportion of these patients, and in some studies up to one-third of the patients with LGV had syphilis. In Alergant's series, 11 of 55 patients had other venereal diseases, and Sonck recorded 192 of 810 cases with two or more venereal diseases (6, 13).

The clinical course of LGV may be divided into three stages: primary, involving the initial infection and early lesions; secondary, involving the regional lymph nodes; and tertiary, with late sequelae.

Initial Infection: the Primary Lesion

After an incubation period that may be extremely variable (ranging from 3 to 30 days), a primary lesion may develop on the genitals. In males, the lesion is commonly on the penis, often on the glans; in women it is often on the vaginal wall or on the labia, and occasionally on the cervix. The primary lesion is usually described as a transient, evanescent lesion. It is painless and may take one of four forms: a papule, a shallow ulcer or erosion, a herpetiform lesion, or a nongonococcal urethritis (possibly due to an intraurethral primary lesion). Occasionally, primary lesions are found on extragenital sites, which indicate the primary site of infection. For example, lesions have been described on the fingers and the tongue.

Histopathologically, no specific finding is associated with the primary lesion. There is a complete loss of epithelial cells in the ulcerated form, with a relatively nonspecific inflammatory response in the surrounding tissue. However, the primary lesion may heal quite rapidly, and in some cases scars may be found only after careful examination of the genitalia. It should be noted that in recent times the primary lesion has usually not been found. For example, Alergant (6) elicited a history of primary lesions in only 23.6% (13 of 55) of his cases in Liverpool, England, and Sigel (7) elicited such a history in 26.5% (39 of 147); Abrams (8) saw them in 40% (8 of 20) of his patients in San Francisco, Calif. Only two of Alergant's patients had primary lesions at their initial examination, and, similarly, Erskine (9) found only 4 of 61 patients with a primary lesion. Obviously, the primary lesion, if present, is not considered serious enough by the patient to require medical attention. It may have completely healed before the sequential development of the disease brings the patient to the physician.

Secondary Lesions: Lymphadenitis or Lymphadenopathy

Most patients who are subsequently diagnosed as having LGV come to venereal disease clinics seeking medical attention because of their enlarging buboes, some because of pain (although the bubo can be painless).

Lymphadenopathy is unilateral by a ratio of approximately 2:1 over bilateral and follows the primary lesion by a period of a few days to weeks. Since the primary lesion is not observed in many cases, those attempting to establish an incubation period have usually relied on the patient's recollection of the time of the presumed infecting sexual contact. Most patients develop lymphadenopathy in 2 to 6 weeks after the presumed infection, although incubation periods have ranged from a few days up to 6 months or longer. In general, many series of lymphadenopathy show a

9:1 or greater ratio of males to females. This does not mean that lymphadenopathy actually occurs at this ratio, but it does reflect a possible sex difference in anatomic sites of lymph node involvement. One (not wholly satisfactory) explanation is based on the difference in distribution of pelvic lymphatics in the male and the female. This theory presumes that a primary penile lesion in male patients or an external primary lesion in female patients, drained by inguinal lymph nodes, will result in inguinal lymphadenopathy, but that women whose primary lesions are intravaginal will have posterior drainage of the lesion, thus involving the retroperitoneal lymph nodes.

In the bubonic form the involved lymph nodes are usually inguinal, although several lymph node systems may be involved. When both inguinal and femoral lymph nodes are involved, they may be separated by the so-called groove sign, which corresponds with the separation of these two lymph node systems by the inguinal ligament. The groove sign is considered pathognomonic of LGV (10), but only occurs in 15 to 20% of the cases. In instances where primary lesions are extragenital, other sites of lymph node involvement (such as axillary or subaxillary) may be seen.

Early in the secondary stage, a single lymph node may be involved, but inflammation may extend to an entire chain that then becomes matted with considerable periadenitis. Epithelial and endothelial cells proliferate, and clusters of large mononuclear cells and small abscesses are found in the lymph nodes. The abscesses expand to become necrotic foci which may vary, depending upon the lymph nodes involved. Suppuration may follow; the nodes may become fluctuant and then harden. The course of lymph node involvement is highly variable: An acute lymphadenopathy in some patients is followed by spontaneous remission; others undergo spontaneous rupture of the suppurative lymph node and develop draining fistulae; and in a relatively small percentage of patients (approximately 5%) this lesion becomes a chronic lymphadenopathy, which may persist for many years. Occasionally, patients with buboes progress, in approximately 5 to 10 years, to the tertiary stage of LGV.

If deep iliac lymph nodes are involved, an initial presumptive diagnosis of appendicitis may be made, although in our experience a more common misdiagnosis is inguinal hernia. We have tested specimens from four patients whose buboes were recognized at a presumed herniorrhaphy. This diagnostic confusion probably reflects the decreasing clinical experience with LGV.

Although it can be beneficial to aspirate fluctuant buboes in order to reduce risk of spontaneous rupture, surgical intervention (incision of nodes) is contraindicated.

In the acutely ill patient, fever, headache, and myalgia are common. White cell counts are often normal, although occasionally mild leukocytosis with normal differential is seen.

Tertiary Stage or the Genito-Ano-Rectal Syndrome

The term "tertiary" is slightly misleading because this chronic LGV manifestation usually has no recognized antecedent lymphadenopathy. The tertiary stage covers a variety of conditions, generally resulting from progressive spread of the disease, with corresponding changes and destruction of tissue in the involved areas. This condition often has a predominately female distribution, perhaps reflecting the involvement of retroperitoneal lymphatics as well as the proximity of the rectal and vaginal tissue. A variety of rectal strictures are found, involving small areas just above the rectum or occasionally involving extensive areas of the lower intestinal tract. Proctitis without strictures, which may occur early, is probably the first objective sign in the anorectal syndrome and usually persists as strictures develop.

Esthiomene (Gr.: eating away) is the misnomer applied to the hypertrophic, chronic granulomatous condition of the vulva occurring in LGV where the organ is quite swollen and disfigured, often with associated fistulae and ulcers. Elephantiasis may occur in both the male and the female.

Systemic Complications and Extragenital Infection

There are a number of secondary complications as well as extragenital LGV infections. These sometimes include a conjunctival infection causing blocked lymphatics and regional adenopathy that may mimic Parinaud's oculoglandular syndrome. Buboes may form in lymph nodes draining any primary site of infection, such as axillary buboes after infection of the fingers and submaxillary buboes after oral infection. Arthritis, meningitis, and cutaneous manifestations have been reported (11).

We have observed some of these complications in a series of 63 patients presumed to have LGV. However, only in one case, that of a febrile woman with pneumonitis where we isolated an LGV agent from the blood, did we feel secure in attributing the "complication" to LGV infection. In the past, when more LGV cases were seen, there was greater opportunity to observe the uncommon complication. This is particularly true of the tertiary stage, which is rarely reported today. Most of the LGV cases in our experience are young men (early twenties, 15:1 sex ratio) who contracted the infection in Southeast Asia or Central America and were usually seeking medical attention for their inguinal lymphadenopathy. Many of the proven LGV cases had constitutional symptoms (fever and chills, headache, myalgias, etc.).

We have seen two proven LGV infections in elderly individuals who had not had sexual contact for many years. Neither had a history of LGV, and both developed acute disease and seroconverted. These observations may suggest persistent subclinical infection with this agent over a long period of time. Some imbalance may activate the infection and result in disease.

We must strongly caution against over attribution of clinical findings to LGV based on positive CF or Frei tests. Our current confirmation of the nonspecificity of these tests due to the prevalence of genital tract chlamydial infections in sexually active individuals urges us to a more critical analysis of some previous reports based on these tests.

CHEMOTHERAPY

It should be noted immediately that a truly specific, highly effective method of treating LGV is not available. One of the problems involved in assessing the effects of chemotherapy is the highly variable natural history of the disease. Spontaneous remission does occur.

No extensive controlled double-blind treatment trials have been done. Only one major effort at such a trial was attempted, which could be faulted only for the relatively small number of patients studied. Greaves et al. in 1957 (12) evaluated three antibiotics (chloramphenicol, oxytetracycline, chlortetracycline), sulfadiazine, or symptomatic therapy with aspirin in a group of patients with LGV. Recognizing the highly variable course of the disease and their inability to predict complications, they decided to measure only progression or remission of the buboes and to follow CF tests on the patient's sera. The duration of the buboes of the 25 patients who finished the course of antimicrobial treatment was significantly less (albeit slightly) than the duration in the control group (31 days versus 69 days). The investigators felt that all the drugs were somewhat beneficial, with an indication that perhaps sulfadiazine or oxytetracycline might be somewhat better than the other two drugs, but the number tested was too small to provide conclusive results. In addition, there appeared to be a somewhat greater rate of complications in the symptomatically treated patients as compared with those receiving antimicrobial drugs (but again the number tested was too small). Antibody titers spontaneously decreased in 7 of 17 patients, whereas all those receiving antimicrobial therapy showed some decline in antibody level 1 year after treatment.

In general, the feeling has been that treatment early in the disease has a better chance of success than treatment initiated at a more chronic stage. Many authors claim that essentially all their patients get better with whatever drug they happen to be using, but careful reading of the literature indicates that in virtually all series a few cases have required two, three, or more courses of therapy.

It seems reasonable to treat LGV with tetracycline (1 to 2 g/day for 4 weeks) rather than sulfonamides because relative sulfa resistance has been recognized for many years.

LABORATORY DIAGNOSIS

Although the Frei test is the classic method of diagnosing LGV, it has been falling into disfavor. The chlamydial group CF test remains a useful serological test. Although there is usually agreement between the CF titer and the Frei test, some workers consider the CF test a more specific and sensitive index of LGV infection. Thus, Alergant found high CF titers in most of his patients (53 of 54), although 16 had negative Frei tests; Sigel found that 97% of his LGV patients had high CF titers. Our experience supports this concept; we have never recovered chlamydiae from a bubo or other site in a patient with a negative CF test but have from patients with negative Frei tests (14). There may be a significant quantitative difference in CF titers in patients with LGV compared to patients with the trachoma or TRIC agent oculogenital infections. For example, in a simple uncomplicated TRIC agent infection, either of the eye or of the eye and genital tract, we have never seen male patients with CF titers above 1:64, whereas at least half of the patients with LGV have CF titers in excess of 1:64, usually 1:256 or higher. Positive CF titers of patients with TRIC agent infections are usually 1:8 and 1:16, and at least 50% of the patients do not have CF titers this high. Patients with LGV also show high titers by micro-immunofluorescence test (15, 16).

LGV agents may be isolated in the yolk sac of embryonated hen eggs (14) or by tissue culture procedures described elsewhere in this monograph. The most productive clinical material is pus aspirated from the fluctuant node.

TRIC AGENT INFECTION IN SITES OTHER THAN THE EYE AND GENITAL TRACT

Although the LGV infections are recognized to be systemic, the TRIC agent infections are generally considered to be superficial and restricted to the epithelium of the conjunctiva, the urethra, the cervix, and the rectum. It has been known for many years that infants with inclusion blennorrhea produce copious numbers of chlamydiae and that passages draining from the eye may yield chlamydial isolates. Thus, Freedman and associates (17) and Scott (18) have been able to recover chlamydiae from the throats and nasal discharges of infants with inclusion blennorrhea. Similar recoveries of organisms have been made from the throats of adults with inclusion conjunctivitis (19). Recently, however, chlamydiae were recovered from the throat of a patient with pharyngitis who did not have any history of conjunctivitis (20). Thus, the possibility that primary chlamydial infections of the throat may occur appears quite probable.

Further evidence of the potential for respiratory spread of chlamydiae has been the recent observation of the development of pneumonitis in an infant after successful therapy for inclusion blennorrhea (21). The recent observations of Beem and Saxon, correlating a distinctive pneumonia syndrome of infants with *C. trachomatis* infection, are presented elsewhere in this volume (p. 153). It must be noted that the throat and the lungs contain cells histologically quite similar to the epithelial cells of the ocular and genital tracts. Therefore, this apparent expansion of the clinical syndromes associated with the TRIC infections does not represent an expansion of the host cell range for this organism. These findings lead to the obvious speculation that TRIC agents may be capable of causing disease in any anatomic site containing the columnar epithelial cells. Thus, if such cells are found in the intestinal tract, one could conceive that chlamydiae, if transmitted to the intestines, might cause infection and disease. It is not known whether chlamydiae or TRIC agents are capable of infecting ciliated columnar epithelium but this, of course, would be extremely pertinent both in terms of the respiratory tree and with relevance to pelvic inflammatory disease. One may speculate that it could be rewarding to seek chlamydial infection and disease involving anatomic sites with similar cell types to those recognized as hosts for TRIC agent infection.

ACKNOWLEDGMENTS

This study was supported by Public Health Service grant AI-10193 from the National Institute of Allergy and Infectious Diseases.

REFERENCES

1. Schachter J, Meyer KF: Lymphogranuloma venereum. II. Characterization of some recently isolated strains. J Bacteriol 99:636–638, 1969
2. Schachter J: A Bedsonia isolated from a patient with clinical lymphogranuloma venereum. Am J Ophthalmol 63:1049–1053, 1967
3. Wang, S-P, Grayston JT: Immunologic relationship between genital TRIC, lymphogranuloma venereum, and related organisms in a new microtiter indirect immunofluorescence test. Am J Ophthalmol 70:367–374, 1970
4. Treharne JD, Davey SJ, Gray SJ, et al: Immunological classification of TRIC agents and of some recently isolated LGV agents by the microimmunofluorescence test. Br J Vener Dis 48:18–25, 1972

5. King AJ, Barwell CF, Catterall RD: Intradermal tests in the diagnosis of lymphogranuloma. Br J Vener Dis 32:209–216, 1956
6. Alergant CD: Lymphogranuloma inguinale in the male in Liverpool, England, 1947 to 1954. Br J Vener Dis 33:47–51, 1957
7. Sigel MM: Lymphogranuloma venereum. University of Miami Press, Coral Gables, Fla., 1962
8. Abrams AJ: Lymphogranuloma venereum. J Am Med Assoc 205:199–202, 1968
9. Erskine D: Lymphogranuloma venereum. A review of 61 cases. Br J Vener Dis 34:163–165, 1958
10. Greenblatt RB, Pund ER, Sanderson ES, et al: Management of chancroid, granuloma inguinale, lymphogranuloma venereum in general practice. US Public Health Serv Publ 255, revised 1964
11. Koteen H: Lymphogranuloma venereum. Medicine 24:1–69, 1945
12. Greaves AB, Hilleman MR, Taggart SR, et al: Chemotherapy in bubonic lymphogranuloma venereum. Bull WHO 16:277–289, 1957
13. Sonck CE: Lymphogranuloma inguinale, Klinische, epidemiologische und immunologische Aspekte. Hautarzt 23:280–286, 1972
14. Schachter J, Smith DE, Dawson CR, et al: Lymphogranuloma venereum. I. Comparison of Frei test, complement fixation test and agent isolation. J Infect Dis 120:372–375, 1969
15. Wang S-P, Grayston JT: Human serology in *Chlamydia trachomatis* infection with microimmunofluorescence. J Infect Dis 130:388–397, 1974
16. Philip RN, Casper EA, Gordon FB, et al: Fluorescent antibody responses to chlamydial infection in patients with lymphogranuloma venereum and urethritis. J Immunol 112:2126–2134, 1974
17. Freedman A, Khalaf Al-Hussaini M, Dunlop EMC, et al: Infection by TRIC agent and other members of the Bedsonia group, with a note on Reiter's disease. II. Ophthalmia neonatorum due to TRIC agent. Trans Ophthalmol Soc UK 86:313–320, 1966
18. Scott JG: TRIC infections (Bedsoniae). Med Proc 14:410–414, 1968
19. Dawson CR, Schachter J: TRIC agent infections of the eye and genital tract. Am J Ophthalmol 63:1288–1298, 1967
20. Schachter J, Atwood G: Chlamydial pharyngitis? J Am Vener Dis Assoc 2:12, 1975
21. Schachter J, Lum L, Gooding CA, et al: Pneumonitis following inclusion blennorrhea. J Pediatr 87:779–780, 1975

Role of the Genital Mycoplasmas in Extragenital Disease

WILLIAM M. McCORMACK

This paper will be devoted to a consideration of the role of the genital mycoplasmas, *Mycoplasma hominis* and *Ureaplasma urealyticum* (T-mycoplasmas), in human extragenital disease. The role of these organisms in nongonococcal urethritis, Reiter's disease, postpartum and postabortal endometritis, pelvic inflammatory disease, and disorders of reproduction has been considered in detail in recent reviews from this laboratory (1–4).

Most studies prior to 1955 and several later papers refer to PPLO (pleuropneumonia-like organisms). Most of these organisms were probably *M. hominis*, although *M. fermentans*, cell wall-deficient bacterial variants, pseudocolonies (artifacts of cultivation), and even *U. urealyticum* may have been included. I shall use the term PPLO in this review if this was the designation used in a cited study. If the organisms were speciated, the appropriate species designation will be used.

URINARY TRACT INFECTIONS

Early reports (5–7) described the isolation of PPLO from the urine of patients of both sexes with abacterial pyuria and/or acute hemorrhagic cystitis. Claus et al. (8) failed to isolate PPLO from patients with similar symptoms. Berg et al. (9) described 15 men with acute hemmorhagic cystitis. PPLO, but not bacteria, were isolated from the urine. Streptomycin apparently eliminated the organisms and produced clinical improvement. More recently, Pachas (10) described a young woman with hemorrhagic cystitis. *M. hominis* was isolated from urine, and when tetracycline was given there was symptomatic improvement.

Witzleb et al. (11) found that 12 of 75 patients with urinary tract infections and none of 71 healthy persons had 100,000 or more colony-forming units of *M. hominis* or *U. urealyticum* per ml of urine. Mycoplasmas were isolated from urine obtained by bladder puncture and directly from the renal pelvis in some cases. Other investigators (12) have also isolated *M. hominis* and *U. urealyticum* from urine specimens obtained by suprapubic aspiration.

More recently, Thompsen (13) isolated *M. hominis* from the upper urinary tract of three women and *U. urealyticum* from the renal pelvis of two women with chronic pyelonephritis. Mycoplasmas could not be cultivated from the upper urinary tracts of a control group of patients with noninfectious urinary tract disease. All three patients who harbored *M. hominis* in their upper urinary tracts showed signs of an acute exacerbation of pyelonephritis, and two of them developed a fourfold or greater change in antibody titer against *M. hominis*. Only one of the three had associated bacteriuria.

Pachas (10) studied a man with hematuria and pyuria. Renal biopsy showed focal glomerulonephritis and chronic focal pyelonephritis. Cultures of the kidney were negative for bacteria. *M. salivarium*, an oropharyngeal commensal, was isolated from cultures of the renal cortex. Cultures of a biopsy obtained 2 years later were negative for mycoplasmas. Tetracycline treatment was without noticeable benefit. A serological response to *M. salivarium* was not demonstrated.

In a recent study, Freidlander and Braude (14) were able to produce bladder stones in male (but not female) rats by injecting a human strain of *U. urealyticum* into the bladder or renal pelvis. Killed organisms also induced bladder stones in male rats.

Marshall et al. (15) described two women with abacterial urethritis characterized by dysuria and by leukocytes in the urethral washings ("first glass") but not in the midstream ("second glass") urine. Both patients responded to treatment with tetracycline, and the authors suggested an etiological role for mycoplasmas, although the patients they reported were apparently not cultured for mycoplasmas.

Since mycoplasmas can be found about the urethral orifices of normal men and women, the presence of these organisms in voided or catheterized urine and even in urine obtained by percutaneous aspiration of the bladder is difficult to interpret. Counting of colony-forming units has not been systematically evaluated and related to urinary tract infection. The data presently available suggest that *M. hominis* may cause some cases of hemorrhagic cystitis and may be associated with exacerbations of chronic pyelonephritis. The role of *U. urealyticum* in urinary infection is even less well defined. The significance of the isolation of *M. salivarium* from renal tissue by Pachas (10) is unclear, as is the relation of *U. urealyticum* to the formation of urinary calculi in humans.

NEONATAL INFECTIONS

Hoogendijk et al. (16) described a neonate who, during the first days of life, developed an abscess of the neck from which *M. hominis* was isolated. Sacker et al. (17) reported a similar case. Presumably, *M. hominis* from the maternal birth canal was inoculated into abraded skin during delivery. These reports led us to suggest that erythema toxicum neonatorum might be due to *M. hominis*. Accordingly, we cultured 30 typical lesions from infants with erythema toxicum neonatorum. All were negative for *M. hominis* (2).

Laughton (18) recovered *M. hominis* and a corynebacterium from the lung of a premature infant who died of respiratory distress syndrome. These organisms were used experimentally to produce similar lesions in the lungs of chick embryos. Brunell et al. (19) isolated *M. hominis* in pure culture from amniotic fluid obtained at caesarean section performed because of maternal fever 3 days after rupture of the membranes. The 2,000-g infant developed respiratory distress syndrome and died. Pathological examination of the lungs showed evidence of interstitial pneumonia; hyaline membranes were not seen. Postmortem cultures were not obtained. Both Laughton (18) and Brunell et al. (19) suggested that infection with mycoplasmas might play a role in some instances of respiratory distress syndrome. Further work will be needed to define the role, if any, of mycoplasmas in this condition.

WOUND INFECTIONS

Stokes (20) described a patient who developed a severe wound infection after hysterectomy. PPLO were isolated from the blood, and a significant rise in antibody titer was noted. The patient was treated with tetracycline and recovered. Lemcke and Csonka (21) noted that their H34 strain, later identified as *M. hominis* (22), was isolated by Stokes from an infected hysterectomy incision. Russell and Fallon (23) isolated *M. hominis* in pure culture from infected caesarean section wounds, and in one case antibody production was documented. Lee et al. (24) isolated *M. hominis* from two patients who developed wound infections after open reduction of mandibular fractures. Systematic studies by Freundt (25) of 53 cultures of pus of wounds and by Hayflick and Stanbridge (26) of 13 wounds and ulcers were negative for mycoplasmas.

RESPIRATORY TRACT INFECTIONS

U. urealyticum has been recovered from throat cultures of 2 to 9% of normal adults (27–30). *M. hominis* has also been isolated from throat cultures of normal adults (30–35; W. A. Clyde, Jr., Fed. Proc. 23:579, 1964). The demonstration (36) that *M. pneumoniae* causes cold agglutinin-positive primary atypical pneumonia led to a search for mycoplasmas in other respiratory infections. Mufson et al. (37) studied hospitalized adults and isolated *M. hominis* from 36 (12.9%) of 278 patients with pneumonia and from 51 (7.7%) of 702 patients without respiratory tract disease. They also found that 3.2% of patients with pneumonia and only 0.3% of patients without respiratory tract disease developed a serological response to *M. hominis*. Although other studies have not supported *M. hominis* as an important cause of upper or lower respiratory tract infection (38–41), the data of Mufson et al. suggest that *M. hominis* may be involved in some cases of pneumonia in adults.

Mufson et al. (42) inoculated volunteers with a strain of *M. hominis* which they had isolated from a patient with pneumonia. Of 50 volunteers, 45 were infected with *M. hominis* and 21 developed a mild afebrile exudative pharyngitis, some with cervical lymphadenopathy. Preexisting antibody to *M. hominis* protected against illness. Thus, *M. hominis* has the potential to cause exudative pharyngitis. However, several studies have failed to implicate this agent in naturally occurring pharyngitis (28, 34, 43, 44; Y. Crawford and J. P. Griffin, Bacteriol. Proc., M61, p. 48, 1966). Mufson has recently reviewed the role of *M. hominis* in pneumonia and pharyngitis (45).

Individuals who practice fellatio are more likely to complain of a sore throat than those who do not practice fellatio. Symptoms of pharyngitis, however, cannot be correlated with the isolation of *N. gonorrhoeae* from the pharynx (46). These observations, coupled with the report of Mufson et al. (42) that *M. hominis* caused pharyngitis when inoculated into volunteers, led us to suggest that *M. hominis* might be responsible for the symptoms of pharyngitis reported by patients who practice fellatio. In a study which is as yet unpublished, pharyngeal cultures from over 900 subjects were examined for genital mycoplasmas. Although pharyngeal colonization with both *M. hominis* and *U. urealyticum* was more prevalent among those who practiced fellatio than among those who did not, we were unable to associate the isolation of either species with symptoms of a sore throat.

Stokes (47) reported a 15-year-old girl who developed empyema following resection of a portion of the right lung. *M. hominis* was isolated in pure culture from the fluid aspirated from the empyema cavity. Therapy with chlortetracycline and drainage was effective, and the patient developed a significant rise in antibody titer to *M. hominis*. There have been no similar cases reported; in the only recorded systematic evaluation, 86 pleural exudates were negative when cultured for PPLO (25).

GASTROINTESTINAL DISEASE

Mycoplasmas have been isolated from rectal cultures of patients with and without proctitis (48). Dunlop et al. (49) found proctitis somewhat more frequently among women whose rectal cultures contained *M. hominis* than among women whose rectal cultures did not contain *M. hominis*. However, most authors believe that mycoplasmas are of no significance in rectal disease and that their presence in rectal cultures merely reflects transfer from the genital tract (25, 50, 51).

Jori et al. (52) isolated *M. hominis* from cultures of the colon of 1 of 10 patients with ulcerative colitis. None of the 10 had complement-fixing antibody against *M. hominis*. The authors concluded that mycoplasmas are of no etiological significance in ulcerative colitis.

Altucci et al. (53) isolated no mycoplasmas when they cultured the upper small intestine of 65 patients, some of whom were normal and some of whom had either biliary tract disorders or malabsorptive states. Lev et al. (54) isolated mycoplasmas (not speciated) from cultures of the small intestine of 6 of 12 patients with tropical sprue. Mycoplasmas were isolated from only 2 of 12 control individuals. Both of the control subjects had visited tropical areas and had morphological abnormalities of the jejunal mucosa. In contrast, Bhat et al. (55) were unable to isolate mycoplasmas from jejunal secretions of 23 patients with tropical sprue or from 28 normal subjects who served as controls.

OCULAR INFECTIONS

Holland (56) and Arm et al. (57) isolated PPLO from patients with acute conjunctivitis. Jones and Tobin (58) isolated *M. hominis* from 8 of 250 newborn infants with conjunctivitis and concluded that the organisms had been acquired from the maternal birth canal during delivery and were responsible for the conjunctivitis. This study has been criticized because of the lack of cultures from a control group of newborn infants without conjunctivitis (59). In other studies mycoplasmas have not been isolated from the eyes of normal individuals or from the eyes of patients with conjunctivitis (25, 60, 61). Further study is needed to define the role, if any, of mycoplasmas in conjunctivitis.

Holland and Worlton (62) isolated PPLO from genital cultures from five patients with nongranulomatous uveitis. Treatment with oxytetracycline was associated with elimination of the organisms and with symptomatic improvement. Although the control group had a comparable rate of isolation of PPLO from genital cultures, the authors postulated that PPLO caused the uveitis through a mechanism involving hypersensitivity. No further data have appeared to support this hypothesis; in one study, cultures from the eyes of 10 patients with uveitis yielded no mycoplasmas (61). Unfortunately, there are available no directly relevant data obtained by use of contemporary techniques for assessing response to mycoplasmal antigens.

CENTRAL NERVOUS SYSTEM INFECTIONS

Paine et al. (63) described a brain abscess which developed after a pipe stem had penetrated the patient's orbit. PPLO were recovered from spinal fluid and from the abscess. Carlson et al. (64) reported a child with cyanotic congenital heart disease who developed a brain abscess. PPLO were isolated from blood and spinal fluid and from postmortem cultures of ventricular fluid. Routine cultures of spinal fluid have been negative for mycoplasmas (25, 26).

Meningitis with the isolation of PPLO (65) and *M. hominis* (66) from cerebrospinal fluid has been noted in infants with meningomyeloceles, and Boe et al. (67) reported a premature infant who developed purulent meningitis and hydrocephalus 2 weeks after delivery. *M. hominis* was isolated from the cerebrospinal fluid. In addition, we have observed a neonate who developed meningitis and a brain abscess. *M. hominis* was isolated from an aspirate of the abscess and, on several occasions, from the spinal fluid (68). In the infants who had meningomyeloceles, organisms acquired from the maternal birth canal presumably entered the central nervous system through the meningomyelocele. The portal of entry in the other neonates who developed meningitis and/or brain abscess due to *M. hominis* is unclear.

BLOOD STREAM INFECTION

Carlson et al. (64) in 1951 described the isolation of PPLO from the blood of three children with a clinical picture suggestive of Henoch-Schonlein purpura. Multiple isolations were made from the blood of two of these children, one of whom had positive blood cultures for 18 months. Other patients with Henoch-Schonlein purpura did not have mycoplasmas in their blood cultures. No further reports have appeared associating mycoplasmas with purpura.

Herschberger et al. (69) described a 23-year-old man with endocarditis. A peculiar organism was isolated from his blood, but the description of the organism suggests that it was not a mycoplasma. In another study, mycoplasmas were not isolated from the blood of 31 patients with endocarditis (26).

ACKNOWLEDGMENTS

This investigation was supported by Public Health Service research grants HD 03693 from the National Institute of Child Health and Human Development and AI 12381 from the National Institute of Allergy and Infectious Diseases.

REFERENCES

1. McCormack WM, Braun P, Lee Y-H, et al: The genital mycoplasmas. N Engl J Med 288:78–89, 1973
2. Lee Y-H, McCormack WM, Marcy SM, et al: The genital mycoplasmas: their role in disorders of reproduction and in pediatric infections. Pediatr Clin North Am 21:457–466, 1974
3. Lee Y-H, McCormack WM: Clinical implication of the genital mycoplasmas in obstetrics and gynecology. J Reprod Med 13:123–127, 1974
4. McCormack WM: Genital mycoplasmal infections: their relation to prematurity and other abnormalities of reproduction. J Clin Pathol 29(Suppl):95–98, 1976
5. Harkness AH, Henderson-Begg A: The significance of pleuropneumonia-like or "L" organisms in non-gonococcal urethritis, Reiter's disease, and abacterial pyuria. Br J Vener Dis 24:50–58, 1948
6. Dienes L, Ropes MW, Smith WE, et al: The role of pleuropneumonia-like organisms in genitourinary and joint diseases. N Engl J Med 238:509–515, 563–567, 1948
7. Dienes L, Smith WE: Studies of the incidence and pathogenicity of pleuropneumonia-like organisms in humans. J Clin Invest 25:911–912, 1946

8. Claus G, McEwen C, Brunner T, et al: Microbiological studies of Reiter's disease. Br J Vener Dis 40:170–180, 1964
9. Berg RL, Weinberger H, Dienes L: Acute hemorrhagic cystitis. An infection associated with pleuropneumonia-like organisms and related to urethritis and prostatitis. Am J Med 22:848–864, 1957
10. Pachas WN: The role of mycoplasma in some unusual conditions of the kidney and the urinary tract. Ann NY Acad Sci 174:786–793, 1970
11. Witzleb W, Farber I, Thieler H, et al: Nachweis von Mykoplasmen in hoheren Abschnitten der ableitenden Harnwege. Zentralbl Bakteriol Orig 208:427–430, 1968
12. Mardh P-A, Lohi A, Fritz H: Mycoplasma in urine collected by suprapubic aspiration. Acta Med Scand 191:91–95, 1972
13. Thompsen AC: The occurrence of mycoplasmas in the urinary tract of patients with chronic pyelonephritis. Acta Pathol Microbiol Scand 83:10–16, 1975
14. Friedlander AM, Braude AI: Production of bladder stones by human T-mycoplasmas. Nature (London) 247:67–69, 1974
15. Marshall S, Lyon RP, Schieble J: Nonspecific urethritis in females. Calif Med 112:9–10, 1970
16. Hoogendijk JL, DeBruijne JI, Herderschee D: Infectie door *Mycoplasma hominis* bij een pasgeborene. Ned Tijdschr Geneeskd 109:1433–1434, 1965
17. Sacker I, Walker M, Brunell PA: Abscess in newborn infants caused by mycoplasma. Pediatrics 46:303–304, 1970
18. Laughton N: Pneumonitis due to a pleuropneumonia-like organism and a corynebacterium associated with hyaline membrane disease. J Pathol Bacteriol 85:413–419, 1963
19. Brunell PA, Dische RM, Walker MB: Mycoplasma, amnionitis, and respiratory distress syndrome. J Am Med Assoc 207:2097–2099, 1969
20. Stokes EJ: P.P.L.O. in genital infections. Br Med J 1:510, 1959
21. Lemcke R, Csonka GW: Antibodies against pleuropneumonia-like organisms in patients with salpingitis. Br J Vener Dis 38:212–217, 1962
22. Lemcke RM: The serological differentiation of mycoplasma strains (pleuro-pneumonia-like organisms) from various sources. J Hyg 62:199–219, 1964
23. Russell FE, Fallon RJ: Mycoplasma and the urogenital tract. Lancet i:1295, 1970
24. Lee Y-H, Nersasian RR, Nguyen KL, et al: Wound infection with *Mycoplasma hominis*. J Am Med Assoc 218:252–253, 1971
25. Freundt EA: The Mycoplasmataceae (the pleuropneumonia group of organisms): morphology, biology, and taxonomy. Aarhuus Stiftsbogtrykkerie, Copenhagen, 1958
26. Hayflick L, Stanbridge E: Isolation and identification of mycoplasma from human clinical materials. Ann NY Acad Sci 143:608–621, 1967
27. Ford DK: Relationships between mycoplasma and the etiology of nongonococcal urethritis and Reiter's syndrome. Ann NY Acad Sci 143:501–504, 1967
28. Purcell RH, Wong D, Chanock RM, et al: Significance of antibody to mycoplasma as measured by metabolic-inhibition techniques. Ann NY Acad Sci 143:664–675, 1967
29. Ford DK: Non-gonococcal urethritis and Reiter's syndrome; personal experience with etiological studies during 15 years. Can Med Assoc J 99:900–910, 1968
30. Hendley JO, Jordan WS: Mycoplasma pharyngeal flora in civilians. Am Rev Respir Dis 97:524–532, 1968
31. Hayflick L, Chanock RM: *Mycoplasma* species of man. Bacteriol Rev 29:185–221, 1965
32. Crawford YE, Kraybill WH: The mixtures of mycoplasma species isolated from the human oropharynx. Ann NY Acad Sci 143:411–421, 1967
33. Crawford YE: Mycoplasma of human derivation, a microbial enigma, p 3–70. *In* C Panos (ed), Mycoplasma and bacterial L-forms. The World Publishing Co., Cleveland, 1967
34. Glezen WP, Clyde WA, Senior RJ, et al: Group A streptococci, mycoplasmas, and viruses associated with acute pharyngitis. J Am Med Assoc 202:455–460, 1967

35. Kundsin RB, Praznik J: Pharyngeal carriage of mycoplasma species in healthy young adults. Am J Epidemiol 86:579–583, 1967
36. Chanock RM, Hayflick L, Barile MF: Growth on artificial medium of an agent associated with atypical pneumonia and its identification as a PPLO. Proc Natl Acad Sci USA 48:41–49, 1962
37. Mufson MA, Chang V, Gill V, et al: The role of viruses, mycoplasmas and bacteria in acute pneumonia in civilian adults. Am J Epidemiol 86:526–544, 1967
38. Glezen WP, Thornburg G, Chin TDY, et al: Significance of mycoplasma infections in children with respiratory disease. J Pediatr 39:516–525, 1967
39. Fawzy KY, Fox JP, Ketler A, et al: The virus watch program: a continuing surveillance of viral infection in metropolitan New York families. V. Observations in employed adults on etiology of acute upper respiratory disease and heterologous antibody response to rhinoviruses. Am J Epidemiol 86:653–672, 1967
40. Stewart SM, Chowdray JE: Isolation of mycoplasmas from the human respiratory tract. J Pathol Bacteriol 95:580–586, 1968
41. Eldridge AE, Jones DM, Tobin BM: Antibody to *Mycoplasma hominis* in patients with respiratory disease. J Clin Pathol 23:664–667, 1970
42. Mufson MA, Ludwig WM, Purcell RH, et al: Exudative pharyngitis following experimental *Mycoplasma hominis* type 1 infection. J Am Med Assoc 192:1146–1152, 1965
43. Moffet HL, Siegel AC, Doyle HK: Nonstreptococcal pharyngitis. J Pediatr 73:51–60, 1968
44. Neu HC, Ellner PD: Role of mycoplasma in exudative pharyngitis. NY State J Med 69:3026–3028, 1969
45. Mufson MA: *Mycoplasma hominis* I in respiratory tract infections. Ann NY Acad Sci 174:798–808, 1970
46. Wiesner PJ, Tronca E, Bonin P, et al: Clinical spectrum of pharyngeal gonococcal infection. N Engl J Med 288:181–185, 1973
47. Stokes EJ: Human infection with pleuropneumonia-like organisms. Lancet i:276–279, 1955
48. Nicol CS, Edward DG$_{ff}$: Role of organisms of the pleuropneumonia group in human genital infections. Br J Vener Dis 29:141–150, 1953
49. Dunlop EMC, Hare MJ, Jones BR, et al: Mycoplasmas and 'nonspecific' genital infection. II. Clinical aspects. Br J Vener Dis 45:274–281, 1969
50. Altucci P, Jori GP, DeVargas F: Mycoplasmas in rectal cultures. Chemotherapia 12:41–46, 1967
51. Berg RL, Daggett W, Madden J, et al: The origin of PPLO found in rectal cultures. Ann NY Acad Sci 79:635–641, 1960
52. Jori GP, DeVargas F, Altucci P: Mycoplasmas in chronic ulcerative colitis: a negative study. Pathol Microbiol 31:209–214, 1968
53. Altucci P, Jori GP, Manguso L, et al: Failure to isolate mycoplasmas from upper small intestine of man. Pathol Microbiol 37:181–184, 1971
54. Lev M, Alexander R, Levenson S, et al: Abnormal microflora in small intestine in tropical sprue. Clin Res 18:384, 1970
55. Bhat P, Boulter EA, Rayan D, et al: Mycoplasma in the upper gastrointestinal tracts of southern Indian control subjects and patients with tropical sprue. Am J Clin Pathol 59:825–828, 1973
56. Holland MC: Uveitis and pleuropneumonialike organisms. Ann NY Acad Sci 79:646–649, 1960
57. Arm HG, Woolridge RL, Cheng K-H, et al: Isolation of mycoplasma organisms (PPLO) from the human eye. Am J Ophthalmol 62:1125–1127, 1966
58. Jones DM, Tobin B: Neonatal eye infections due to *Mycoplasma hominis*. Br Med J 3:467–468, 1968

59. Tyrrell DAJ, Taylor-Robinson D: Sticky eye and *Mycoplasma hominis*. Br Med J 3:801–802, 1968
60. Sachdev KS, Flamm H: Human mycoplasma infections. Ger Med Mon 14:512–516, 1969
61. Millian SJ, Leopold IH, Schneierson SS: Relationship between mycoplasma (PPLO) and eye disease; a negative report. Am J Ophthalmol 64:289–291, 1967
62. Holland MC, Worlton JT: Relationship of pleuropneumonialike organisms and uveitis. Am J Ophthalmol 43:597–606, 1957
63. Paine TF, Jr, Murray R, Perlmutter I, et al: Brain abscess and meningitis associated with a pleuropneumonia-like organism: clinical and bacteriological observations in a case with recovery. Ann Intern Med 32:554–562, 1950
64. Carlson HJ, Spector S, Douglas HG: Possible role of pleuropneumonia-like organisms in etiology of disease in childhood. Am J Dis Child 81:193–206, 1951
65. Davis JH, Arnstein LH: Pleuropneumonia organism meningitis complicating ruptured meningocele. Pediatrics 11:381–384, 1953
66. Wealthall SR: Mycoplasma meningitis in infants with spina bifida. Dev Med Child Neurol 17:535, 117–122, 1975
67. Boe O, Diderichser J, Matre R: Isolation of *Mycoplasma hominis* from cerebrospinal fluid. Scand J Infect Dis 5:285–288, 1973
68. Siber GR, Alpert S, Smith AL, et al: Neonatal central nervous system infection due to *Mycoplasma hominis*. J Pediatr 90:625–627, 1977
69. Herschberger C, Dantes DA, Shwartzman G: A case of subacute bacterial endocarditis caused by an unusual microorganism related to the "pleuropneumonia-like" or grahamella group. J Mt Sinai Hosp NY 12:295–308, 1945

Chairman's Report and Discussion

KING K. HOLMES

William Fair discussed the diagnosis, pathogenesis, and etiology of epididymitis and prostatitis. There is circumstantial evidence in humans and experimental evidence in dogs that reflux of "sterile" urine can produce epididymitis, but more than 50% of cases of epididymitis seen by urologists today cannot be attributed to specific microbial etiology or reflux. Less than 5% are due to gonorrhea. Similarly, the great majority of patients with symptoms of prostatitis (perineal aching, low back pain, irritation during voiding) have no demonstrable microbial infection, and their condition has been termed prostatosis. Those who do have bacterial prostatitis have depressed concentrations of zinc in semen and prostatic fluid (<150 μg/ml). Preliminary data show normal semen zinc levels in men with nonbacterial prostatitis. E. R. Alexander described a Seattle study of 24 men with acute epididymitis (1), which showed that four of six men over 45 years of age had coliform or *Pseudomonas* bacteriuria, whereas of 18 men under 32 years of age urethral cultures yielded *Neisseria gonorrhoeae* in 6, *Chlamydia trachomatis* in 6, and both agents in 1. Several patients also had *Ureaplasma urealyticum* infection of the urethra. The data suggest that in older men epididymitis is secondary to bacterial prostatitis, whereas in younger men it is secondary to urethritis caused by sexually transmissible agents. In a separate study of men with nonbacterial prostatitis, *C. trachomatis* was isolated from the urethra from 1 of 19 men and from the prostatic expressate from 2 of 14 men. Dr. Alexander also noted that undiluted semen was toxic to cell cultures, making semen culture difficult to interpret. S. Razin said he had previously reported that the antibacterial activity of semen was due to the polyamine spermine and asked how Dr. Fair had excluded spermine in his experiments. Dr. Fair replied that spermine is primarily active against gram-positive organisms and also that his method of column separation of the zinc-peptide antibacterial compound left both spermine and spermadine on the column. Dr. Holmes asked if nonbacterial prostatitis was temporally related to prior or current attacks of urethritis in most cases. Dr. Fair replied that most patients gave a history of urethritis, but it was usually not closely related historically to the onset of prostatitis.

D. Ford discussed the relationship of Reiter's syndrome to histocompatibility antigen HLA-B27 and evidence that *Shigella* and *Yersinia* infections precipitate Reiter's syndrome in a proportion of patients who are B27 positive. Urethritis is a feature of Reiter's syndrome, even in those cases apparently precipitated by enteric infection. This might suggest reactivation of a latent urethral infection. Data are needed concerning the rate of isolation of *C. trachomatis* and *U. urealyticum* from the urethra of untreated cases of acute Reiter's syndrome, on the efficacy of tetracycline for urethritis in Reiter's syndrome, and on serum antibody to *C. trachomatis* and *U. urealyticum* in Reiter's syndrome. J. Schachter responded that *C. trachomatis* can be recovered from about 30% of untreated men with acute Reiter's syndrome who have urethritis. The serum antibody titer against *C. trachomatis* in such individuals is higher than that seen in matched controls with uncomplicated chlamydial urethritis, but this is difficult to interpret, since Reiter's patients are reported to have a hyperreactive serum antibody response nonspecifically to other antigens, such as influenza virus. B. Jones and Dr. Holmes agreed that *C. trachomatis* would be isolated from the urethra of about one-third of men with untreated, active Reiter's syndrome. In view of the fact that some of Dr. Schachter's isolates from joints have been subgroup B chlamydia, it was of interest to Dr. Jones that the reactivity of serum from Reiter's patients was impressive against subgroup B chlamydiae. W. R. Bowie reported that E. D. Thomas and colleagues had done HLA typing on 95 of his nongonococcal urethritis (NGU) patients. Twelve of these men were HLA-B27 positive, of whom five had *C. trachomatis* and eight had ureaplasma infection. Thus far, none has gone on to develop Reiter's syndrome or spondylitis. S. Darougar described a married couple in which the wife had conjunctival and cervical infection with *C. trachomatis*, serotype D, and antibody titer against serotype D of 1:1,000. The husband had Reiter's syndrome, and his serum also had an antibody titer of 1:1,000 against serotype D.

E. Rees then described the clinical manifestations of genital *C. trachomatis* infection in women. Of particular interest was a study of female contacts of men with NGU, in whom hypertrophic erosion of the cervix and endocervical mucopus were significantly correlated with *C. trachomatis* infection. Oxytetracycline treatment of such cases usually resulted in disappearance of the hypertrophic (edematous) component of the erosion, in clearing of the mucopus, and in negative post-treatment cultures for *C. trachomatis*. Reisolation after treatment was associated with a high probability of reinfection.

P. A. Mårdh and colleagues discussed the epidemiology, etiology, and complications of pelvic inflammatory disease (PID) in the south of Sweden. Complications include recurrent PID in 27% of cases, and involuntary childlessness is observed in up to 75% of women who have had three or more episodes of PID. From 1965 to 1975, the total number of cases of PID increased, as did the percentage which were nongonococcal and the percentage using an intrauterine device. In one group of 134 women with laparoscopically confirmed PID associated with cervical gonorrhea, only 13% had *N. gonorrhoeae* recovered from the fallopian tubes. Dr. Mårdh also reported isolation of *C. trachomatis* from fallopian tubes from two patients with PID.

Dr. Schachter discussed the clinical manifestations, diagnosis, and therapy of lymphogranuloma venereum (LGV), a systemic *C. trachomatis* infection, and the evidence for *C. trachomatis* involvement of mucosal columnar epithelium of sites other than the eye and genitalia (e.g., eustachian tube, nasopharynx, lower respiratory tract, and gut). The prevalence of antibody to *C. trachomatis* is surprisingly high in children, and there is a great need to define the clinical spectrum of *C. trachomatis* infection, particularly in children. C. C. Kuo discussed a collaborative study with P. Perine of 30 patients in Ethiopia with clinically suspected LGV. Micro-immunofluorescence (micro-IF) antibody to *C. trachomatis* was detected in 21 of 30 patients, in very high peak titers, ranging from 256 to 16,000 (mean, 3,000). The remaining patients had no detectable antibody and presumably did not have LGV. Antibody to the purified *C. trachomatis* species-specific protein antigen (2) were detected by Dr. Caldwell, using counter-immunoelectrophoresis, in 20 of 20 of the micro-IF positive LGV cases and in none of a group of controls with chlamydial urethritis (3). *C. trachomatis* was isolated from bubo, urethra, or cervix from 6 of the 21 seropositive cases, and five of the six isolates were successfully immunotyped as L_1, L_2, or L_3. Dr. Kuo described five biological markers which differentiate LGV strains from other strains of *C. trachomatis*. (i) *Diethylaminoethyl (DEAE) enhancement*: DEAE pretreatment of HeLa 229 cells enhances infectivity of trachoma organisms, but not of LGV strains. (ii) *Intracerebral mouse virulence test*: The LD_{50} of LGV strains inoculated intracerebrally in mice is usually 10^7 organisms, whereas up to 10^9 trachoma organisms show no mouse virulence. (iii) *Neuraminidase inhibition of infectivity for HeLa 229 cells*: Pretreatment of cells with neuraminidase inhibited infectivity of trachoma organisms, but not of LGV strains. (iv) *Inhibition of infectivity of live organisms by heat-inactivated organisms*: With an inoculum containing 100 heat-inactivated to 1 live organism, infectivity of trachoma strains for HeLa 229 cells was inhibited by heat-inactivated trachoma organisms but not by heat-inactivated LGV organisms; infectivity of LGV was not inhibited by heat-inactivated trachoma or heat-inactivated LGV. (v) *Monkey eye infection*: Conjunctival inoculation of trachoma organisms into Taiwan monkeys produces follicular conjunctivitis, whereas LGV organisms do not produce follicular conjunctivitis.

D. Graham discussed studies, conducted by a colleague, J. Praszker, of three LGV strains isolated in Australia. The three Australian LGV strains were similar to one another with respect to ID_{50} for eggs, number of days until chick embryo death after a standardized lethal dose, number of particles per intracerebral mouse infectious dose, and number of particles per inclusion-forming unit in tissue culture after centrifugation (approximately 1:1 for the LGV strains, and at least 100:1 for ordinary trachoma-inclusion conjunctivitis strains). The three LGV strains were all sent to S.-P. Wang and typed as L_2, but analysis of proteins of these organisms by polyacrylamide gel electrophoresis showed distinctive differences between strains. Concerning *C. trachomatis* pharyngeal infection, Dr. Bowie commented that he had failed to recover *C. trachomatis* from the pharynx of any of 40 female sex partners of men with NGU. Most had had fellatio with their contacts, but few had sore throat.

There have been several recent reviews of the role of mycoplasmas in genital infection. W. McCormack discussed the role of *Mycoplasma hominis* and *U. urealyticum* in extragenital diseases. These agents have been variously associated with hemorrhagic cystitis and abacterial urethritis, neonatal respiratory distress syndrome, obstetrical-gynecological wound infections, pharyngitis and pneumonitis, conjunctivitis, and central nervous system and blood stream infections. D. Taylor-Robinson suggested that the possible etiological role of *M. hominis* in NGU should be reconsidered. Dr. McCormack noted that *M. hominis* seemed more invasive than *U. urealyticum* in terms of uterine infection and bacteremia. However, Drs. Holmes and Bowie said that the ability to distinguish chlamydial from nonchlamydial urethritis showed that *M. hominis* was not particularly associated with nonchlamydial urethritis. Dr. Ford pointed out that *M. hominis* was resistant to erythromycin, whereas NGU was responsive to that drug. Dr. Mårdh asked whether mycoplasmemia was connected with postpartum fever. Dr. McCormack defined postpartum fever as oral temperature of over 100°F (37.8°C) at 24 or more hours after delivery. Most cases who lack urinary tract infection or other obvious extragenital causes of fever have endometritis, and in such patients *M. hominis* bacteremia probably reflects endometritis. *U. urealyticum* can be recovered from blood from about 5% of women during the first few minutes after vaginal delivery, but *U. urealyticum* bacteremia has not been associated with postpartum fever in Boston. G. Kenny has observed a high prevalence of *U. urealyticum* and *M. hominis* in the uterus of patients undergoing caesarean section, but a large spectrum of other aerobes and anaerobes can also be recovered. He and H. Foy have seen two patients who had only *U. urealyticum* recovered from the uterus. One of them had ureaplasma recovered from the blood at 24 h and again at 5 days after delivery.

REFERENCES

1. Harnisch JP, et al: Etiology of acute epididymitis. Lancet i:819, 1977
2. Caldwell HD, Kuo C-C, Kenny GE: Antigenic analysis of chlamydiae by two-dimensional immunoelectrophoresis. II. A trachoma-LGV-specific antigen. J Immunol 115:969–975, 1975
3. Caldwell HD, Kuo C-C: Serologic diagnosis of lymphogranuloma venereum by counterimmunoelectrophoresis with a *Chlamydia trachomatis* antigen. J Immunol 118:442–445, 1977

Session III

OCULOGENITAL *CHLAMYDIA TRACHOMATIS* INFECTIONS

Pathogenesis of Ocular *Chlamydia trachomatis* Infections in Humans

J. THOMAS GRAYSTON, LUNG-JUI YEH, SAN-PIN WANG, CHO-CHOU KUO, R. PALMER BEASLEY, AND JAMES L. GALE

Departments of Epidemiology and Pathobiology, School of Public Health and Community Medicine, University of Washington, Seattle, Washington 98195; Trachoma Control Program, Provincial Health Department, Taiwan; and United States Naval Medical Research Unit no. 2, Taipei, Taiwan

Trachoma eye disease has been recognized since antiquity. There have been innumerable published descriptions of the natural history of the eye disease. These descriptions have not presented a consistent picture. For example, the onset of the disease has been described both as acute and insidious. The various reasons for progression of the disease have included hot, dry climate, blowing sand, and secondary bacterial infection.

Following our first isolation of trachoma organisms on Taiwan in 1958, we undertook a number of studies directed toward prevention or control of trachoma eye disease. The usual concept of the pathogenesis of trachoma, namely, transmission of the organism to the eye during youth with progressive development of disease through the various stages, leading eventually to corneal and conjunctival scarring and inactive eye disease in adulthood, did not fit the observations we were making. Although prevalence studies revealed children with active disease and older children and adults with healing or healed trachoma, such progression was often not observed in individuals. We carried out a number of field observational studies of the natural history of trachoma in Taiwan children in association with trachoma vaccine and antibiotic treatment studies (1). In a given group over time, the incidence of new disease was higher than could be accounted for by the prevalence rate. Both conversion to new disease and reversion to normal eyes were observed.

At the same time, we were finding in an experimental monkey eye infection model that pannus formation, the pathognomonic clinical sign of trachoma, and progressive disease with conjunctival and limbal scarring occurred only in monkeys who had been infected more than one time or had received immunological stimulation through vaccine before infection (2).

Our trachoma vaccine studies showed that eye disease could be prevented for 1 to 2 years, but the disease in the immunized group gradually caught up with that in the controls (3). It was also shown in the monkey eye infection model and in human volunteer infections that trachoma vaccine with inadequate antigenic strength to prevent infection was associated with a deleterious effect, with more severe and often more frequent infections (4, 5). Such deleterious effects were also seen if challenge was with a trachoma immunotype separate from that in the vaccine. Since trachoma, lymphogranuloma venereum, and psittacosis are characteristically prolonged chronic diseases, the idea arose that this is due to the fact that they are "poor antigens" that fail to stimulate the body's immunological system adequately. However, our findings with trachoma eye disease suggested to us that the progression of the chronic disease was associated with the immunological response and that pannus and scar formation followed reinfection disease and the development of immunopathology (2, 3, 6).

An understanding of the pathogenesis of trachoma eye disease is important to effective control measures. With the development of a serological test for *Chlamydia trachomatis* that was both specific and sensitive, it was possible to undertake studies of the pathogenesis and natural history of trachoma in which the clinical observations could be supported by meaningful laboratory studies (7). Cell culture techniques for isolation of trachoma organisms also assisted in these studies.

MATERIALS AND METHODS

In November 1970, we placed a group of family households near the sea coast in southwestern Taiwan under observation (8). In February 1972, we added another group of families from an adjacent county (9). Except for one segment of the 1970 families that was dropped, the families in the study were chosen to have at least one child with active trachoma eye disease and at least one child with no evidence of eye infection. There were a total of 32 such families kept under observation, 17 from the first area and 15 from the second. Examinations were carried out four times in 1971 and three times in 1972. Subsequently, most families have been examined twice a year. In addition to a clinical eye examination utilizing hand-magnifying lenses, at each examination a tear secretion specimen was collected by placing a standard filter-paper strip in the lower conjunctiva. The saturated filter-paper strip was eluted in normal saline, frozen, and later examined for antibody. Other specimens, venous blood for serum antibody and conjunctival swab for culture, were collected less frequently. All clinical examinations were performed by one ophthalmologist (L. J. Yeh) without reference to or knowledge of previous results.

Culture of conjunctival swabs for trachoma organisms was carried out first in eggs and then in McCoy cells, and since 1974 in HeLa 229 cell culture (Kuo et al., this volume, p. 328). Isolates were immunotyped as previously described (10). Both sera and tears were examined with the micro-immunofluorescence technique for *C. trachomatis* antibodies. The standard test with 12 antigens and twofold serum dilutions was used during the first part of the study. The simplified test using pooled antigens and fourfold serum and tear dilutions has been used for the last 2 years (Wang et al., this volume, p. 237). Originally tear specimens were tested only at the approximate 1:10 dilution of the eluate and graded ± to +++ by intensity of fluorescence. Since 1974, tears have been handled similarly to serum, with the eluate diluted fourfold. Some of the earlier tear specimens have been retested with dilutions.

A total of 219 persons in the 32 families were originally examined. This includes 80 adults (older than 15 years at the time of original examination) and 139 children. Of these, 41 adults and 118 children have been followed through at least 1974, with approximately equal numbers from the families placed under observation in late 1970 and those added in 1972.

Examples of spontaneous healing. The most striking finding from the total group

TABLE 1. *Demonstration of spontaneous remission of trachoma eye disease in members of a family living in one household followed for 5 years: results of representative clinical and laboratory examinations*

Age (yrs) and sex	Clinical diagnosis[a]		Micro-IF antibody[b]				Isolation type[c]	
			Serum[d]		Tear[e]			
	1970–1973	1974–1976	1970–1973	1974–1976	1970–1973	1974–1976	1970–1973	1974–1976
39F	IV	IV	16	—	0	0	—	0
37M	IV	IV	128	—	++	160	—	—
14F	III	IV	32	8	++	10	0	0
12M	IV	IV	32	16	40	40	0	0
10F	II++	NP	128	32	160	0	0	0
9F	II++	NP	512	128	160	40	C	0
6F	II++	NP	16	16	160	160	0	0
4M	ND	NP	256	0	160	0	0	0

[a] Clinical diagnosis is by the World Health Organization classification, where I is trachoma at onset, II is active trachoma, III is active and scarring trachoma, and IV is inactive trachoma with residual scars and/or pannus. Abbreviations and symbols: NP, Nothing particular (no clinical findings of trachoma); ++, papillae (indicates acute relatively severe infections); ND, not determined (clinical findings not definitely normal or trachoma).

[b] All micro-immunofluorescent (micro-IF) antibody results show a pattern specific for type C.

[c] Conjunctival swab, type isolated in egg yolk sac or tissue culture. 0, Culture negative; —, not tested.

[d] Reciprocal of serum dilution, $0 = <8$.

[e] Reactivity of positive tear specimens shown as reciprocal of highest dilution of tear specimen which showed fluorescence, or as ±, +, ++, +++ = intensity of fluorescence at original 1:10 dilution. $0 = <10$.

was the tendency for persons with active trachoma on original observation to spontaneously heal their disease or revert to normal eyes. This was despite the fact that those under observation lived in areas of Taiwan previously known for high incidence of severe trachoma, and the families had been specifically chosen for inclusion in the study on the basis of the presence of active disease in children.

Table 1 presents an example of a family with severe trachoma in several children on original examination in which there has been spontaneous remission or healing during the period of observation. No antibiotic treatment for trachoma was received by members of the family. Three children with extensive trachoma stage II on original examination reverted to normal eyes. Isolation of type C organism was accomplished from one of the children. All members of the family had serum micro-immunofluorescent antibody with a pattern specific for immunotype C, and all but the father showed similar tear secretion antibodies. Recent examination has shown fall-off in the antibody titers.

In 18 of the 32 families there was no evidence of new eye disease, and all disease activity ceased during observation. Including two families that had new cases, all members of 20 of the 32 families had no evidence of active trachoma disease at the time of last examination.

Examples of new cases of active trachoma infection. There were 20 persons in 9 families that developed clinical and laboratory evidence of new trachoma eye infec-

TABLE 2. *Results of representative clinical and laboratory examinations over 4 years of a family household with many new trachoma eye infections*

Age (yrs) and sex	Clinical diagnosis[a]			Micro-IF antibody[b] Serum[d]			Micro-IF antibody[b] Tear[e]			Isolation type[c]	
	1972	1974	1975–1976	1972	1974	1975–1976	1972	1974	1975–1976	1974	1975–1976
75M	IV	IV	IV	32	—	—	++	40	10	—	—
67F	—	—	IV	0	—	—	—	—	0	—	—
41M	IV	III	IV	64	—	—	0	160	160	—	C
38F	IV	III	III	32[f]	—	—	+	±	±	—	0
14M	III	III	III++	—	—	512	40	40	160	—	0
11M	IV	—	II++	256	—	128	160	—	160	—	C
9M	IV	I++	II++	256	128	256	20	80	160	C	0
6F	NP	I++	II++	32	128	512	0	40	>160	C	C
4M	NP	I	I++	0	128	1,024	0	40	>160	0	C
1F	NP	NP++	I++	0	128	32	0	40	10	C	0

[a–e] See Table 1.

[f] Serum antibody of 38F shows pattern of both type C and type B. Tear antibody of 38F is type C only. All other antibody results are specific for type C.

tion. Results from the family with the most new infections can be seen in Table 2. In 1972, when this family was placed under observation, the three youngest children had normal eyes. Inactive healed trachoma was found in two older children and three adult members of the family. One older child was diagnosed as trachoma stage III. Type C serum antibody was found in most family members, including the 6-year-old girl with normal eyes. Tear antibody was less prominent. Isolation attempts unfortunately were not made in 1972. By 1974, clinical and laboratory evidence of new trachoma infection was observed in five family members (41-, 9-, 6-, and 4-year-old males, 1-year-old female). Subsequent observa-

TABLE 3. *Additional detail on examination results of the three youngest children of the family shown in Table 2* [a]

		1972	1973	1974	1975	1976
1F	Dx	NP	NP	NP++	I++	NP
	Isol	—	—	C	0	0
	Serum	0	0	128	32	—
	Tear	0	0	40	10	0
4M	Dx	NP	NP	I	II++	I+
	I	—	—	0	C	0
	S	0	—	128	1,024	—
	T	0	0	40	>160	>160
6F	Dx	NP	I	I++	II++	II++
	I	—	—	C	C	0
	S	32	—	128	512	—
	T	0	0	40	>160	160

[a] All serum and tear antibody are type C pattern. See footnotes of Table 1.

tions have shown that a sixth, the 11-year-old boy, also reverted to active disease and that there was increasing severity of the eye disease throughout the family members. The serum and tear antibody titers continued to rise, and isolation of type C organisms was accomplished from all new infections.

The 9- and 11-year-old boys had the most definite reversion to clinically active disease. In each it was supported by tear antibody titer rise and isolation of organism. The 41-year-old father had less definite clinical illness, but a high-titer tear antibody conversion and isolation of organism.

Table 3 shows more detail on the findings from the three youngest children of this family. The 1- and 4-year-olds had no evidence of antibody before development of their type C infection. In the most recent examination of the youngest child, the eye disease has disappeared as has the tear antibody. The eye disease and high-titer antibody persist in the two older children after 3 years, and pannus formation has been observed in both. The 6-year-old girl had serum antibody against type C before evidence of the new infection.

Table 4 presents another family in which the three youngest children converted to active disease. On original examination, the 9- and 11-year-old children had active disease from which type B organism was isolated. The serum of all family members except the two youngest showed antibody of the type B pattern. Both the 9- and 11-year-olds have tended to heal their disease since the original observation and have not yielded organisms on culture since 1971. The three younger children converted to active disease, with laboratory evidence of type B infection. All three have now lost their active disease. The 35-year-old mother of this family has shown trachoma stage III throughout the period of observation. A recent isolation attempt from her yielded a type B organism. Except for this finding, it appeared that trachoma disease was disappearing from this family.

Table 5 presents in greater detail the conversion observed in the three youngest members of this family. In the two younger children, there was relatively brief

TABLE 4. *Results of representative clinical and laboratory examinations over 5 years of a family household with three new trachoma eye infections that have run their course*

Age (yrs) and sex	Clinical diagnosis[a]			Micro-IF antibody[b]: Serum[d]			Micro-IF antibody[b]: Tear[e]			Isolation type[c]		
	1970–1971	1972–1974	1975–1976	1970–1971	1972–1974	1975–1976	1970–1971	1972–1974	1975–1976	1970–1971	1972–1974	1975–1976
77F	IV	IV	—	64	—	—	0	0	—	—	—	—
37M	IV	IV	IV	8	—	—	0	0	0	—	—	—
35F	III	III	III	64	—	—	++	++	80	—	—	B
11M	II++	—	IV	64	—	—	++	—	10	B	—	0
9F	II++	III++	III	256	32	32	+++	10	20	B	0	0
6F	NP	I++	IV	16	512	128	0	160	40	0	B	0
3M	NP	NP++	NP	0	128	—	0	10	0	0	0	0
1F	NP	I++	NP	0	128	8	0	80	0	0	B	0

[a–e] See Table 1. All serum and tear antibody are of type B pattern.

TABLE 5. *Additional detail on examination results of the three youngest children of family shown in Table 4* [a]

		1970	1971	1972	1973	1974	1975	1976
1F	Dx	NP	NP++	I++	NP++	NP	NP	NP
	Isol	0	B	+	0	0	0	0
	Serum	0	128	—	128	32	8	—
	Tear	0	120	40	80	0	0	0
3M	Dx	NP	NP	NP	NP++	NP	NP	NP
	I	0	0	0	0	0	0	0
	S	0	0	—	128	8	—	—
	T	0	0	0	10	0	0	0
6F	Dx	NP	NP	I++	II	I++	I	IV
	I	0	0	0	B	0	0	0
	S	16	8	—	—	512	128	—
	T	0	0	160	40	40	40	40

[a] All serum and tear antibody are type B pattern. See footnotes of Table 1.

clinical disease and persistence of antibody. In the case of the 3-year-old boy, a definite clinical diagnosis of trachoma was never made, but infection was demonstrated serologically in association with the appearance of acute papillary hypertrophy in the conjunctiva. Neither of the younger children developed pannus. The 6-year-old girl had serum antibody before her new infection which persisted for 4 years, showed pannus formation, and healed to trachoma IV in 1976.

One of the fascinating findings from this study is the type specificity of trachoma infections within a household. In 31 families there was evidence of the infecting type from serum and/or tear antibody (in 11 there was additional evidence of type from trachoma strain isolation). In 28 of these 31 families, the infecting type was either B (13 families) or C (15 families), but not both. In these 28 families, all children that had antibody had it against either type B or type C, but not against both. Further, within a single family, all children had antibody against the same type. The same was true for the adults, with four exceptions. They consisted of two persons from C families showing both B and C antibody and two from B families showing C antibody only. Three of these adults were females, including the two with C antibody in B families. It would seem reasonable to assume that their type represented infection in their youth and with their own family type, rather than their husband's family type.

Examples of families with both immunotype B and C infections and one with type D infection. The three family exceptions were two families with serological evidence of both type B and type C infection and one family in which the infecting type was D.

Table 6 presents a family with both type B and C infection. On original examination, this appeared to be a typical type B family in which the mother had antibody for both types B and C. The father and four children showed type B antibody only. The 17-year-old girl, who had active trachoma stage III, showed type B antibody in both serum and eye secretion. However, on two separate attempts 8 months apart during 1971, a type C trachoma organism was isolated from the girl's con-

TABLE 6. *Selected representative examination results from 5 years of observations of a family in which both trachoma type B and type C infections have occurred*

Age (yrs) and sex	Clinical diagnosis[a]			Micro-IF antibody[b]							Isolation type[c]		
				Type	Serum[d]			Tear[e]					
	1970–1972	1973–1974	1975–1976		1970–1972	1973–1974	1975–1976	1970–1972	1973–1974	1975–1976	1971–1972	1973–1974	1975–1976
46M	IV		IV	B	32	—	—	+	—	—	—	—	—
41F	IV	IV	IV	C and B	64	—	—	0	±	40	—	—	—
19F	NP			B	32	—	—	0	—	—	—	—	—
17F	III	III++	IV	B	64	16	—	++	40	—	C	B	0
13M	NP	NP	NP		0	—	—	0	0	0	0	—	—
10F	I++	NP	IV	B	256	0	—	++	0	0	0	—	0
8F	ND	I++	II++	C	0	256	128	0	320	160	0	C	C
5M	NP	NP	I++	C	0	0	128	0	0	160	0	0	C

[a–e] See Table 1.

junctivae. In 1973 her conjunctival swab yielded a type B organism. The activity of the clinical disease subsequently subsided. This girl, whose last examination was in 1974, never developed any type C antibody.

Evidence for transmission of type C organisms in this family was obtained when the two youngest children converted. On original examination they had no clinical or laboratory evidence of trachoma infection. The 8-year-old girl converted to active trachoma in 1973, and the 5-year-old boy converted in 1975. Both conversions to type C were supported by all laboratory tests. The clinical disease in both has included development of pannus.

Table 7 shows results from the family with type D infections. Members of this family had antibody against types B, D, and E, types that have been shown to be related (7). The titers were higher with type D (or DE) than B except in the case of the 7-year-old girl. However, we were reluctant to identify this as a type D family because only types C and B had previously been associated with trachoma in Taiwan. We had immunotyped strains isolated from 84 persons in Taiwan with trachoma. Sixty-two of these persons were shown to have type C, and 22 had type B. The first immunotype D strain, G-1, was isolated from a child with trachoma living in Gambia, Africa, a trachoma-endemic country (11). However, most type D strains that have been identified have been associated with genital tract infection or with ocular infection in areas without endemic trachoma (10). It was not until the youngest child converted to active trachoma and yielded an isolate that was proven to be type D that confirmatory evidence was obtained of trachoma due to type D in Taiwan. The child was 6 years old at the onset of disease, which showed pannus formation and healed to trachoma IV after 2 years.

During the observation of this family, the 7-year-old girl, who originally had relatively mild inactive trachoma IV, developed active disease of about 1.5 years' duration. This was associated with an antibody rise in the tear secretion, but isolation was not accomplished. The antibody pattern in her first serum and eye secretion was clearly higher titered against type B than DE, which suggests that this child was originally infected with the common ocular type B rather than with the family type D. However, the reinfection, which occurred about the same time as the new infection in the 3-year-old, resulted in a fourfold rise in type D serum antibody, suggesting that the reinfection was with type D.

Relationship of laboratory evidence of infection to disease activity. In most of the persons converting to new trachoma disease during the period of observation, the laboratory evidence of infection tended to occur simultaneously with the clinical changes. When specimens were available, the first clinical changes of disease were often accompanied by rises in both tear and serum antibody and isolation of trachoma organism. However, two children in Table 8 showed the unusual finding of isolation of trachoma organism before the development of clinical disease or antibody. This was a type C family in which there was no evidence of active disease in the parents or older children. Among the three younger children, when first observed in 1972, trachoma eye disease was found only in the 8-year-old girl. The 4-year-old boy was diagnosed as having nontrachomatous follicles. The 2-year-old had normal eyes. Routine conjunctival swab cultures in June 1972 on these three children all cultured type C organisms. Only the 8-year-old with active disease had antibody. Repeated isolation attempts were mostly positive on all

TABLE 7. *Selected representative examination results from 5 years of observations of a family with trachoma eye disease due to type D*

Age (yrs) and sex	Clinical diagnosis[a]		Micro-IF antibody[b]					Isolation type[c]	
			Type	Serum[d]		Tear[e]			
	1970–1972	1974–1975		1970–1972	1974–1975	1970–1972	1974–1975	1971	1974–1975
62M	IV	IV	D	64	—	++	40	—	—
59F	IV	IV	DE	128	—	0	0	—	—
37M	IV	IV	DE	—	—	—	20	—	—
35F	IV	IV	DE	64	—	0	0	—	—
15F	IV		DE	256	—	+	—	0	—
11M	III	IV	D	256	128	0	40	0	0
8F	III	IV	D	128	—	0	0	0	0
7F	IV	III	B and D	128	128[f]	0	160	0	0
5F	NP	NP		0	0	0	0	0	0
3M	NP	I++	D	0	256	0	160	0	D

[a–e] See Table 1.
[f] Serum antibody to type B antigen in 1971 and 1974 was 1:128, but to type D antigen it was 1:16 in 1971 and 1:128 in 1974.

three children. Low-titer serum antibody was demonstrated against type C antigen in January 1974 in the 4-year-old boy, along with development of clinical disease. Well-established clinical disease along with high-titer antibody in serum and tears was not demonstrated in either of the younger children until 1975, 3 years after the organism was first isolated. The 4-year-old developed pannus, but the 2-year-old did not. The antibody titer rises were more associated with the development of clinically obvious disease than with isolation of the trachoma organism. At last

TABLE 8. *Results of representative examinations of the three youngest children of a family (not shown) followed for 4 years in which isolations of type C trachoma organisms were made prior to appearance of clinical disease and development of antibody* [a]

		1972	1974	1975	1976
8F	Dx	III++	III++	III++	III
	Isol	C	C	C	—
	Serum	64	128	—	—
	Tear	160	160	160	160
4M	Dx	NP	I	I++	I
	I	C	0	C	C
	S	0	8	64	—
	T	0	0	40	40
2M	Dx	NP	NP	I++	NP
	I	C	C	C	0
	S	0	0	128	—
	T	0	0	160	10

[a] All serum and tear antibody reactions are specific for type C. See footnotes for Table 1.

examination, the youngest child has shown return to normal eyes clinically, rapid fall in tear antibody, and for the first time absence of organism in conjunctival swab.

We do not know the significance of isolation from normal eyes in the absence of antibody formation. In experimental monkey eye inoculation, we have a few times isolated trachoma organisms from the conjunctiva, although no disease developed. The findings suggest that the isolation technique is sensitive and that in families where one child is excreting organism, it is very likely to get into the eyes of the other family members. It would appear that, at least occasionally, the presence of trachoma organisms in the conjunctivae may not be sufficient to initiate disease or even an asymptomatic infection with antibody response.

The findings of clinical diagnosis of eye disease are quantitated under the McCallum-World Health Organization classification scheme. In addition to the quantitative grading of follicles, papillae, and scars on a +, ++, +++ basis, the qualitative acuteness of the papillae has been determined in this study on each clinical examination. We have analyzed the various parts of the clinical diagnosis to determine their relationship to positive laboratory findings of trachoma organism infection. The presence of papillae, especially of the acute type, was found to correlate best with laboratory evidence of infection (9).

DISCUSSION

Study of the clinical course and laboratory findings of the 20 cases of new trachoma infection has shown that they fall readily into three groups which are suggestive of primary, secondary, and tertiary infection. The five primary infections occurred in the youngest children (average age, 3.8 years; range, 2 to 5 years at onset of disease), both clinical disease and antibody response was for the shortest time, and no vascularization of the cornea (pannus) developed. Healing was by disappearance of eye disease, with return to normal eyes. Nine secondary infections occurred in somewhat older children (average age, 7.2 years; range, 5 to 10 years). All had normal eyes, but four had serum antibody and one tear antibody before the new disease. At the onset of secondary infection, clinical trachoma was diagnosed with follicle formation, whereas in primary infections the onset was often (three of five) with papillary hypertrophy only. Secondary infections had higher-titered and more prolonged antibody response. All showed pannus formation, and the few that have become inactive have residual corneal changes and are diagnosed as trachoma stage IV. The six tertiary infections all had inactive trachoma IV with pannus before the new disease. All had prior serum antibody, and four of six had prior tear antibody. Four of the infections have occurred in children (average age, 11.3 years; range, 10 to 14 years) and two in adults (40 and 43 years old). The course of disease in the adults has been relatively brief, but in the children it has been at least as long as the secondary infections.

Tables 3, 5, and 8 show many of the new trachoma infections, with examples of primary, secondary, and tertiary infections. The youngest child in each family (Table 3, 1-year-old female; Table 5, 1-year-old female; Table 8, 2-year-old male) underwent a very similar course. They were 2 to 5 years old when infected. The clinical disease diagnosed as trachoma did not last over 1 year, although acute papillary hypertrophy was observed in two, associated with laboratory conversion, before trachoma was diagnosed clinically. Tear antibody disappeared in 2 years or less after disease activity ceased, but serum antibody persisted longer. In no case did the clinical disease progress beyond initial trachoma stage I. Corneal vascularization (pannus) was not found. The clinical picture was typical of what has been called "inclusion conjunctivitis" and would not have been diagnosed as trachoma in a nonendemic country. The brief mild course in the 3-year-old boy in Table 5 shows that infection with less substantial disease may occur. His disease with acute papillary hypertrophy and no follicular formation was similar to the syndrome inclusion conjunctivitis of the newborn. These four infections and one other still active (not shown) we have classified as primary infections.

The somewhat older children of these families who had onset of new disease at 5 to 7 years of age (Table 3, 4-year-old male and 6-year-old female; Table 5, 6-year-old female) developed more-established trachoma disease with mature follicles (stage II) and beginning pannus formation (vascularization, 1 mm). The clinically diagnosable trachoma eye disease coincided with conversion of one or more laboratory tests. The antibody response was higher titered and more prolonged. Although one child did not have previous serum antibody, all three appeared to be undergoing reinfection with a secondary response. Only the eye disease of the child of Table 5 has ceased activity after 5 years, leaving inactive stage IV trachoma.

The converter in Table 7 (3-year-old male) is similar to the three just described with secondary trachoma IV at the 1976 exam. The two converters in Table 6 (8-year-old female and 5-year-old male) are also considered to have secondary infections, although neither had antibody before the appearance of new disease. This is clear with the girl whose disease has been active for 4 years, with the development of mature follicles and pannus. The boy has had disease for less than 1 year, but pannus has been seen. Pannus has also developed in the 4-year-old boy of Table 8 during 3 years of active disease.

Examples of tertiary infection are shown in Table 2 (9-, 11-, and 41-year-old males) and on Table 7 (7-year-old female). In each case, new active disease has appeared in persons with healed trachoma stage IV. All had substantial serum antibody, and several had tear antibody before the new disease.

The recurrence of active trachoma disease raises the question of whether it represents relapse or reinfection. In some individuals with trachoma IV it may be relapse. However, in five of six of the cases of laboratory-proven tertiary infection, there was evidence of disease transmission in the family members. Further, our associates have shown that when Sikh Indians with previous trachoma infection moved from the trachoma-endemic Punjab to nonendemic British Columbia, new disease (either relapse or reinfection) ceased (12).

The great usefulness of tear secretion to study antibody response in ocular trachoma infection is demonstrated in this study. The procedure for securing the specimen, placing the strip of filter paper in the lower conjunctiva, is easily tolerated, whereas attempts to obtain repeated blood specimens result in loss of cooperation. Although tear and serum antibody tend to develop about simultaneously with appearance of clinical disease, the tear antibody is more closely associated with active disease in that it disappears sooner after cessation of activity.

We believe that scarring and then aggravated scarring that lead to functional disability of the eye from severe trachoma is caused at least in part by repeated reinfections. Thus, as the number of acute cases in a population decreases and the transmission of organisms decreases, the first effect is a decrease in the severity of disease. The prevalence of disease responds later, decreasing sharply, especially in the younger age group. We realize that trachoma eye disease in Taiwan and other Far Eastern countries has always been milder (as measured by blindness rates) than that seen in the Middle East. For example, in our families only the grandparents have entropian and other functional distortions. In the middle-aged adults, + or ++ conjunctival scars are all that are observed in their nearly universal trachoma IV. Younger persons with stage IV usually show only 1 mm of inactive vascularization and a few Herbert's pits.

The epidemiological studies of trachoma in the Middle East by Nichols and associates (13) and more recently Jones (14) have resulted in findings in concert with ours in Taiwan. Jones has used the excellent term "multicyclic" to describe the repeated episodes of trachoma eye disease seen in an individual. He has also spoken of "ocular promiscuity" to dramatize the pool of organisms available for reinfection in trachoma-endemic areas, similar to our expression "pressure for reinfection." Jones' earlier studies in London described eye disease syndromes of inclusion conjunctivitis, intermediate disease with keratitis, and typical trachoma, all associated with genital transmission of the organism (15). He, however, did not conclude that these diseases were different stages of the same process based on the hosts' immunological response to the organisms.

ACKNOWLEDGMENT

This investigation was supported by Public Health Service research grant EY 00219 from the National Eye Institute.

REFERENCES

1. Woolridge RL, Grayston JT, Perrin EB, et al: Natural history of trachoma in Taiwan school children. Am J Ophthalmol 63:1313–1320, 1967
2. Wang SP, Grayston JT: Pannus with experimental trachoma and inclusion conjunctivitis agent infection of Taiwan monkeys. Am J Ophthalmol 63:1133–1145, 1967
3. Grayston JT: Immunization against trachoma, p 546–559. *In* First International Conference on Vaccine Against Viral and Rickettsial Diseases of Man. Pan American Health Organization Scientific Publ. no. 147. Pan American Health Organization, Washington, D.C., 1967
4. Wang SP, Grayston JT, Alexander ER: Trachoma vaccine studies in monkeys. Am J Ophthalmol 63:1615–1630, 1967
5. Grayston JT, Wang SP, Lin HM, et al: Trachoma vaccine studies in volunteer students of the National Defense Medical Center. II. Response to challenge eye inoculation of egg grown trachoma virus. Chin Med J 8:312–318, 1961
6. Grayston JT: Symposium on trachoma: biology of the virus. J Invest Ophthalmol 2:460–470, 1963
7. Wang SP: A micro immunofluorescence method. Study of antibody response to TRIC organisms in mice, p 273–288. *In* RL Nichols (ed), Trachoma and related disorders caused by chlamydial agents. Excerpta Medica, Amsterdam, 1971
8. Wang SP, Kuo CC, Grayston JT: Simplified method for immunological typing of trachoma-inclusion conjunctivitis–lymphogranuloma venereum organisms. Infect Immun 7:356–360, 1973

9. Yeh LJ, Yang CY, Beasley RP, et al: Studies of trachoma in families on Taiwan. Chin J Microbiol 8:120–132, 1975
10. Grayston JT, Gale JL, Yeh LJ, et al: Pathogenesis and immunology of trachoma. Trans Assoc Am Physicians 85:203–211, 1972
11. Collier LH, Sowa J: Isolation of trachoma virus in embryonated eggs. Lancet i:993–996, 1958
12. Detels R, Alexander ER, Dhir SP: Trachoma in Punjabi Indians in British Columbia: a preliminary study with comparisons to India. Am J Epidemiol 84:81–91, 1966
13. Nichols RL, Bobb AA, Haddad NA, et al: Immunofluorescent studies of the microbiologic epidemiology of trachoma in Saudi Arabia. Am J Ophthalmol 63:1372–1408, 1967
14. Jones BR: The prevention of blindness from trachoma. Trans Ophthalmol Soc UK 95:16–33, 1975
15. Jones BR: Ocular syndromes of TRIC virus infection and their possible genital significance. Br J Vener Dis 40:3–15, 1964

Clinical Epidemiology of Oculogenital *Chlamydia* Infection

CARL H. MORDHORST

Trachoma is defined classically as a chronic inflammatory disease of the eye which eventually may lead to scarring of the conjunctiva and cornea. From a public health viewpoint, trachoma has been classified increasingly as either blinding or nonblinding. In both cases, the causative agent has been shown to be the *Chlamydia trachomatis* organisms. In addition, it is now recognized that *C. trachomatis* is one of the most prevalent genital pathogens causing sexually transmitted diseases such as urethritis, cervicitis, proctitis, and ophthalmia neonatorum.

Whereas the prevalence of trachoma has decreased significantly in many areas of the world during the last decades as economic conditions have improved, blinding trachoma is still a major public health problem in some countries of Africa, the Middle East, and Asia. However, the milder nonblinding trachoma, which includes a low prevalence of potentially blinding lesions but does not cause substantial visual loss, also constitutes a public health problem. This is true in trachoma-endemic areas and also in nonendemic areas with the existing potential for genital tract-to-eye *C. trachomatis* transmission. Limited knowledge is available concerning the occurrence of nonblinding trachoma in areas with a high prevalence of genital chlamydial infections.

Since 1963, a prospective study of *C. trachomatis* infections has been carried out in Denmark. The aim of the study has been primarily to determine the association with *C. trachomatis* of various clinical eye syndromes among patients referred to an outpatient clinic with an inflamed eye of uncertain etiology. Infection by *C. trachomatis* organisms was either proved or very suggestive among 187 patients referred to the eye clinic. *C. trachomatis* eye infections were associated with a rather wide range of ocular clinical manifestations, including cases of classical trachoma. Furthermore, it became apparent that such sporadic cases of *Chlamydia* eye infections in most instances depended upon genital tract-to-eye transmission.

Some results from these studies have previously been reported (1–5). The present communication summarizes the clinical and epidemiological findings from the study during the period 1963–1973.

MATERIALS AND METHODS

Eye examination. Most of the patients were seen at the Out-patients' Eye Clinic, Rigshospitalet, University of Copenhagen. They were examined by conventional biomicroscopy. The observations were recorded in detail by the same person (C.H.M.). The pathological signs suggestive of *C. trachomatis* infection of the conjunctiva and the cornea were graded, and trachoma-stage diagnosis was made according to the criteria recommended by the World Health Organization (6).

Laboratory studies. Epithelial scrapings and/or swabs of conjunctiva and genital tract were collected from pertinent patients for isolation of *C. trachomatis* and other bacterial organisms and for microscopic examinations.

For the demonstration of *C. trachomatis* inclusions, smears of the epithelial cell scrapings were stained by the Giemsa method (3).

Isolation of *C. trachomatis* organisms was carried out by inoculation of epithelial scrapings into embryonated eggs (3). Immunotyping of the isolates was by the mouse toxicity prevention test (7) and/or by the micro-immunofluorescence (micro-IF) method (8, 9).

Serum specimens collected from patients were tested by the complement fixation (CF) test using a *Chlamydia* group antigen derived from *C. trachomatis* strain E/DK-1/ON (5). Sera were also tested by either the standard (10) or simplified micro-IF method (11).

RESULTS

During the period of 1963–1973, a total of 1,001 individuals, mostly from the Copenhagen area, were examined for an external eye disease. In addition to the routine clinical examination, a conjunctival cell scraping was smeared, stained with Giemsa, and studied for the presence of inclusion bodies and cell types. Various additional laboratory studies were carried out on selected cases. The distribution of the overall clinical categories among the patients, based on these examinations, is given in Table 1. There were 187 patients who were suspected of having an ocular *C. trachomatis* infection. Along with their contacts, the persons with suspected *C. trachomatis* infection comprised almost half of all those examined.

During the initial period of the study, a wide variety of external eye diseases were referred to the clinic. Neonatal ocular infection comprised the largest number of *C. trachomatis* eye patients seen. Accordingly, a special effort was made to trace *C. trachomatis* infection among newborns and their families during 1967–1969. Trachoma-like eye infections in increasing numbers were referred to the clinic during the period 1970–1973. The medical community had become increasingly aware of *C. trachomatis* as an oculogenital pathogenic agent.

Survey of mothers and newborns. In 1964 and again in 1969, studies were undertaken of consecutive admissions over a 3-month period to the maternity department of Rigshospitalet, University of Copenhagen. The results are summarized in Table 2. The mothers and their newborn infants had a blood specimen taken at delivery (cord blood on infant) and a conjunctival scraping (1969 only) collected 2 to 6 days after delivery. Despite a two- to threefold increase in the

TABLE 1. *Distribution of patients examined for external eye diseases in 1963–1973 by presumptive clinical categories*

Clinical category	No. of patients
Suspected ocular *Chlamydia trachomatis* infection	187
Contacts of suspected *C. trachomatis* patients	271
Bacteria-like conjunctivitis	211
Viral and atypical conjunctivitis	76
Allergic conjunctivitis or keratoconjunctivitis	42
Various corneal manifestations	144
Normal eye findings	70
Total	1,001

TABLE 2. *Results of laboratory tests for Chlamydia complement fixation (CF) antibodies in sera and inclusion bodies (IB) in conjunctiva cells in mothers and their newborn infants*

Group	No. of patients	*C. trachomatis*	
		CF positive	IB (Giemsa)
1964			
Mothers	120	16 (13%)	NT[a]
Infants	120	15 (13%)[b]	NT
1969			
Mothers	191	25 (13%)	0
Infants	191	24 (13%)[b]	4

[a] Not tested.
[b] Cord blood.

rate of gonorrhea in the Copenhagen area between 1964 and 1969, the level of *Chlamydia* CF antibodies remained constant and was similar to that found in a control group of females. The mothers showed no eye disease. Although the 1969 mothers' Giemsa-stained conjunctival cell smears showed no *Chlamydia* inclusion bodies, four had a very suspicious trachoma-like cytology. Eight infants with a purulent discharge from the conjunctiva had a similar trachoma-like cytology, and in four cases inclusions could be demonstrated. *C. trachomatis* organisms were isolated from three.

Collection of specimens from infants 2 to 6 days after birth was employed for practical reasons, but was poorly timed to demonstrate neonatal *C. trachomatis* eye infections. One of the four *Chlamydia*-positive infants was first diagnosed 3 weeks after delivery and 2 weeks after discharge from the Maternity Department. All newborn infants received silver nitrate eye drops right after birth, and in no instance was a gonorrheal eye infection found during the described survey.

***C. trachomatis* eye disease after the neonatal period.** Among the 187 patients with proven or suspected *C. trachomatis* eye infection were 133 over 1 year of age. The eye disease in these patients fell into three diagnostic categories: inclusion conjunctivitis, subepithelial keratitis, and trachoma. Figure 1 shows the distribution of these cases by age, sex, and clinical diagnosis. The age and sex distribution for each diagnosis was roughly the same. Most cases fell in the 20- to 29-year age group of greatest sexual activity.

The relationship between inclusion conjunctivitis and trachoma, both caused by *C. trachomatis*, has been the subject of a longstanding controversy. Inclusion conjunctivitis has been defined as self-limited, follicular conjunctivitis without keratitis or visual impairment. It has primarily been observed in developed countries in regions with a temporate climate. It is associated with a genital tract-to-eye transmission often linked to cases of nongonorrheal urethritis or cervicitis. In contrast, trachoma has been defined as a chronic follicular keratoconjunctivitis followed by scar formation and is therefore a potentially blinding eye disease. Trachoma is prevalent in underdeveloped countries and is associated with eye-to-eye transmission. Some of the patients summarized in Fig. 1 could not be labeled as having either trachoma or inclusion conjunctivitis. They showed a follicular keratoconjunctivitis without pannus or scarring, but with sub-

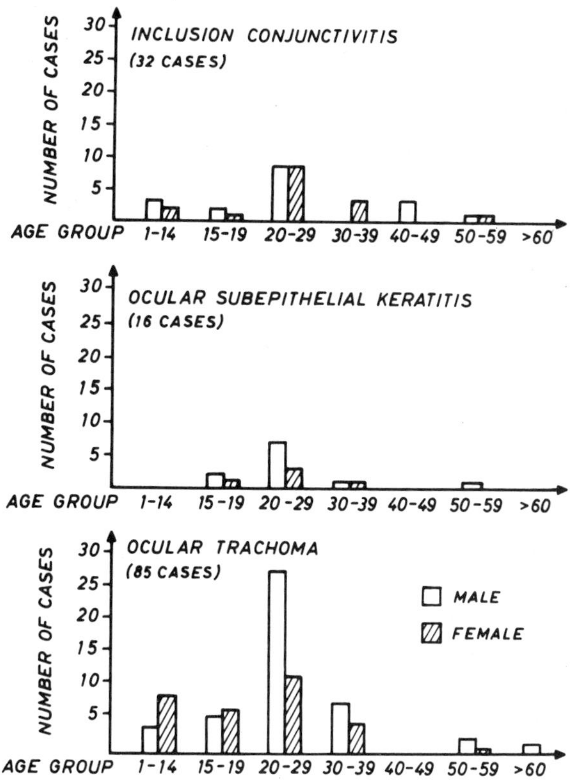

FIG. 1. *Distribution by age, sex, and clinical diagnosis of 133 patients (over 1 year of age) with proven or suspected C. trachomatis eye infection.*

epithelial keratitis. This clinical picture has similarities to epidemic keratoconjunctivitis caused by certain adenovirus types. The patients with subepithelial keratitis were seen earlier in the course of eye disease than most of the patients with inclusion conjunctivitis and trachoma.

The majority of patients were characterized clinically as having trachoma, because they fulfilled the World Health Organization diagnostic criteria, e.g., invasion of vessels into the upper part of the cornea, follicles present on the upper surface of the conjunctiva, and conjunctival scarring. The trachoma cases included both those that were more acute as well as many with protracted courses (up to several years or more).

All patients had at least one Giemsa-stained conjunctival swab smear examined for the presence of chlamydial inclusion bodies. In addition, 103 of the 187 patients suspected of *C. trachomatis* eye infection had at least one conjunctival swab cultured in embryonated egg yolk sac, and 31 of these patients had a genital tract swab similarly cultured. The results of these efforts are shown in Table 3. The organism was demonstrated in most of the newborns but much less frequently in older children and adults. No matter what the syndrome, those with the most acute eye disease were most frequently positive. Culture-positive cases, with four

TABLE 3. *Demonstration of C. trachomatis organisms in 187 patients clinically suspected of having C. trachomatis eye infection*

Clinical diagnosis	No. of patients	No. of patients positive/no. examined			
		Smear for inclusions		Culture of *C. trachomatis*	
		Conjunctiva	Genital tract	Conjunctiva	Genital tract
Neonatal inclusion conjunctivitis	54	50/54	0/2	17/44	0/3
Inclusion conjunctivitis	32	7/32	0/12	5/13	1/6
Ocular subepithelial keratitis	16	8/16	0/12	3/11	0/5
Ocular trachoma	85	21/85	1/48	16/35	4/17
Total	187	86/187	1/74	41/103	5/31

exceptions, had typical inclusion bodies demonstrated on smear. During the last few years of the study, only very few patients were cultured. The genital tract smears and cultures were carried out mostly on patients with symptoms or signs of inflammation. Although few positive genital tract swabs were found, culture was more sensitive than smears.

Growth of potential pathogenic bacteria from conjunctival swabs was uncommon whatever the clinical syndrome. Apart from cultures of *Staphylococcus aureus*, nonhemolytic streptococci, and occasional strains of other species, a few *Haemophilus* strains were isolated and further tested by Kilian et al. (12). From these studies it was concluded that the presence of the *Haemophilus* strains examined was most likely of little significance.

Upper respiratory tract swabs were obtained from a few patients, but none of 26 specimens yielded an isolate. Cultures of 160 conjunctival swabs from 98 patients classified clinically as non-*C. trachomatis* infections were negative.

Laboratory studies of the 271 contacts of *C. trachomatis* patients showed 7 with inclusion bodies in conjunctival smears, but none of 52 had isolates on culture of the conjunctiva. However, nine isolates (from 62 cultured) were obtained from genital tract specimens, all but one from parents of infants with ophthalmia neonatorum.

C. trachomatis was isolated from 55 persons in this study; 41 isolates came from the conjunctiva and 14 were of genital origin. Table 4 shows 50 of the strains according to immunotype and origin. In no instance was immunotype A, B, or C isolated. These have been the dominant types found in endemic trachoma by Wang and Grayston (8). Instead, the types found were those associated with genital

TABLE 4. *Distribution of C. trachomatis immunotypes according to type and origin*

Origin	*C. trachomatis* immunotype[a]								No. of isolates[b]
	Ba	E	D	G	F	K	H	J	
Ocular	4	15	8	5	2	1	2	1	38
Genital		6	2	2	1		1		12
Total	4	21	10	7	3	1	3	1	50

[a] In no instance was immunotype A, B, C, I, L_1, L_2, or L_3 isolated.
[b] Isolates obtained from 47 patients.

TABLE 5. *Distribution of micro-IF antibodies in sera among clinically suspected ocular C. trachomatis-infected patients and their contacts*

Clinical category	No. of sera	Micro-IF serotype										
		C/J	A	B	Ba	ED	GF	K	H	I	Broad	Negative
Index eye patients	158	5[a]	<1	<1	3	42	13	3	4	0	1	27
Contacts of index patients .	230	5	2	0	3	31	9	3	<1	<1	1	43

[a] Percentage of patients.

tract infection (with the exception of Ba) and ocular infection in nonendemic trachoma countries. The types isolated from the genital tract were similar.

Table 5 shows the distribution of micro-IF antibodies among the eye disease patients and their contacts. The distribution of antibodies by serotype was similar among the index patients and their contacts and corresponded to the immunotypes obtained by culture.

Sera from 147 of the 187 patients with proven or suspected *C. trachomatis* eye infection were tested by the CF and micro-IF methods with chlamydial antigens. All 48 patients for whom the organism had been demonstrated in smear or culture had micro-IF antibody, and 24 had CF antibody. In those patients from whom the organism was not demonstrated, about 63% had micro-IF antibody (62 of 99) and 40% had CF antibody (39 of 99).

C. trachomatis serum antibody among contacts of the patients was found about as frequently as in the eye patients. This was especially true for sexual contacts, more than 90% of whom had micro-IF serum antibody. The same trachoma immunotype specificity was found in the serum antibody from the patient and the contact. Similar findings were observed among parents (47 mothers and 35 fathers tested) of newborn infants with proven *C. trachomatis* eye infection. More than 90% of the mothers had micro-IF antibodies at the time of delivery. The immunotype specificity of serum antibody from the mother corresponded to that of her infant. The antibody remained demonstrable for 6 to 7 years, with the same immunotype persisting. Among the fathers, 80% showed micro-IF antibodies, most of which (25 of 28) were the same immunotype as those of the infant and the mother. The fathers' antibody titers were usually lower and tended to disappear earlier than those of the mothers.

In a previous publication (5), follow-up examination of newborns with *Chlamydia* eye infection, and of their parents, has been reported. In that study more than 50% of the infants demonstrated signs of subsequent development of mild ocular trachoma, even though they had initially been treated with tetracycline eye ointment for 2 to 3 weeks. Many of the mothers gave a history of symptoms of pelvic inflammation months or years after delivery. These observations were confirmed in 1967 to 1968, when all mothers of babies with neonatal *Chlamydia* infection were closely followed over a 15-month period. During that period, 16 infants were diagnosed and 15 mothers were followed in the clinic. Paired sera were obtained from 12 mothers, and 7 of these demonstrated a rise in micro-IF antibody titer in the months following delivery. Seven mothers also showed a rise in *Chlamydia* CF antibody titers during the same period, including six of the seven with a micro-IF titer rise. The mothers with increase in antibody titers were more apt to have both symptoms and signs of prolonged genital tract infection. Although the infants seemed to react satisfactorily to local tetracycline treatment given over 3 to 4 weeks, relapses at a later time were found in a few. The potential of neonatal *Chlamydia* infection developing into more serious trachoma-like disease with visual impairment has been suggested by observation of cases of childhood trachoma among Danish eye patients (C. H. Mordhorst, S. P. Wang, and J. T. Grayston, in preparation).

Figure 2 shows the frequency of the principal clinical manifestations in adults with proven ocular *Chlamydia* infection. The high rate of both upper conjunctival follicles and keratitis including pannus and scarring may well indicate that the patients in the present study were those with more severe disease among a larger pool of *Chlamydia* eye infections. Their histories suggest that they responded unfavorably to treatment of the eye disease, usually as a result of inadequate or improper therapy. Only after prolonged, relapsing, or especially severe eye disease were they referred to this study. The high rate of symptoms from the genital tract among the present series of referred eye patients may or may not be representative for all *Chlamydia* eye patients, but it reaffirms the need for examination for possible genital disease due to *C. trachomatis* organisms.

Upon completion of treatment with specific antibiotics and with the disappearance of oculogenital symptoms, many patients reported that upper respiratory symptoms, which had been present for months or even years, disappeared.

When contacts of the *C. trachomatis* patients were examined, eye disease was only rarely found, but symptoms of genital tract infection in the past or present were noted in almost 75% of these individuals.

DISCUSSION

During the past half century, trachoma has not been regarded as a public health problem in Denmark. Trachoma could be diagnosed only by clinical observation

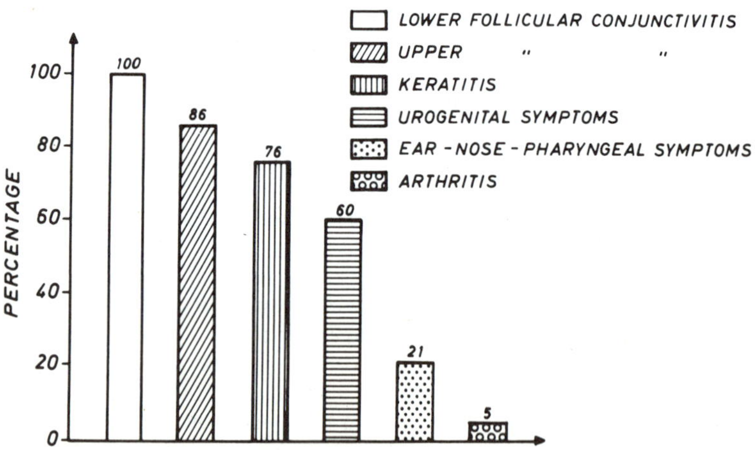

FIG. 2. *Prevalence of the principal clinical manifestations in 42 patients (over 1 year of age) with proven C. trachomatis eye infection.*

until 1957, when *C. trachomatis* organisms were first propagated in chick egg yolk sac (13). We have used this technique since 1963 along with cytological and serological methods for diagnosis. The patients referred for these studies were a very select group. The eye patients presented with an external eye disease of unknown etiology which had not responded to treatment. The special intensive laboratory studies for *Chlamydia* infection were concentrated on the 187 patients suspected clinically of *C. trachomatis* eye infection and their contacts, and on a special study among mothers and their newborn infants. The latter study resulted in the finding that at least 2% of consecutively admitted newborns had a severe neonatal blennorrhea due to *C. trachomatis*. That finding, along with the finding of many indigenous trachoma cases, demonstrates that these organisms do cause a public health problem in Denmark and suggests that diagnostic procedures should be available and preventive measures should be implemented.

The long-term observations of this study yielded several important findings. Chronic eye disease with mild trachoma clinical changes was observed as a sequela to inclusion conjunctivitis of the newborn. Some of the severe relapsing childhood trachoma cases may have had their origin in neonatal conjunctivitis. Three syndromes (inclusion conjunctivitis, ocular subepithelial keratitis, and ocular trachoma) of eye disease were seen in association with *C. trachomatis* infection in persons more than 1 year of age. The largest number were diagnosed as trachoma. Long-term observation and careful review of earlier ophthalmological examinations suggested a unity in these syndromes with progression to signs diagnosable as trachoma (pannus and typical scar formation). In addition to the cases demonstrating subepithelial keratitis without pannus formation, many of those diagnosed as trachoma had a history of keratitis diagnosed previously. Many of the patients had been diagnosed as having epidemic keratoconjunctivitis of the type due to adenoviruses. Ophthalmologists have great concern over the communicability of these infections, and they were much more apt to diagnose adenovirus than trachoma infection in patients with subepithelial keratitis.

It is necessary to emphasize that this study was not set up to provide similar laboratory and clinical examinations on all patients. Some of the patients that were of interest for special reasons received much more thorough workups and were much more apt to have a specific etiological diagnosis confirmed. This, along with the special nature of those patients that were referred for examination, makes it impossible to determine any percentages or rates of *C. trachomatis* infections in Denmark. The conclusion here is that such infections with several different clinical syndromes are appearing with some frequency in Denmark, and it is estimated that if they were vigorously sought, many more could be identified.

ACKNOWLEDGMENTS

I am most grateful to San-Pin Wang, who carried out all the immunotyping of isolates and tests for micro-IF serum antibodies, and to J. Thomas Grayston for his helpful comments on the manuscript. I also wish to thank Ellen Christensen for valuable technical assistance throughout the study.

REFERENCES

1. Mordhorst CH: Isolation of the TRIC agent from 3 cases of neonatal inclusion blennorrhoea and from the cervix of one of the mothers. Acta Pathol Microbiol Scand 62:301–302, 1964

2. Mordhorst CH: Isolation of the TRIC agent from eye and urogenital tract of trachoma patients in Denmark. Acta Pathol Microbiol Scand 63:637–638, 1965
3. Mordhorst CH: TRIC agents isolated in Denmark. I. Isolation of the TRIC agent from inclusion blennorrhea. Acta Pathol Microbiol Scand 64:277–285, 1965
4. Mordhorst CH: Studies on oculogenital TRIC agents isolated in Denmark. Am J Ophthalmol 63:1282–1288, 1967
5. Mordhorst CH, Dawson CD: Sequelae of neonatal inclusion conjunctivitis and associated disease in parents. Am J Ophthalmol 71:861–867, 1971
6. Tarizzo ML: Field methods for the control of trachoma. World Health Organization, Geneva, 1973
7. Alexander ER, Wang SP, Grayston JT: Further classification of TRIC strains from ocular trachoma and other sources by the mouse toxicity prevention test. Am J Ophthalmol 63:1469–1478, 1967
8. Wang SP, Grayston JT: Classification of TRIC and related strains with microimmunofluorescence, p 305–321. *In* RL Nichols (ed), Trachoma and related disorders caused by chlamydial agents. Excerpta Medica, Amsterdam, 1971
9. Wang SP, Kuo CC, Grayston JT: A simplified method for immunological typing of trachoma-inclusion conjunctivitis–lymphogranuloma venereum organisms. Infect Immun 7:356–360, 1973
10. Wang SP, Grayston JT: Human serology in *Chlamydia trachomatis* infection with microimmunofluorescence. J Infect Dis 120:388–397, 1974
11. Wang SP, Grayston JT, Alexander ER, et al: Simplified microimmunofluorescence test with trachoma-lymphogranuloma venereum (*Chlamydia trachomatis*) antigens for use as screening test for antibody. J Clin Microbiol 1:250–255, 1975
12. Kilian M, Mordhorst CH, Dawson CR, et al: The taxonomy of haemophili isolated from conjunctivae. Acta Pathol Microbiol Scand Sect B 84:132–138, 1976
13. T'ang FF, Chang HL, Huang YT, et al: Studies on the etiology of trachoma with special reference to isolation of the virus in chick embryo. Chin Med J 75:429–446, 1957

Response to Treatment in Ocular Chlamydial Infections (Trachoma and Inclusion Conjunctivitis): Analogies with Nongonococcal Urethritis

C. R. DAWSON, J. P. WHITCHER, C. LYON, AND J. SCHACHTER

Severe endemic trachoma is analogous to nongonococcal urethritis (NGU) in a number of ways: the disease process is a chronic inflammation of mucous membranes, endemic trachoma and NGU are multifactorial in origin but chlamydiae play an important role, and the response to antibiotic treatment is often unsatisfactory. The complex microbiology and treatment of endemic trachoma appear to have parallels in the disease process in NGU.

PATHOGENESIS OF TRACHOMA

Trachoma is a chronic inflammatory disease of the conjunctiva which leads to scarring of the conjunctiva and cornea. In regions where trachoma leads to a substantial amount of blindness, the conjunctival inflammation is a result of infection with *Chlamydia* in association with a number of bacterial pathogens, commonly *Haemophilus* sp. (*H. aegyptius* and *H. influenzae*) and the pneumococcus, occasionally *Moraxella* sp. and *Staphylococcus aureus*, and rarely the gonococcus (as epidemic conjunctivitis) or beta-hemolytic streptococcus (1, 2). The inflammatory response consists of infiltration of the subepithelial tissues with inflammatory cells (particularly lymphocytes and plasma cells) and the formation of lymphoid germinal centers that are observed clinically as "lymphoid follicles." Giemsa-stained conjunctival scrapings show both polymorphonuclear leukocytes and lymphocytes in large numbers and, in more severe cases, plasma cells, macrophages with ingested debris ("Leber cells"), and immature lymphocytes and stem cells ("follicle cells") (3).

Scarring of the conjunctiva results from the inflammation, and the degree of scarring depends on the intensity and duration of the inflammation. We have recently described four degrees of intensity of inflammation in trachoma, based on clinical finding, as: severe, moderate, mild, and inactive or insignificant (4–7). There is a clear-cut correlation between chlamydial infection and clinical intensity; in one study we found trachoma inclusions by Giemsa or fluorescent-antibody (FA) staining of conjunctival smears in 56% of severe, 20% of moderate, and 4% of mild cases of endemic trachoma in southern Tunisia (5), and this correlation has been confirmed by Jones (6, 7).

Although endemic trachoma is always associated with chlamydial infection, other pathogenic bacteria contribute to the intensity of the disease. The dual role of chlamydial and bacterial infection in determining the intensity of chronic trachoma is illustrated in Table 1. Severe disease is clearly associated with the presence of trachoma agent in conjunctival smears; yet, in a substantial proportion of severe

TABLE 1. *Chlamydial agents and Koch-Weeks bacillus as factors in trachoma intensity in Tunisia, October 1972*

Inclusions[a]	K-W[b]	No.	Trachoma intensity		
			Severe	Moderate	Mild
+[c]	+	11	82[d]	9	9
+	−	19	84	11	5
−	+	28	68	25	7
−	−	80	28	60	12
Total		138	48	42	10

[a] Trachoma inclusions in Giemsa- and immunofluorescent-stained smears.
[b] Koch-Weeks bacillus in Giemsa-stained smears.
[c] +, Organism found; −, organism not found.
[d] Percentage of patients.

intensity cases trachoma inclusions were not found. Among these inclusion-negative cases there is a significant association of disease severity with *H. aegyptius* (Koch-Weeks bacillus) infection. Corneal involvement (keratitis) had a similar pattern (Table 2).

In Egypt we have observed a similar relationship with *Neisseria gonorrhoeae* infections of the eye which occur in self-limited outbreaks in young village children but do not lead to the corneal ulceration seen with the occasional gonococcal eye infections in developed countries. Other bacterial pathogens which appear to contribute to trachoma intensity are the pneumococcus (most common in winter months), *Moraxella* sp., and, less commonly, *S. aureus*, beta-hemolytic streptococcus, and gram-negative rods (1, 8). The bacterial flora also includes organisms such as Viridans group streptococci and diphtheroids, which are present in abundance but are not associated with conjunctival inflammation (1). Ocular adenovirus infections have been reported in children under the age of 2 years (9), and, although self-limited, they can produce marked conjunctival scarring (10).

EPIDEMIOLOGY AND NATURAL HISTORY OF TRACHOMA

In those rural areas of developing countries where trachoma is still a blinding disease, almost all children acquire the disease by the second year of life (5, 7). The prevalence of active disease begins to decrease between 5 and 6 years of age, and there is little active disease in persons after age 15 years. In such communities, the intensity of inflammatory disease in any one individual is relatively constant and is a

TABLE 2. *Keratitis in endemic trachoma (1): relation of inclusions and Koch-Weeks bacillus (K-W) to corneal infiltrates (Tunisia, October 1972)*

Inclusions[a]	K-W[a]	No.	Infiltrates (%)
+	+	10	90
+	−	18	83
−	+	28	75
−	−	77	47

[a] Detected in Giemsa-stained conjunctival smears.

result not only of eye infection (chlamydial and bacterial) but also of environmental factors (crowding, lack of water, eye-seeking flies, poor housing) and host factors (being less than 10 years old) (7). There is a clear-cut association of severe trachoma with impoverished rural economic systems, and economic development leads to the amelioration and then the disappearance of active trachoma in affected communities.

In some patients, the conjunctival scarring produced by chronic inflammation causes the eyelids to turn inward so that the lashes constantly abrade the cornea. In addition, the scarring reduces the amount of tears, so that abrasions of the corneal epithelium, whether produced by the inturned lashes or by minor ocular injuries, are frequently followed by secondary infection, corneal opacity, and loss of vision.

Only in patients with conjunctival inflammation that reaches a minimal degree or "threshold" of intensity is the conjunctival scarring so marked that there is a high risk of loss of vision. Unexpectedly, the prevalence of potentially disabling lesions of the eye (distortion of the lids and inturned eyelashes) continues to increase for 10 to 20 years after the peak of inflammatory disease (5, 11). This apparent progression of scarring may result either from recurrent bouts of active trachoma or, more likely, from slow shrinkage of the conjunctival scars. Thus, the active inflammatory disease in the first 10 years of life may lead to visual disability as long as 20 to 40 years after the inflammatory disease has subsided.

RESPONSE OF ACTIVE TRACHOMA TO TREATMENT

Since endemic trachoma is a chronic, multifactorial disease, the response to treatment has some pertinence to the treatment of NGU. In a series of treatment trials of active trachoma in Tunisia, we compared 10 weeks of treatment with topical antibiotics (tetracyclines, erythromycin, or rifampin) with topical boric acid application (4, 8). Although there was a significant decrease in trachoma intensity in antibiotic-treated groups, active disease persisted in some patients, and disease intensity increased again in the absence of continued treatment. For example, among 42 patients with severe disease treated with either tetracycline or rifampin ointment, trachoma of severe intensity was found in 7% at 5 weeks after treatment, in 24% at 19 weeks after treatment, and in 38% at 39 weeks. In contrast, among 20 severe intensity cases treated with boric acid, severe disease persisted at 5 weeks after treatment in 80%, at 19 weeks in 70%, and at 39 weeks in 60% (8). Since this therapeutic trial was limited to schoolchildren, we assumed that the recurrence of disease activity was due primarily to reinfection in the home environment. The decline in disease intensity in the boric acid-treated group reflects the rapid improvement that is occurring naturally in this age group.

Another important feature of these trials is that the chlamydial agent was suppressed to a greater extent than would be expected by the clinical response. Thus, among children with severe disease, 56% were inclusion positive before treatment, but 20% of severe cases had inclusions after treatment; among those with moderate disease, 20% were inclusion positive before treatment, but only 5% of moderate cases were inclusion positive after treatment (5).

In adult oculogenital chlamydial infection, the effect of treatment on isolation of chlamydiae is even more dramatic; after one or two applications of sulfonamides or antibiotic drops to the eye, chlamydiae are rarely isolated, even when inclusions (albeit somewhat abnormal ones) are still present in the smears.

POSSIBLE IMPLICATIONS FOR THE STUDY OF NGU

There is still considerable controversy about the etiology of NGU, but it is probable that chlamydial infection is one of the most important causes. Even in oculogenital cases with proven infection of the eye, we have not been able to demonstrate genital chlamydial infection in some cases (12). Extrapolating from the data on trachoma and adult oculogenital infections, it would be expected that chlamydial infection could not be demonstrated in every case of NGU caused by that organism and that chlamydial organisms may account for a much higher proportion of NGU cases than would be indicated by agent isolation alone.

Studies of viral and chlamydial infection of the female genital tract have shown that multiple infections are relatively common, with cytomegalovirus, herpes simplex virus, and *Chlamydia* all being found frequently in the same patient (13). In both the male and the female, bacterial pathogens may contribute to the intensity of inflammation in the genital tract, as they do in endemic trachoma. Although some of these organisms (e.g., *Haemophilus* sp.) may not ordinarily be considered as pathogens in the genital tract, they might well enhance the intensity of inflammation in patients with chlamydial urethritis.

As in trachoma, the intensity of inflammation in NGU may be worse in poor or lower class patients with less access to medical care. We are not aware that chronic NGU leads to scarring of the urethra, but such scarring might be expected in severe cases that persist for some length of time. If complications were to occur from such scarring, they would be seen 10 to 20 years after the initial episodes as a result of slow shrinkage and contraction of the scars.

Although tetracycline reduces the symptoms of most NGU caused by chlamydial infection, it would be expected that some chronic cases might not respond quickly to antibiotics. Even relatively small amounts of antibiotics, however, appear to suppress the infection and render negative the laboratory tests for the demonstration of chlamydiae. In our treatment studies of trachoma, disease activity probably recurred because of continued exposure to infection in the home. Similarly, some cases of NGU would not be expected to respond to treatment because of continued exposure to untreated sexual consorts.

ACKNOWLEDGMENTS

This investigation was supported by Public Health Service grants EY.00427 and EY.53797 from the National Eye Institute and AI.01093 from the National Institute of Allergy and Infectious Diseases, and also by PL.480 grant 07-075.

REFERENCES

1. Vastine DW, Dawson CR, Daghfous T, et al: Severe endemic trachoma in Tunisia. I. The effect of topical chemotherapy on conjunctivitis and ocular bacteria. Br J Ophthalmol 58:833–842, 1974
2. Vastine DW, Dawson CR, Hoshiwara I, et al: Comparison of media for the isolation of *Haemophilus* species from cases of seasonal conjunctivitis associated with severe endemic trachoma. Appl Microbiol 28:688–690, 1974
3. Yoneda C, Dawson CR, Daghfous T, et al: Cytology as a guide to the presence of chlamydial inclusions in Giemsa stained conjunctival smears in severe endemic trachoma. Br J Ophthalmol 59:116–124, 1975
4. Dawson CR, Daghfous T, Messadi M, et al: Severe endemic trachoma in Tunisia. II. A

controlled therapy trial of topical chlortetracycline and erythromycin. Arch Ophthalmol 92:198–203, 1974
5. Dawson CR, Daghfous T, Messadi M, et al: Severe endemic trachoma in Tunisia. Br J Ophthalmol 60:245–252, 1976
6. Jones BR: Laboratory tests for chlamydial infection. Their role in epidemiological studies of trachoma and its control. Br J Ophthalmol 58:438–454, 1974
7. Jones BR: The prevention of blindness from trachoma. Trans Ophthalmol Soc UK 95(Pt 1):16–33, 1975
8. Dawson CR, Hoshiwara I, Daghfous T, et al: Topical tetracycline and rifampicin therapy of endemic trachoma in Tunisia. Am J Ophthalmol 79:803–811, 1975
9. Bell SD, McComb DE, Murray E, et al: Adenoviruses isolated from Saudi Arabia. I. Epidemiologic features. Am J Trop Med Hyg 8:492–500, 1959
10. Dawson CR, Hanna L, Togni B: Adenovirus type 8 infections in the U.S. IV. Observations on the pathogenesis of lesions in severe eye disease. Arch Ophthalmol 87:258–268, 1972
11. Reinhard J, Weber A, Nizetic B, et al: Studies of the epidemiology and control of seasonal conjunctivitis and trachoma in Southern Morocco. Bull WHO 39:497–545, 1968
12. Dawson CR, Schachter J: TRIC agent infections of the eye and genital tract. Am J Ophthalmol 63(Pt 2):1289–1298, 1967
13. Wentworth BB, Bonin P, Holmes KK, et al: Isolation of viruses, bacteria and other organisms from venereal disease clinic patients; methodology and problems associated with multiple isolations. Health Lab Sci 10:75–81, 1973

Perinatal Chlamydial Infection

ELISABETH REES, I. ANNE TAIT, DEREK HOBSON
AND F. WILLIAM A. JOHNSON

The presence of an inflammatory process in the cervix may affect the fetus in the prenatal, intrapartum, or neonatal period. In the prenatal period, the fetal environment may be compromised through bacteremic spread in the mother or ascending infection from the cervix, resulting in chorioamnionitis. The latter commonly develops after prolonged rupture of the membranes, but infection across intact membranes also occurs, and, on examination of placentas, limitation of the chorioamnionitis to the portion of the membranes overlying the os is often found (1). Premature birth is likely, possibly due to rupture of the weakened and inflamed membranes at this site (2).

The most important factor influencing the health of the infant is premature birth. In the neonatal period, 60% of deaths occur in the 10% of infants born prematurely. In addition to increased mortality, neurological sequelae are more frequent (3).

In considering chlamydial cervicitis in pregnancy, therefore, the relative incidence of prematurity not associated with general factors affecting maternal health, such as toxemia, may be of importance. Ascending infection due to *Chlamydia trachomatis*, in either the pregnant or nonpregnant woman, has not yet been established. Chlamydial conjunctivitis in an infant delivered by caesarean section is suggestive, but postnatal transmission by fingers of the attendants cannot with certainty be excluded, although the chances of this are more remote if the infant is transferred to a special-care unit immediately after birth.

Neonatal conjunctivitis and associated infection in the parents has been recognized since 1907, when Halberstaedter and von Prowazek (4) described inclusions in eye and genital exudate. Since the development of sensitive tissue culture techniques, the epidemiological pattern of the disease and the clinical manifestations and sequelae of infection in infants referred to departments of ophthalmology have been defined (5, 6).

An investigation into the presentation of chlamydial conjunctivitis in maternity units was undertaken to examine the perinatal factors influencing the clinical picture, the response to treatment, and the pattern of infection in the parents.

MATERIALS AND METHODS

Of the 81 babies examined between October 1973 and March 1976, 74 were referred by pediatricians, 3 were referred by ophthalmologists, and 4 were examined because their mothers had been treated for chlamydial cervicitis (3 during the pregnancy and 1 before pregnancy). Four attended as outpatients. The group forms a selected series of cases of conjunctivitis referred mainly because of failure to respond to routine treatment or because no bacterial growth was reported on bacteriological investigation. A few cases were seen because of very early development of signs, the youngest being 6 h. In the United Kingdom, cases of gonococcal conjunctivitis are usually referred to the venereologist.

Specimens were obtained with cotton wool-tipped swabs from the mucosal surface of the lower lids in all cases, and from the upper lids when possible, and were examined microscopically and by culture for *Neisseria gonorrhoeae* and other bacteria. Stuart's medium was used for transport, but most gonococcus swabs were inoculated directly onto modified Thayer-Martin medium. Swabs for *C. trachomatis* were put into transport medium for culture in untreated replicating McCoy cells (7). Specimens for herpes simplex virus identification by electron microscopy and culture were taken in some cases.

Mothers were investigated in the maternity unit when possible, but mothers of premature babies had usually been discharged and were requested to attend as outpatients. All were investigated for syphilis, gonorrhea, chlamydial cervicitis, trichomoniasis, candidiasis, and, in a few cases, genital herpes. When possible, fathers were similarly investigated if the infant and mother had gonococcal, chlamydial, or syphilitic infection. *Chlamydia*-positive mothers and infants were requested to attend once during treatment and then 2, 5, 8, 16, 28, 40, and 52 weeks post-treatment. Swabs for *C. trachomatis* were taken from the infant on each occasion. Swabs from the mother were taken for *C. trachomatis*, gonococcus, *Trichomonas vaginalis*, and *Candida* species at each visit after treatment.

RESULTS

Babies. *C. trachomatis* was isolated from 29 babies, and *N. gonorrhoeae* was isolated from 9. Delivery was by caesarean section in four cases; *N. gonorrhoeae* was isolated from one of these babies, and *Chlamydia* was isolated from one.

In the 29 *Chlamydia*-positive babies, the onset of discharge varied between 3 and 13 days (mean, 6.9 days) (Table 1). Early onset was associated with concurrent bacterial infection, which was not common. *N. gonorrhoeae* was isolated in three cases, *Staphylococcus aureus* in two, *S. albus* in two, *N. pharyngis* in one, diphtheroids in one, and *Streptococcus faecalis* in one. No swab was taken in one case treated by the general practitioner after discharge from hospital. In the remainder, the laboratory reported no bacterial growth or no pathogens grown. Late onset at 9, 11, and 13 days occurred in three babies treated at birth with intramuscular cloxacillin and ampicillin for prematurity associated with prolonged rupture of the membranes.

C. trachomatis was isolated from both eyes in 18 cases, the right eye in 7, and the left eye in 4. The inclusion count per whole cover slip (IFU/WC) varied between 1

TABLE 1. *Day of onset of discharge in 29 cases of neonatal chlamydial conjunctivitis*

Day of onset	No. of babies
3	3
5	3
6	8
7	7
8	3
9	2
11	2
13	1

and 86,000, and was low in cases which had received prior antibiotic therapy other than topical neomycin.

All babies developed edema of the lids and a purulent discharge, which was often thin and watery in the absence of concurrent bacterial infection. In two cases it was blood stained. Localized "bubbling" of the mucosa was a marked feature in the more severe cases and in those in which the request for investigation was late (Fig. 1).

Twenty-two cases had received prior topical treatment as follows: neomycin, 13; chloramphenicol, 5; neomycin and chloramphenicol, 2; gentamicin and chloramphenicol, 1; penicillin and tetracycline, 1. Neomycin had no effect on the development of the conjunctivitis, but babies receiving chloramphenicol at the time of taking swabs usually had no discharge and were said to be improving. Slight to moderate lid edema and slight to severe conjunctivitis and mucosal edema were found on examination, depending on the duration of therapy (Fig. 2).

Diagnosis was delayed until 30 days in one ill premature baby (30-week gestation) who developed conjunctivitis at 3 days and was treated with intramuscular penicillin and kanamycin for 5 days followed by gentamicin for 5 days for *Escherichia coli* infection. *Streptococcus faecalis* was then isolated, and chloramphenicol eye drops were given for 5 days. The conjunctivitis subsided, but recurred and was retreated twice. *Chlamydia* was isolated on the second and fifth days of the third course of chloramphenicol. Herpesvirus particles were identified on electron microscopy. Only slight edema of the lids and slight conjunctivitis were present. No discharge was seen. The mucosa was edematous, and "bubbling" was present.

One baby (35-week gestation) was delivered by caesarean section because of fetal distress 3 h 15 min after spontaneous rupture of the membranes. The baby was nursed in the special-care unit where, at 11 days, slight edema of the lids of one eye, conjunctivitis, and mucopurulent discharge developed. This failed to respond to neomycin eye drops, and only scanty *Staphylococcus albus* was grown on culture. *C. trachomatis* was isolated from swabs taken on the 13th day (Fig. 3). No other baby was under treatment for chlamydia conjunctivitis in the unit. The mother developed pyrexia on the third day and was treated with ampicillin, 500 mg every 6 h for 7 days.

Treatment and follow-up. Tetracycline eye ointment five times daily for 4 weeks was prescribed. Babies attended for follow-up for periods up to 11 months. *Chlamydia* was reisolated from four infants 1 to 6 weeks after cessation of treatment. One had received treatment for 2 weeks, one for 3 weeks, and two for 4 weeks. Isolates were obtained from both eyes of the caesarean baby on relapse, although only one eye had shown clinical or bacteriological evidence of infection initially, and both eyes had been treated. Two mothers reported slight intermittent swelling of the lids and discharge. Slight conjunctivitis was present in only one.

Prematurity. Twelve of the 29 babies were premature (41.4%). The gestation period was 30 to 32 weeks in five, 33 to 34 weeks in four, and 35 to 37 weeks in three. None of the mothers of these babies had toxemia of pregnancy. Two babies had a ventricular septal defect. In addition, one *Chlamydia*-positive mother had a baby who died of cardiac failure at 3 days and was found at postmortem examination to have a hypoplastic left ventricle, patent ductus arteriosus, and preductal co-arctation.

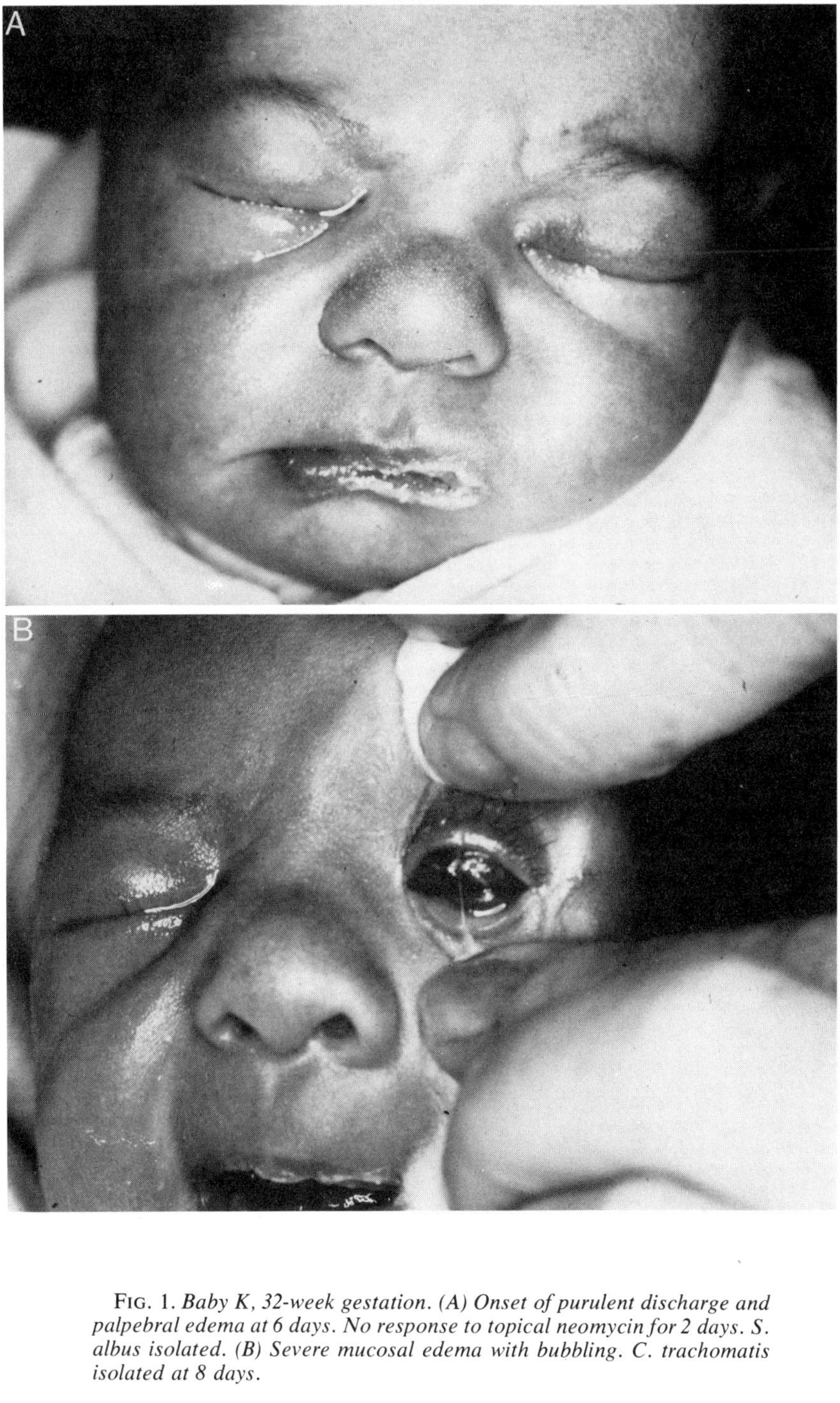

FIG. 1. *Baby K, 32-week gestation. (A) Onset of purulent discharge and palpebral edema at 6 days. No response to topical neomycin for 2 days. S. albus isolated. (B) Severe mucosal edema with bubbling. C. trachomatis isolated at 8 days.*

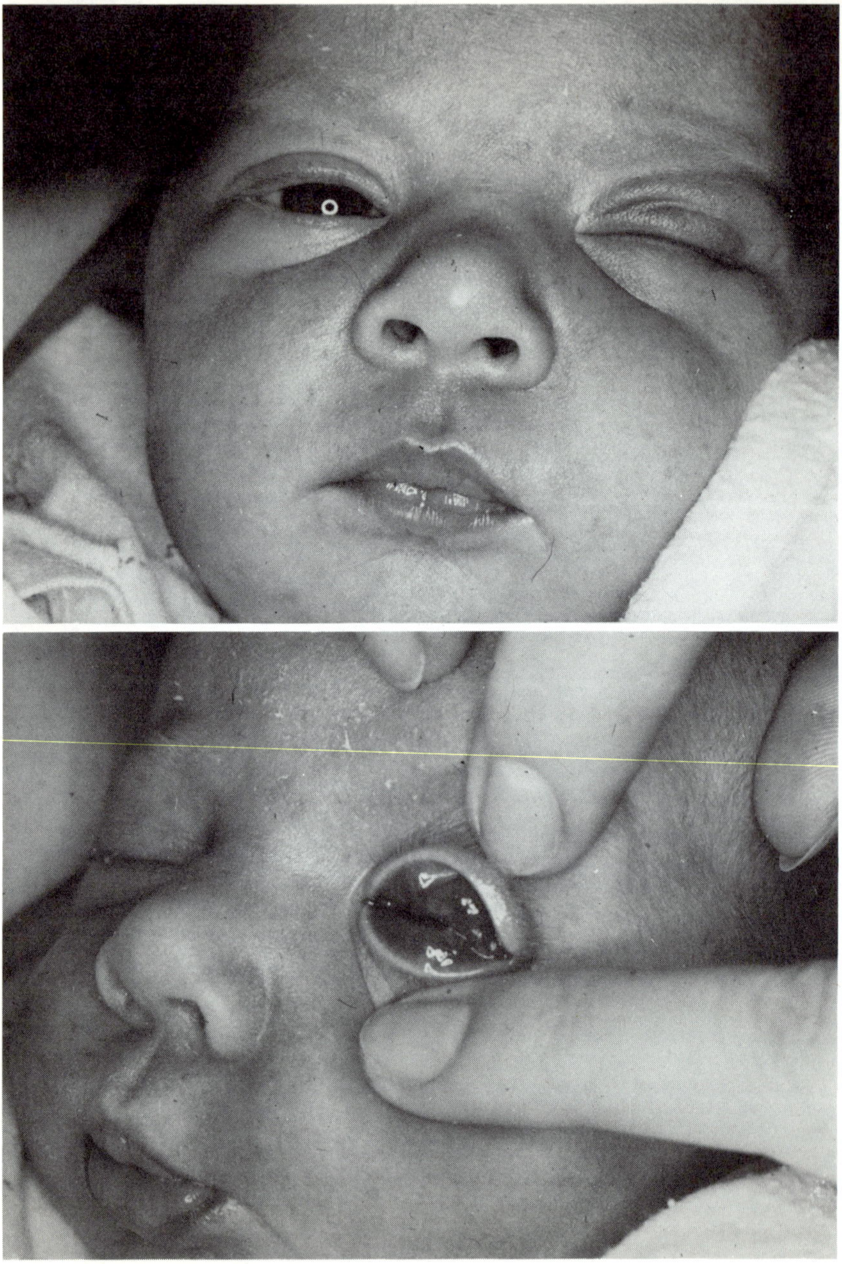

FIG. 2. *Baby R, 40-week gestation. (A) Onset of purulent discharge and palpebral edema at 5 days. Treated with topical neomycin for 2 days (no response) and topical chloramphenicol for 5 days with improvement. (B) Mucosal edema with bubbling, no discharge. C. trachomatis isolated at 12 days.*

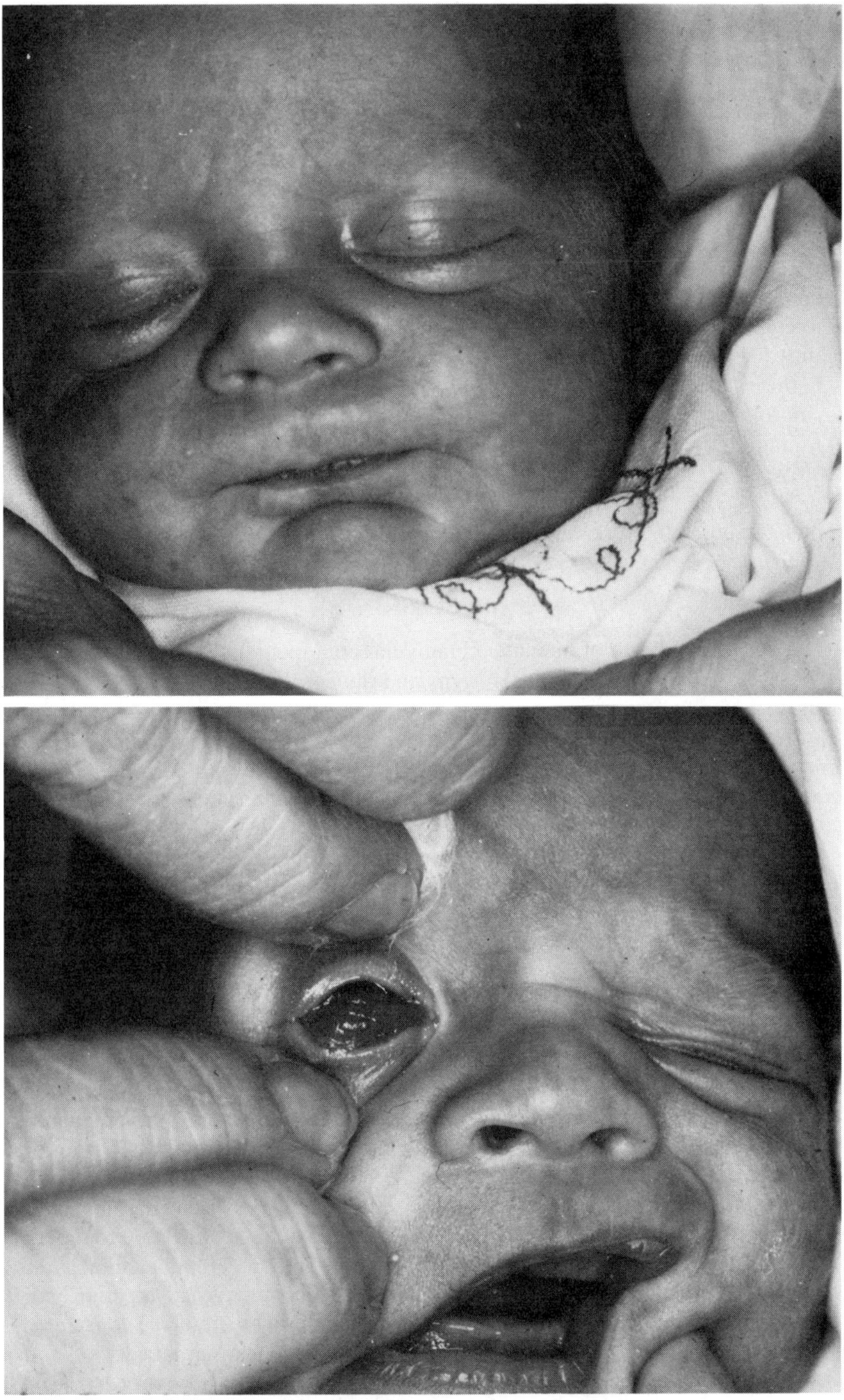

FIG. 3. *Baby C, 35-week gestation, delivered by caesarean section. (A) "Sticky eye" at 11 days. Early palpebral edema of right eye. No response to topical neomycin for 2 days. (B) Typical mucosal edema, scanty mucopurulent discharge in right eye. C. trachomatis isolated at 13 days.*

Mothers. Twenty-four mothers, aged 17 to 36 years, were examined, of whom 14 were married and 10 were single. Seven had received antibiotics prior to examination. The cervical specimen from one mother was cytotoxic. Among the 16 remaining patients, *C. trachomatis* was isolated from the cervix of 15 and *N. gonorrhoeae* from 3. Postpartum infection occurred in 16 (66.6%). Seven mothers received antibiotics in the maternity unit for pyrexia (3) or offensive lochia (4). High vaginal swabs were reported as follows: no pathogens, three; *Staphylococcus aureus,* one; gonococci, one; alpha-hemolytic streptococci, one; and anaerobic streptococci, one. Nine mothers developed pelvic infection after discharge from hospital (13 to 38 days postpartum), and details of this group are reported elsewhere in this volume (Rees et al., p. 67). One mother was treated with antibiotics for a wound infection.

Fathers. Eleven fathers of *Chlamydia*-positive babies were examined. All were asymptomatic. Two were gonococcus positive, and one of these admitted extramarital contact 3 months earlier. Of 10 examined for *C. trachomatis*, 4 were positive 24, 36, 59, and 84 days after the birth of the baby. The IFU/WC counts were low, being 3, 2, 10, and 420, respectively. No discharge was found, but shreds containing pus cells were present in the urine. One gonococcus-positive father was *Chlamydia* negative.

DISCUSSION

The actual incidence of neonatal chlamydial conjunctivitis is difficult to assess because of the late onset of clinical signs and the early discharge from hospital of healthy babies.

In the United Kingdom, treatment of conjunctivitis is usually undertaken by the general practitioner unless the condition is refractory, in which case the infant is referred to an ophthalmologist. Investigation of parents is then unlikely to be carried out.

Nine of 24 mothers (37.5%) of infants with neonatal chlamydial infection developed pelvic infection 13 to 38 days postpartum, one having been treated with ampicillin for infected lochia prior to discharge from hospital. Six additional mothers (25.0%) were treated for pyrexia and infected lochia in the maternity unit. This confirms the association of neonatal conjunctivitis with maternal pelvic infection reported in several studies (8, 9). Mordhorst and Dawson (9) found pelvic infection before and after the birth of the infected child in 4 of 16 mothers. They suggest that the physiological stress of childbirth may reactivate a subclinical genital infection. In our group of 81 infants, four of the mothers had been treated for chlamydial cervicitis, three during pregnancy and one 2 months prior to pregnancy. *Chlamydia* was not reisolated during or after the pregnancy, and swabs from the babies were all negative. The four pregnancies continued to term.

The high incidence of prematurity in this series, 12 of 26 babies with chlamydial infection having gestation periods of 30 to 37 weeks, could be attributed to the fact that their longer stay in hospital allows conjunctivitis to develop and an accurate diagnosis to be made. However, the possibility of ascending cervical infection being a contributory factor to premature rupture of the membranes should be considered. The possible infection in utero of one baby delivered by caesarean section lends support to the hypothesis that chlamydial infection may ascend from the cervix and, therefore, may be directly involved not only in chorioamnionitis but also in the development of nonspecific salpingitis.

The asymptomatic nature of both the gonococcal and chlamydial infections in the fathers underlines the importance of investigating them to prevent reinfection of the mother.

It is suggested that the treatment of pregnant women infected with *C. trachomatis* is strongly indicated. Tetracycline should be avoided in pregnancy because of its effect on the bones and teeth of the fetus. Erythromycin is not contraindicated and has given satisfactory results in the small number of cases in which we have used it.

In summary, in a selected series of 81 babies *C. trachomatis* was isolated from 29 in the neonatal period. In the *Chlamydia*-positive group, concurrent gonococcal conjunctivitis was present in three and herpesvirus particles were identified by electron microscopy in one. One baby was delivered by caesarean section. The onset of conjunctival discharge varied between 3 and 13 days postpartum (mean, 6 to 8 days). Concurrent bacterial infection and prior antibiotic therapy influenced the clinical presentation. After treatment with oculentum tetracycline prescribed for 4 weeks, *C. trachomatis* was reisolated in four cases during follow-up extending to 11 months. A high incidence of postpartum maternal infection occurred. Four fathers with chlamydial infection and two with gonococcal infection were asymptomatic. Accurate diagnosis of neonatal conjunctivitis is important in assessing the need for family investigation. The treatment of pregnant women with chlamydial cervicitis is strongly indicated.

ACKNOWLEDGMENTS

We are indebted to F. Harris and Anne E. McCandless for referring babies in their units, to Alan Percival and R. E. Rewell for bacteriological investigations, to Bruce White for viral studies, and to G. C. Turner for serological tests for syphilis.

This investigation was supported by a project grant from the Medical Research Council of Great Britain.

REFERENCES

1. Reid DE: Amniotic sac infection syndrome—chorio-amnionitis, p 756–757. *In* DE Reid, KJ Ryan, and K Bernirschke (ed), Principles and management of human reproduction, 1st ed. W. B. Saunders Co., Philadelphia, 1972
2. Gamsu H: Intrauterine bacterial infections. Ciba Found Symp 10:135–149, 1973
3. Shaffer TE: Preventive pediatrics, p 138–139. *In* WE Nelson (ed), Textbook of pediatrics, 7th ed. W. B. Saunders Co., Philadelphia, 1959
4. Halberstaedter L, von Prowazek S: Uber Zellemschlusse parasitarer Natur beim Trachom. Arb Kais Gesundh Amt Berl 26:44, 1907
5. Freedman A, Al-Hussaini MK, Dunlop EMC, et al: Infection by TRIC agent and other members of the Bedsonia group. Ophthalmia neonatorum due to TRIC agent. Trans Ophthalmol Soc UK 86:313–320, 1966
6. Forster RK, Dawson CR, Schachter J: Late follow-up of patients with neonatal inclusion conjunctivitis. Am J Ophthalmol 69:467–472, 1970
7. Hobson D, Johnson FWA, Rees E, et al: Simplified method of diagnosis of genital and ocular infections with chlamydia. Lancet 2:555–556, 1974
8. Dunlop EMC, Al-Hussaini MK, Freedman A, et al: Infection by TRIC agent and other members of the Bedsonia group. Genital infection and disease of the eye. Trans Ophthalmol Soc UK 86:321–334, 1966
9. Mordhorst CH, Dawson C: Sequelae of neonatal inclusion conjunctivitis and associated disease in parents. Am J Ophthalmol 71:861–867, 1971

Prospective Study of Perinatal *Chlamydia trachomatis* Infection

E. RUSSELL ALEXANDER, JOHN CHANDLER, TERRENCE A. PHEIFER, SAN-PIN WANG, MARILYN ENGLISH, AND KING K. HOLMES

This study was undertaken as it became apparent that the prevalence of chlamydial infection was increasing. It is well established that the incidence of nongonococcal urethritis (NGU) has been increasing since the 1960s and that NGU due to chlamydial infection has become extremely common in the United States (1; Wiesner, p. 9, this volume). In a study of normal pregnant women conducted by Foy et al. in Seattle at University Hospital nearly 10 years ago, the prevalence of cervical chlamydial infection was 1.3% (as measured by isolation of *Chlamydia trachomatis* in eggs) (2). In the past few years, the high prevalence rates for *Chlamydia* in a variety of studies of disease states (3, 4) made us suspect that the prevalence in normal pregnant women was probably higher.

It was decided, as a pilot study, to determine the prevalence of cervical chlamydial infection in a sample of pregnant women and to follow the offspring of infected women to assess the risk of neonatal infection and disease.

STUDY POPULATION

Cervical cultures were obtained on 142 unselected pregnant women attending a prenatal clinic at University Hospital, Seattle, Wash., who were in their 36th to 40th week of gestation. The only limitation was that they were recruited on certain days when our study nurse was in attendance. After the closing of the obstetric service at the King County (Harborview) Hospital, the obstetric clinic at University Hospital assumed care for many of the County Hospital patients, so that University Hospital now serves a wider socioeconomic spectrum than in the years of the Foy study (2).

Examination of pregnant women included a swab culture obtained from the cervix for isolation of *C. trachomatis*, as described below. All 142 infants were given 1% silver nitrate ophthalmic prophylaxis at birth, the standard procedure at University Hospital. Infants of mothers with cervical chlamydia were seen in the newborn nursery, again at 10 to 14 days of age (home visit), and at 6 weeks and 3 months whenever possible. At each examination, conjunctival cultures were obtained, and the eyes were examined for gross conjunctivitis by a trained public health nurse. A history of eye disease was sought. In the latter part of the study, tear antibody was collected on a filter-paper strip.

If an infant developed conjunctivitis, the parents were asked to bring the infant back to be examined by an ophthalmologist (John Chandler). Conjunctival cultures were obtained at that point, and cytological smears were made for Giemsa and fluorescent-antibody staining. In the last half of the study, tear antibody was obtained on filter-paper strips. Infants with inclusion conjunctivitis were treated with sulfonamide ophthalmic solution applied topically, for the most part, although treatment was not a part of the study. When initial therapy was unsuccessful, sulfonamide or tetracycline ointments were used.

LABORATORY METHODS

Specimens for chlamydial isolation were frozen at −70°C until tested. 5-Iodo-2′-deoxyuridine-treated McCoy cells were utilized for isolation of *C. trachomatis* as previously described (5). Isolates were serotyped by the simplified micro-immunofluorescence technique (6).

Tear antibody was measured in the following manner. Tears were obtained by applying a strip (5 by 20 mm) of sterile filter paper (Whatman no. 41) along the conjunctival margins and were eluted in 0.2 ml of 0.01 M phosphate-buffered saline (pH 7.0). Tear eluate was then tested in the micro-immunofluorescence test by methods previously described (7; Wang et al., p. 237, this volume).

RESULTS

Of the 142 women examined in this pilot study, 18 (12.7%) had *C. trachomatis* recovered from their cervix. The infected women differed significantly from the uninfected women in a number of ways (Table 1). They were considerably younger than the uninfected (a 3-year mean difference: 20.3 years versus 23.1 years). They were further characterized as having a higher proportion of non-whites (50% versus 17%), a lower proportion who were married at the time of conception (33% versus 76%), and a more frequent history of gonorrhea (50% versus 9%). In addition, they were clearly of lower socioeconomic status than the uninfected. One reflection of this was the proportion who were on maternal and infant care governmental support (61% versus 31%). A number more were on various other welfare programs.

Infected women were offered a reculture of the cervix at 6 weeks postpartum. Fifteen were tested and only three had remained culture positive. A urethral culture was offered for the sexual partner at the same time. Four of five males tested were positive. All culture-positive individuals and their sex partners were offered tetracycline therapy at that point.

There were no striking differences between the infants of culture-positive and -negative mothers in birth weight, gestation, duration of labor, rupture of membranes, Apgar of infant, or any characteristic of the birth history. However, such comparisons were difficult in this pilot study, because many of the infected and uninfected mothers were in the high-risk category (e.g., four of the positive mothers were preeclamptic and one was toxemic).

Nine infants suffered conjunctivitis (Fig. 1). Only two cases were severe enough that a physician was consulted by the mother. For four more, our public health

TABLE 1. *Correlates of chlamydial cervical infection in pregnant women, University Hospital, Seattle, 1975*

Factor	Noninfected (n = 124)	Infected (n = 18)	P
Age (years)	23.1 ± 4.4	20.3 ± 4.1	<0.01
Race (white)	103 (83%)	9 (50%)	<0.01
Married at conception	94 (76%)	6 (33%)	<0.001
Gonorrhea history	11 (9%)	9 (50%)	<0.001
Maternal and infant care program	39 (31%)	11 (61%)	<0.05

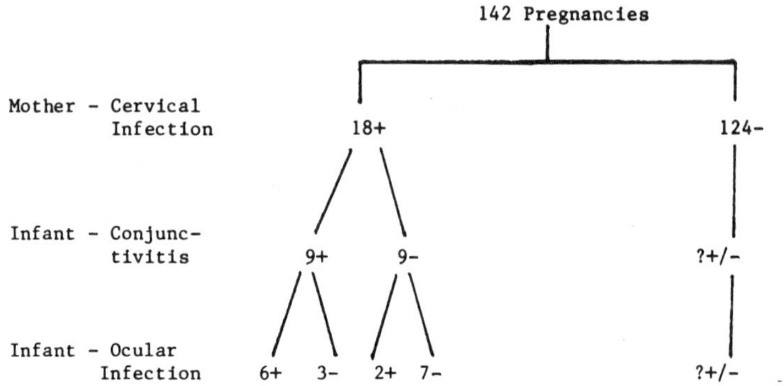

FIG. 1. *Occurrence of neonatal conjunctivitis and of ocular infection due to C. trachomatis in infants of 18 mothers with cervical chlamydial infection near term.*

nurse was notified by the family. The initial severity was mild in six, but only three remained mild throughout the course. Six of nine had onset in the first 2 weeks (none in the hospital nursery). Three of the six had recurrences at 5 to 8 weeks of age, after treatment. In three more instances, the initial onset was 6 to 7 weeks of age.

At least 8 of the 18 infants born to culture-positive mothers were believed to have been infected with *C. trachomatis*. Six were among those with conjunctivitis and two were in the remaining nine without conjunctivitis. Only three were culture positive (all three developed recurrent conjunctivitis). Isolates recovered from infants and their mothers were consonant, and were of ED, GF, and H immunotypes. All three showed inclusions by one or both methods employed. A fourth had a very few inclusions seen on one smear. That case and the remaining four positive cases were considered to be infected with *C. trachomatis* on the basis of tear antibody results (Table 2). Evidence that tear antibody reflects actual eye infection is presented elsewhere (7; Wang et al., p. 237, this volume). It should be noted that this estimate of chlamydial infection is considered a minimal one, since chlamydial infection was not firmly ruled out in 6 of the 10 chlamydia-negative infants. Two were not tested for tear antibody and another four did not have tear

TABLE 2. *Micro-immunofluorescence tear antibody results in 16 infants of culture-positive mothers, tested at various intervals after birth*

Age (weeks)	No. of specimens from 16 infants tested		No. of specimens from 7 Tab-positive infants[a]	
	Positive	Negative	Positive	Negative
0–8	0	13	0	5
9–16	5	7	5[b]	1
17–24	5	4	5	2[c]
25–32	2	3	2	0

[a] Tab = Tear antibody.
[b] Five of five tested at 10 to 14 weeks of age were Tab positive.
[c] Two infants who were initially Tab positive become Tab negative at 19 to 23 weeks of age, after treatment.

antibody specimens collected at what we now believe to be the optimal time. Furthermore, in this pilot study there was no measurement of how much conjunctivitis or chlamydial infection occurred in the infants born to mothers who had negative cervical cultures for *C. trachomatis*.

The tear antibody results are incomplete, because collection was begun after the start of the study, and we were not sure of the optimal time to obtain specimens. A summary of the results can be seen in Table 2. This refers to specimens rather than infants. All positive results were seen at 9 to 30 weeks of age. Of seven who ultimately were tear antibody positive, five of five who were tested at 0 to 8 weeks had no tear antibody at that time. However, five of five tested at 10 to 14 weeks of age were positive, and two lost antibody at 19 to 23 weeks, even though others remained positive to 30 weeks of age. On this basis, it appears that 10 to 14 weeks of age is an optimal time for tear antibody to become apparent (6 to 10 weeks after onset of clinical conjunctivitis).

DISCUSSION

This pilot study represents a preliminary look to see whether a problem exists. Clearly it does. Mothers were cultured for *C. trachomatis* only once before delivery. Tear antibody and serum antibody were not systematically obtained at the optimal time from all exposed infants, and infants of culture-negative mothers were not followed. Nonetheless, an infection rate of 13% in an obstetric clinic population is serious enough. This is a much higher prevalence of chlamydia than was observed 10 years ago in the same clinic (2), although the method of culturing, the stage of pregnancy when cultures were obtained, and the composition of the University Hospital prenatal population were different in the two studies. It will be important to examine the prevalence in other clinics and other cities.

Eight (44%) of 18 infants born to infected mothers had evidence of chlamydial infection. This is approximately 6% of the total group of 142 infants. These figures are higher than the older estimate of Thygeson and Stone (8) or the newer estimate of Armstrong et al. (9), both of which were less than 1%. This may reflect a change in prevalence of the organism as well as a difference in severity of the conjunctivitis measured, with less severe cases being detected in the present prospective study. Most studies of inclusion conjunctivitis concern more severe disease, which comes to the attention of a physician. However, it is no longer clear that chlamydial infection of the infant is benign. In the first place, Mordhorst and Dawson (10) and others have shown that sequelae to mild inclusion conjunctivitis may occur despite treatment. Furthermore, initially asymptomatic chlamydial infection of infants may precede pneumonia, as evidenced by the report of Beem and Saxon (p. 153, this volume).

This report does confirm that inapparent infection with *C. trachomatis* can occur, but probably gives only a minimal estimate of its frequency. Such inapparent infection had been previously demonstrated in a primate model, which suggested that it might be relatively common (11).

The use of tear antibody as a measure of prior infection is well illustrated here. In all instances where early specimens were obtained in this study, they were negative. The immunotype specificity of tear antibody remained constant while antibody was detectable, and it matched the immunotype of the mother or infant isolate. It is somewhat surprising that early tear specimens were all negative. This may be attributable, in part, to the mildness of the clinical disease in the majority of cases.

This pilot study suggests that there is a problem of chlamydial infection of neonates of considerable magnitude. This is occurring in the face of silver nitrate (Credé) prophylaxis at birth used in every infant in this study. To prevent this infection, mothers could undergo cervical culture and be treated during pregnancy if infected. However, many potentially useful antibiotics should not be used during late pregnancy (e.g., tetracycline, sulfonamides), and, in any event, tissue culture diagnosis of *C. trachomatis* infection is not available for most hospitals. Alternative methods of ophthalmic prophylaxis should be evaluated in controlled field trials. More comprehensive prospective studies of the offspring of infected, as well as uninfected, mothers should be designed on the basis of this study and that of Beem and Saxon (p. 153, this volume).

ACKNOWLEDGMENTS

This research was supported by Public Health Service research grants AI-12192 from the National Institute of Allergy and Infectious Diseases, EY-00219 from the National Eye Institute, and CA-11073 from the National Cancer Institute.

REFERENCES

1. Schachter J, Hanna L, Hill EC, et al: Are chlamydial infections the most prevalent venereal disease? J Am Med Assoc 231:1252–1255, 1972
2. Foy HM, Wang SP, Kenny G, et al: Isolation of TRIC agents and Mycoplasma from the cervix of pregnant women. Am J Ophthalmol 63:27–30, 1967
3. Holmes KK, Handsfield HH, Wang SP, et al: Etiology of nongonococcal urethritis. N Engl J Med 292:1199–1205, 1975
4. Wentworth BB, Bonin P, Holmes KK, et al: Isolation of viruses, bacteria and other organisms from venereal disease clinic patients. Methodology and problems associated with multiple isolations. Health Lab Sci 10:75–81, 1973
5. Wentworth BB, Alexander ER: Isolation of *Chlamydia trachomatis* by use of 5-iodo-2-deoxyuridine-treated cells. Appl Microbiol 27:912–916, 1974
6. Wang SP, Kuo CC, Grayston JT: Simplified method for immunological typing of trachoma-inclusion conjunctivitis-lymphogranuloma venereum organisms. Infect Immun 7:356–360, 1973
7. Wang SP, Grayston JT: Local and systemic antibody response to trachoma eye infections in monkeys, p 217–232. *In* RL Nichols (ed), Trachoma and related disorders caused by chlamydial agents. Excerpta Medica, Amsterdam, 1971
8. Thygeson P, Stone W: Epidemiology of inclusion conjunctivitis. Arch Ophthalmol 27:91–119, 1942
9. Armstrong JH, Zacarias F, Rein MF: Ophthalmia neonatorum: a chart review. Pediatrics 57:884–892, 1976
10. Mordhorst CH, Dawson C: Sequelae of neonatal inclusion conjunctivitis and associated disease in parents. Am J Ophthalmol 71:861–867, 1971
11. Alexander ER, Chiang WT: Infection of pregnant monkeys and their offspring with TRIC agents. Am J Ophthalmol 63:1145–1153, 1967

Distinctive Pneumonia Syndrome in Infants Infected with *Chlamydia trachomatis*

MARC O. BEEM AND EVELYN M. SAXON

During etiologic studies of acute respiratory disease, we have become aware of a distinctive interstitial pneumonia of young infants that cannot be attributed to presently recognized bacteria, viruses, or *Mycoplasma pneumoniae*. Two infants recently seen with this syndrome had histories of postnatal inclusion conjunctivitis, leading us to wonder if this unusual, but by no means rare, pneumonia might be caused by *Chlamydia trachomatis*. Pursuit of this was encouraged by observations made by those working with oculogenital aspects of chlamydial infection that rhinitis, nasopharyngitis, tracheitis, ear pain, and pneumonia may sometimes be associated with *C. trachomatis* infections of the eye. There have been several reports of isolation of this organism from the nasopharynx, and Schachter et al. (1) reported isolation of *C. trachomatis* from sputum from an infant with pneumonia.

The following studies suggest that this pneumonia syndrome is, indeed, associated with chlamydial infection of the respiratory tract. In addition, there is evidence that involvement of the respiratory tract is not the exception, but the rule, in infants infected by *C. trachomatis*.

CLINICAL MANIFESTATIONS

A Distinctive Pneumonia Syndrome

The salient features of the pneumonia syndrome are summarized in Table 1. It occurs in early infancy, with gradual onset, little systemic reaction, and respiratory findings of distinctive cough (a series of closely spaced staccato coughs, each separated by a brief inspiration) and diffuse pneumonia. Elevated blood eosinophil and immunoglobulin values are noted, and the course is protracted. This syndrome is similar to that described initially by Botsztejn in 1941 as pertussoid eosinophilic pneumonia of infants (2, 3).

Retrospective Study of Eight Infants with the Distinctive Pneumonia Syndrome

In our first effort to associate *Chlamydia* with this illness, we attempted isolation from the respiratory specimens which had been freezer stored after collection from eight infants with the syndrome who had previously been studied for virus shedding. We used 5-iodo-2′-deoxyuridine-treated McCoy cells according to the method of Wentworth and Alexander (4). The results of these retrospective isolation studies are shown in Table 2. *C. trachomatis* was isolated from the nasopharynx and/or tracheal aspirate from six of eight infants. This number of isolates seemed particularly impressive since factors possibly unfavorable to chlamydia isolation, such as prolonged freezer storage and antimicrobial therapy, pertained to most specimens. Viruses isolated from three of these infants were

TABLE 1. *"Distinctive" pneumonia syndrome of infants*

Age	1–3 months
Onset	Gradual, over many days
Systemic manifestations	Absent; fever = 0, malaise = ±
Respiratory	Distinctive cough Tachypnea X-ray: Interstitial infiltrates Hyperexpansion Pleural thickening Auscultation: Good breath sounds Inspiratory crepitant rales Absent or minimal expiratory wheezing
Laboratory	Eosinophilia Low PaO_2, normal $PaCO_2$ High immunoglobulins G and M; sometimes A
Course	Protracted Clinical illness lasts weeks X-ray and physical signs last months

shed transiently and were considered to be superinfecting agents. Of interest are the 7:1 male preponderance, the mean age of 9 weeks, and the fact that only five of these eight infants had conjunctivitis by history or examination.

Prospective Study of Infants with the Distinctive Pneumonia Syndrome, Inclusion Conjunctivitis, or Other Illnesses

We turned next to prospective studies of this condition and also did similar studies in a second cohort comprised of infants with inclusion conjunctivitis alone

TABLE 2. *Retrospective studies of eight infants with distinctive pneumonia syndrome*[a]

Patient	Sex	Age at admission (weeks)	Conjunctivitis	Chlamydiae		Viruses	
				NP	Trach	NP	Trach
Nel	M	15	+	0	+	Rhino	0
Web	M	10	0	+	+	Adeno	0
Hoo	M	9	+	+	0	0	
Stu	M	7	0	+		0	
Ken	M	7	0	+		Rhino	
Broo	M	6	+	+		0	
Clar	M	6	+	0		0	
Adam	F	5	+	0		0	
Summary	F = 1 M = 7	$\bar{x} = 9$	5/8	6/8		3/8	

[a] NP, nasopharyngeal aspirate; Trach, tracheal aspirate; +, present or isolated; 0, absent or not isolated.

TABLE 3. *Prospective studies of 12 infants with distinctive pneumonia syndrome*[a]

Patient	Sex	Age at admission (weeks)	Conjunctivitis	Chlamydiae			Viruses		
				Conj	NP	Trach	Conj	NP	Trach
Saw	F	12	+	+	+		0	CMV	
Bark	M	10	0	+	+	+	0	0	0
Har	F	10	0	0	+		0	CMV	
Han	M	9	0	+	+		0	RS	
Barn	F	7	0	0	+	+	0	Rhino	Rhino
Stub	M	7	+	+	+	+	0	0	0
Bog	M	7	0	0	+	+	0	0	0
Port	F	7	+	0	+	+	0	CMV	CMV
Chat	M	6	+	+	+	+	0	CMV	CMV
Shi	M	6	+	+	+	+	0	0	0
Ruw	M	6	0	0	+	+	0	0	0
Wash	M	6	+	+	+	+	0	0	0
Summary	F = 4 M = 8	$\bar{x} = 8$	6/12	7/12	12/12	9/9	0/12	6/12	3/9

[a] Conj, conjunctiva; NP, nasopharyngeal aspirate; Trach, tracheal aspirate; +, present or isolated; 0, absent or not isolated; CMV, cytomegalovirus; RS, respiratory syncytial virus; Rhino, rhinovirus.

and a third cohort comprised of infants with a variety of other acute problems. These studies consisted of attempts to isolate bacteria, chlamydiae, *M. pneumoniae*, and viruses from conjunctival swabs, nasopharyngeal aspirates, tracheal aspirates (from some of the pneumonia infants), and urines. No consistent or unique species of bacteria were identified; eight nasopharyngeal aspirates of pneumonia infants cultured for *M. pneumoniae* were negative. Chlamydia and virus isolation results from these prospectively studied infants were as follows (Tables 3–6):

Considering first infants with the pneumonia syndrome, it might be commented that, once this syndrome was perceived, 12 additional patients were identified in about 6 months (Table 3). As in the retrospective group, most were males, and mean age was 8 weeks. Conjunctivitis was present by history or examination in only six. Chlamydial infection was evidenced by positive conjunctival culture in 7 and by positive nasopharyngeal culture in all 12. The nine tracheal aspirates tested were all *Chlamydia* positive by isolation. Thus, 100% of the prospectively studied infants with the distinctive pneumonia syndrome had respiratory tract shedding of chlamydiae.

These isolates, like all others reported here, formed inclusions that stained with iodine and had the morphology of subgroup A organisms on Giemsa staining. On this basis, they were considered to be strains of *C. trachomatis*. Of the six virus-positive infants, respiratory syncytial virus (RS) and rhinovirus infections were considered to be coincidental. The occurrence of four cytomegalovirus (CMV) infections suggests at least an epidemiological and possibly a synergistic relation between CMV and chlamydiae.

To learn whether nasopharyngeal shedding or chlamydiae might be a unique marker of the pneumonia syndrome, we looked for chlamydiae in the respiratory secretions of infants with chlamydial infection who did not have pneumonia (Table 4). Twelve infants with culture-proven acute postnatal inclusion conjunctivitis

TABLE 4. *Results in 12 infants with postnatal inclusion conjunctivitis sampled at ≥4 weeks of age*[a]

Patient	Sex	Age at sampling (weeks)	Conjunctivitis	Chlamydiae		Viruses	
				Conj	NP	Conj	NP
Phil	F	19	0	0	0	0	RS
Bord	M	16	+	+	+	0	0
Ell	F	15	+	+	+	0	0
Town	F	10	+	+	+	0	0
Tur	F	9	+	+	+	0	0
Spee	F	8	+	+	0	0	0
Enn	F	8	+	+	+	0	0
Brow	M	6	+	+	+	0	0
Under	M	6	+	+	+	0	0
Harm	M	5	+	+	+	0	RS
Pow	M	5	+	+	+	0	0
Alex	M	4	+	+	+	0	0
Summary	F = 6 M = 6	$\bar{x} = 9$	11/12	11/12	10/12	0/12	2/12

[a] Conj, conjunctiva; NP, nasopharyngeal aspirate; +, present or isolated; 0, absent or not isolated; RS, respiratory syncytial virus.

were examined and cultured again when they were about 9 weeks of age. All had been treated with topical sulfacetamide or tetracycline. At the time of reexamination, none had pneumonia, although two had nondescript acute respiratory illnesses and were shedding RS virus. Eleven still had conjunctivitis, and the conjunctival cultures from all 11 were *Chlamydia* positive. Moreover, 10 of the 11 with persisting conjunctivitis also had nasopharyngeal aspirates that were *Chlamydia* positive. Thus, it is evident that nasopharyngeal shedding is not unique to infants with the pneumonia syndrome, but is common among infants in the chronic phase of inclusion conjunctivitis. Indeed, for the total group of 30 prospectively studied infants, nasopharyngeal shedding (present in 24) was a more comprehensive marker of chlamydial infection than conjunctival shedding (present in 18).

Finally, to estimate the prevalence of chlamydia shedding in the general population from which the pneumonia patients were drawn, we sampled 15 infants hospitalized for various acute problems—chiefly, enteritis, RS disease, and acute respiratory illness. Like the pneumonia cohort, all were black and most came from households on public aid, but they differed in being somewhat older and in having similar numbers of males and females. As shown in Table 5, none had conjunctivitis by history or examination, and none yielded chlamydiae from conjunctival culture. However, nasopharyngeal aspirates of two were positive for chlamydiae, for an isolation rate of 13%. Although the small number studied gives only a rough estimate of the frequency of nasopharyngeal chlamydia shedding in the general infant population, it is clearly less than the 90% found in the pneumonia group as a whole, or the 100% found in those studied prospectively. We believe, therefore, that chlamydial infection is significantly related to the distinctive pneumonia syndrome. Table 6 summarizes the results of the prospective studies.

TABLE 5. *Recovery of C. trachomatis and viruses from conjunctiva and nasopharyngeal aspirates from 15 infants with various illnesses other than pneumonia*[a]

Patient	Diagnosis	Sex	Age at admission (weeks)	Conjunctivitis	Chlamydiae		Viruses	
					Conj	NP	Conj	NP
Hen	Enteritis	M	4	0	0	0	0	0
Alex	Enteritis	M	6	0	0	0	0	0
Joi	Enteritis	M	16	0	0	0	0	0
Bro	Enteritis	F	16	0	0	0	0	0
Coch	RS disease	M	24	0	0	+	0	RS
McC	RS disease	M	8	0	0	0	0	RS
Full	RS disease	M	8	0	0	0	0	0
Cun	RS disease	F	8	0	0	+	0	RS, CMV
Sim	RS disease	F	5	0	0	0	0	RS, Adeno
Dick	ARI	M	16	0	0	0	0	0
Jon	ARI	F	14	0	0	0	0	RS
Mon	ARI	M	13	0	0	0	0	RS
Brun	ARI	F	13	0	0	0	0	RS
Hud	ARI	F	9	0	0	0	0	0
Lov	Child abuse	F	10	0	0	0	0	0
Summary		M = 8 F = 7	$\bar{x}$ = 11	0/15	0/15	2/15	0/15	7/15

[a] Conj, conjunctiva; NP, nasopharyngeal aspirate; +, present or isolated; 0, absent or not isolated; RS, respiratory syncytial virus; ARI, acute respiratory illness; CMV, cytomegalovirus; Adeno, adenovirus.

DISCUSSION

Two findings suggest that chlamydiae are indeed replicating in the respiratory tract and are not simply being carried there by secretions from infected conjunctiva. One is that six pneumonia patients with abundant infectious material in their nasopharyngeal aspirates had negative conjunctival cultures. The other is that

TABLE 6. *Summary of prospective chlamydial studies*

Diagnosis	No. in group	Sex ratio, M:F	Age (weeks)		No. with conjunctivitis	No. with isolation of:[a]				
						Chlamydiae			Viruses	
			Range	Mean		Conj	NP	Trach	Conj	NP
"Distinctive pneumonia syndrome"	12	2:1	6–10	8	6 (50%)	7 (58%)	12 (100%)	9 (100%)	0	6 (50%): CMV, 4; RS, 1; Rhino, 1
Inclusion conjunctivitis	12	1:1	4–19	9	11 (92%)	11 (92%)	10 (83%)	ND[b]	0	2 (17%): RS, 2
Other illness[c]	15	8:7	4–24	11	0	0	2 (13%)	ND	0	7 (47%): RS, 5; RS + Adeno, 1; RS + CMV, 1

[a] Conj, conjunctiva; NP, nasopharyngeal aspirate; Trach, tracheal aspirate; CMV, cytomegalovirus; RS, respiratory syncytial virus; Rhino, rhinovirus; Adeno, adenovirus.

[b] Not done.

[c] Enteritis, 4; respiratory syncytial virus disease, 5; acute respiratory infection, 5; child abuse, 1.

tracheal secretions obtained by direct bronchoscopy or laryngoscopy from nine pneumonia infants were positive for chlamydiae. On the other hand, two observations raise questions about the extent of replication in the lung itself. First, there were four infants who had many chlamydiae in their nasopharyngeal aspirates but only a few in their tracheal aspirates, as determined by the number of inclusions seen on primary isolation. The second is failure to identify chlamydiae in lung biopsies of two pneumonia infants. These were negative by culture as well as by light, immunofluorescence, and electron microscopy. This question and others, such as the apparently high incidence of CMV in this group, are under further study.

We conclude that *C. trachomatis* infection of the young infant commonly involves the respiratory tract. At least one respiratory illness is associated with this infection, a distinctive pneumonia syndrome similar to that described by Botsztejn as pertussoid eosinophilic pneumonia of infants.

REFERENCES

1. Schachter J, Lum L, Gooding CA, et al: Pneumonitis following inclusion blennorrhea. J Pediatr 87:779–780, 1975
2. Botsztejn A: Die pertussoïde, eosinophile Pneumonie des Säuglings. Ann Paediatr 157: 28–46, 1941
3. Biró Z: 12 wietere Fälle von interstitieller pertussoider eosinophiler Pneumonie des Säuglings. Helv Paediatr Acta 15:135–154, 1960
4. Wentworth BB, Alexander ER: Isolation of *Chlamydia trachomatis* by use of 5-iodo-2-deoxyuridine-treated cells. Appl Microbiol 27:912–916, 1974

Chairman's Report and Discussion

J. THOMAS GRAYSTON

This session began with recognition of the recent death of Francis Gordon. He made far-reaching contributions to knowledge of *Chlamydia trachomatis*, and he played an important role in the previous three international symposiums on *C. trachomatis*. He was a kind, thoughtful, quiet person who will be remembered and his friendship missed by scientists throughout the world interested in these organisms.

Roger Nichols presented a review of the clinical, microbiological, immunological, and epidemiological aspects of *C. trachomatis* infections of the human eye. In response to a question from Chandler Dawson concerning immunity to *C. trachomatis* eye infection, Dr. Nichols states his opinion that "in humans and animals we have pretty good evidence that an infection with trachoma does indeed elicit a response which has a reasonable degree of resistance to reinfection and, perhaps, to prolongation of infections. It is the bed in which trachoma lies which permits the infection to continue. I am convinced that if we inoculated everyone in this room with chlamydia in the eye we would all have severe infections, but only one or two of us would continue after 3 or 4 months of time with the classical ocular infection. Those of us who continue would have other predisposing factors which would permit this disease to become prominent."

J. T. Grayston presented a paper on the pathogenesis of ocular *C. trachomatis* infections in humans based on studies of a series of families living in an endemic area of Taiwan. Teresa Rota asked about impairment of the immune system in the children in the family where *C. trachomatis* was isolated from the eye before the development of disease and antibody response. Dr. Grayston responded that no evidence of deficiency was found in the family under discussion. The children were normal to physical examination and had responded to other infections normally. David Taylor-Robinson further inquired whether aberrations in cell-mediated response have been looked for in relation to pathogenesis and recovery from trachoma. Another participant stated, "We have obtained isolations from the eyes of children who are normal; however, they usually rapidly go on to become clinically infected. We have interpreted this as simply the latency period during which time the infection is building up. We have looked very closely at the children who have eye infections which do not seem to be clearing, and we as yet have not been able to find an immunologic difference in the children who are obviously severely infected and are getting into difficult or blinding sequelae of the disease as opposed to those who are clearing their infections." In response to a question from Doris Graham about reversion of trachoma II organisms to normal eyes, Barrie Jones stated, "This was described by McCallum himself, who introduced this classification. He pointed out that it could happen, and I think it's been observed practically everywhere it's been looked for." King Holmes asked whether intercurrent eye infection or other inflammation in an eye that is latently infected could recall active trachoma. Dr. Grayston had noted no such phenomenon in his subjects. Dr. Holmes then pointed out that Dr. Nichols (see above) and Dr. Grayston were proposing different mechanisms for the continuing activity of chronic trachoma. Dr. Grayston agreed and again stated his opinion that in most cases the development of chronic trachoma depends on reinfection and that pannus and conjunctival scarring are an immunopathological reaction to reinfection. In response to Dr. Holmes' question about the exact nature of the aberrant immunological response, Dr. Grayston stated that it had not been demonstrated but that he thought it would be similar to the "allergic" reaction in tuberculosis responsible for cavitation. Thomas Hatch stated that if reinfection played a prominent role in development of chronic trachoma, he would expect, in areas where there is more than one serotype prevalent, to get isolations of different serotypes very frequently in the more serious trachoma. Dr. Grayston pointed out that one of the striking findings in the data he presented was the tendency for all members of a family to have infection with the same immunotype. This strongly suggested that the usual mode of transmission is within the family group. Under these circumstances reinfection would most likely be with the family immunotype. The appearance of some exceptions noted in the presentation suggests reinfection from outside the family, resulting in two immunotypes in a family and even two from an individual subject. The discussion then brought out the strong tendency for *C. trachomatis* micro-immunofluorescent antibodies to demonstrate the concept of "original antigenic sin." Infection in the family subjects and in controlled experiments in monkeys (S.-P. Wang) resulted in antibodies to the first infecting type. These antibodies persisted, and with a second infection with a new type, antibodies to the first type often boosted higher than antibodies to the second infecting type. Dr. Jones emphasized the importance of examining all areas of the conjunctivae and not just the upper tarsal area before concluding there is no disease. Dr. Nichols discussed his experience with following Saudi Arabia children on an almost weekly basis. He found that the first thing that happened after demonstration of conjunctival inclusions was the appearance of antibodies in eye secretions. These exceeded the titer of antibodies in serum. They were first immunoglobulin G, and later they became immunoglobulin A. Both antibodies neutralized trachoma organism infectivity. Over time the antibodies in eye secretions diminished relative to the serum antibody. Maurice Shepard asked whether there were nonpathogenic strains of *C. trachomatis*. No one was aware of such strains.

Carl Mordhorst presented a summary of his extensive experience with the clinical epidemiology of oculogenital *C. trachomatis* infection in Denmark. His observations were based on studies of over 1,000 persons with eye disease and their contacts seen over 11 years. E. R. Alexander asked for an estimate on the frequency of micropannus development in babies with inclusion conjunctivitis. Dr. Mordhorst and Dr. Dawson agreed on about 50% as the answer. Dr. Holmes asked whether anyone had data on the frequency of eye disease in patients with genital *C. trachomatis* disease. He has not recognized eye disease in his venereal disease patients and wondered whether they should be seen by an ophthalmologist in the absence of overt eye disease. Drs. Mordhorst, Jones, and Dawson agreed there were no data on the frequency of clinically manifest, clinically mild, and subclinical eye disease with *C. trachomatis* genital infection. There was the opinion that clinically manifest eye disease is uncommon but that milder disease and subclinical infection occur with greater frequency and are often not recognized.

Julius Schachter told of reviewing the charts of 173 patients in a walk-in eye clinic with isolation-proven *C. trachomatis* genital infection and finding that four were patients who were also being seen in the eye clinic with follicular conjunctivitis. Unfortunately, no specific diagnosis of the eye disease had been attempted. Dr. Dawson argued for the use of the term "inclusion conjunctivitis" for the *C. trachomatis* eye disease resulting from genital tract transmission and reserving the term "trachoma" for the severe endemic disease. He feels that using the name "trachoma" to describe the eye disease in a developed country like Denmark promotes confusion and misunderstanding among ophthalmologists and other physicians. Dr. Alexander points out that the organisms isolated from the eye disease in both situations are the same. He reminded the group that Dr. Mordhorst had used both terms, inclusion conjunctivitis and trachoma, to classify on the basis of clinical findings the eye disease in Denmark. He supported such classification, with chronic infection with pannus and scar formation being appropriately called trachoma.

Dr. Dawson presented his results of response to treatment of ocular *C. trachomatis* infections. He pointed out analogies between endemic trachoma and nongonococcal urethritis, particularly the complex microbiology and sometimes unsatisfactory antibiotic treatment. Gerald Masover wondered whether there are any deficiencies in the normal eye secretions such as lysozyme that would enable the organisms to become established. Dr. Dawson believes the eyes are normal. He attributes decreased rates of active eye disease after 6 years of age to increasing social distance from younger siblings and better facial hygiene, decreasing the chance for reinfection. William Fair raised several issues from his perspective as an urologist. He was interested in the analogy of eye disease with nongonococcal urethritis and commented on the role of the similar epithelium. He wondered about the frequency of urethral strictures in association with *C. trachomatis* infection both in the human and experimentally. Dr. Dawson and others, while agreeing that no systematic study had been done, pointed out that various urologists had reported follicular urethral disease and even scar formation in association with *C. trachomatis* infection as early as 1910. James Gale remarked that he would be presenting data from a baboon urethral infection later in the meeting. Dr. Fair reported that in cystoscopic studies of persons without knowledge of urethral disease, supposedly with normal urethras, he has seen scarring.

Dr. Jones presented a report on prevention of blindness from trachoma and associated diseases that together constitute communicable ophthalmia. D. Gump asked for comment on the possible role of cervix infection in women in regard to cervical carcinoma. Dr. Alexander briefly reviewed his prospective study of genital infection and carcinoma in Taiwan. There was very little recurrent cervical infection with *Chlamydia* demonstrated, and it was not significantly associated with cancer. In another study involving cancer patients and two types of controls, one age matched and one more tightly matched for other factors, there was no difference in chlamydial antibody. P.-A. Mårdh asked about the comparative effect of bacteria and *Chlamydia* on the eye. Dr. Dawson commented that bacterial infection was easier to eliminate but recurred more easily than *Chlamydia* infection. Dr. Holmes asked whether there are data of a case-control nature, actually showing that there are bacterial pathogens other than *Neisseria gonorrhoeae, C. trachomatis*, and *Staphylococcus aureus*, associated with mucopurulent conjunctivitis. Dr. Dawson responded that in Tunisia he looked at specific pathogens, and there was a very clear-cut correlation between clinical conjunctivitis, mucopurulent discharge, and *Haemophilus*. He found much better correlation of disease with bacteria seen on smears than with cultures. Many people were carrying *Haemophilus* who did not have disease, but when the organism could be seen on smears, they had disease. He also stated that when the gonococcus or pneumococcus appeared on smears, disease was associated. Dr. Nichols agreed from his experience in Saudi Arabia with one exception. He has seen children with *N. gonorrhoeae* isolated from their eyes who have normal eyes and carry the organism for many weeks without developing conjunctivitis. Dr. Nichols then commented on the "very persuasive elegant presentation" by Dr. Jones on the role of flies in eye-to-eye transmission of *C. trachomatis*. Although from his experience, also in the Middle East, he would not deny that flies can carry and transmit trachoma, he believes that contact, especially finger to eye, is by far the most important means of transmission. There followed a discussion by Drs. Graham, Jones, and Dawson about the relation of trachoma to upper respiratory disease and middle ear infection. Very severe purulent infection of eye, ear, and nose in Australian aboriginal children with numerous bacteria, including *Escherichia coli*, are unique in Drs. Graham's and Jones' experience. Similar associations of respiratory infection or ear infection with trachoma in the Middle East and in native Americans have also raised the question of an etiological relationship of these conditions. Rare *C. trachomatis* isolations have been reported from the respiratory tract. It is probable that the chronic changes of trachoma may promote bacterial infection. Dr. Alexander pointed out the frequency of severe middle ear disease in Alaska natives and North American Indians where no trachoma is recognized. Dr. Nichols reported sore throat as a prominent primary symptom when he was accidently infected with trachoma. Dr. Mordhorst stated that respiratory symptoms, often with nasal discharge, accompanied or preceded acute relapsing conjunctivitis of several of the Danish childhood trachoma cases. Dr. Holmes postulated that *Chlamydia* genital tract infection may impair proper anatomic drainage or other bacterial clearance mechanisms, thereby promoting bacterial infection. He used the analogy to gonococcal pelvic inflammatory disease (PID), where gonococcal cervicitis is thought to create the first episode of gonococcal PID. However, subsequent episodes of PID, even in people who initially had gonococcal PID, may be nongonococcal and are thought to be related to the tubal damage produced by the gonococcus. A. Barron asked whether the epidemiological situation could account for the differences in opinion regarding reinfection versus recurrent infection. He pointed out that Dr. Jones' presentation was similar to Dr. Grayston's in emphasizing reinfection in trachoma and that Dr. Dawson also spoke of reinfection, whereas Dr. Nichols emphasized relapse of continuing infection. Dr. Grayston responded to Dr. Jones' comments that the primary cycle of infection he demonstrated in children in Taiwan was clinically what Dr. Jones and Dr. Dawson call inclusion conjunctivitis by saying that interpretation was absolutely correct. One 3-year-old boy had what would be similar to inclusion conjunctivitis of the newborn with only papillary hypertrophy, and others developed follicular disease characteristic of the adult inclusion conjunctivitis syndrome. This emphasized his opinion that it is better not to talk about inclusion conjunctivitis as something different from trachoma, but as the beginning stage of trachoma disease.

Three presentations followed before additional discussion. Elisabeth Rees presented data on perinatal *C. trachomatis* infection; Dr. Alexander a prospective study of neonatal *C. trachomatis* infection; Elizabeth Saxon (for Marc Beem) a report on a "distinctive pneumonia syndrome" in infants associated with *C. trachomatis* infection.

Dr. Rees answered in the negative to D. McComb's question as to whether there was a history of abortion in the 29 women or conjunctivitis in the older siblings of the babies. She answered Dr. Schachter's question on newborn eye prophylaxis by stating that none was used in her hospital.

Dr. Holmes commented on Dr. Alexander's study by saying that although the cervical isolation rate of *Chlamydia* was 13%, it was less than 2% for *N. gonorrhoeae*. This higher *Chlamydia* rate was unexpected. Dr. Taylor-Robinson asked for a rough estimate on how many babies do develop neonatal conjunctivitis, and what proportion of those are due to *Chlamydia*. He had done a study of 100 newborns and found that seven developed conjunctivitis, of whom four had *Chlamydia* agents isolated. Dr. Alexander commented that in the small number he had studied, the overall rate of neonatal conjunctivitis was 6%. Both Dr. Alexander and Dr. Schachter agreed that about 50% of neonatal conjunctivitis is caused by *C. trachomatis*. S. Darougar stated that the rate of newborn *C. trachomatis* infections varies greatly with the population group. In one community he studied, he found only one positive baby out of 500, whereas in a London study he found five positive cases among 30 babies.

Dr. Alexander, commenting on Miss Saxon's presentation, stated that he had received some isolates from Dr. Beem's laboratory and could confirm that they were *C. trachomatis*. He then asked for more detail on how the culture material was obtained. Miss Saxon stated that all tracheal cultures were taken through a bronchoscope and that the nasal specimens were obtained with an infant-feeding tube from the posterior nasopharynx. Williams McCormack asked Miss Saxon whether the infants with pneumonia were treated, and if so, what were the results. Miss Saxon replied that some were treated and some were not. They all recovered. There was a feeling that there was symptomatic response and conversion of cultures to negative within 2 to 3 days after the onset of antimicrobial therapy. Dr. Holmes stated that cultural results from specimens taken by bronchoscopy should be interpreted with caution as far as proving lower respiratory infection. Studies have shown that if you put methylene blue in the nasal pharnyx and pass a bronchoscope into the trachea and the bronchus and then aspirate sputum, the sputum will contain methylene blue. Dr. Grayston then commented that although rare isolations have been reported from the respiratory tract, results of the study by Dr. Beem and Miss Saxon of infant pneumonia represent an important extension of the range of body tissues significantly infected by *C. trachomatis*. Their demonstration of the association of infant pneumonia with these organisms emboldened him to briefly report serological data that he and Dr. Wang have been accumulating in recent years that suggest an association between *C. trachomatis* organisms and respiratory tract infection. They had previously used the micro-immunofluorescent serological test with *C. trachomatis* antigens in serological epidemiological studies in several population groups. In various adult populations, they never found less than 25% of the serum showing some *C. trachomatis* antibodies. These results suggested an unexpectedly high frequency of genital *C. trachomatis* infections in some of the higher socioeconomic groups studied or the occurrence of *C. tracho-*

matis infections of other areas of the body as yet unrecognized. They then sought out sera from persons in age groups least likely to have genital tract infection. The sera examined were from children involved in respiratory disease studies. The sera came both from Seattle and from Copenhagen (Dr. Mordhorst). Several different groups of children's sera have been studied. Over 500 single sera and about 200 paired sera were examined. The results showed that about 20% of the sera from children up to the age of 2 years had *C. trachomatis* antibodies, whereas 10% of children from the ages of 3 through 15 years had antibodies. In a subsequent study of sera from children with a diagnosis of pneumonia in the year 1969, they found a fourfold rise or fall in antibody titer in paired sera from 23 and 7%, respectively, in the two age groups. The distribution of antibody types in these children's sera differed from those found in adults. In adults, the antibodies seen most frequently are types ED and GF, the immunotypes most frequently isolated from genital tract infections. In these children's sera, other types (A, CJ, I, K, and B) appeared with equal and sometimes greater frequency. The investigators have been puzzling over the meaning of these results and at the same time expanding their experience by examining additional sera from other population groups. Dr. Grayston stated, "The report of Beem and Saxon adds to our feeling that *C. trachomatis* organisms cause respiratory tract infections in more than rare cases. The results in children above the age of 2 years suggest that respiratory *C. trachomatis* infections may occur without transmission from the genital tract. If respiratory tract-to-respiratory tract transmission of *C. trachomatis* organisms does indeed occur, this would represent another important extension of our knowledge of the role of these organisms in human disease."

Session IV

CHLAMYDIA: BIOLOGY AND EXPERIMENTAL PATHOLOGY

Biological Characteristics of Chlamydiae

G. PHILIP MANIRE

INTRODUCTION

The genus *Chlamydia* is comprised of two species, *C. trachomatis* and *C. psittaci*, which share certain characteristics, including morphology, a strict requirement for intracellular growth in susceptible cells, a common heat-resistant antigen usually detected by the complement fixation test, susceptibility to tetracycline and certain other antibiotics, and a unique developmental cycle involving two morphological forms adapted to extracellular survival and intracellular multiplication, respectively (1). The two species are distinguished from each other on the basis of the criteria shown in Table 1.

The question of classification has been clearly settled in the past 20 years by the demonstration that chlamydiae have a continuous morphology throughout their growth cycle, that they possess a complex cell envelope resembling those of gram-negative bacteria, that both ribonucleic acid (RNA) and deoxyribonucleic acid (DNA) are present, that they contain ribosomes very similar to those in other bacteria, and that, although they are energy parasites, they do possess numerous metabolic systems. Thus, they are quite correctly classified as bacteria, albeit of a very special sort.

The chlamydiae have numerous biological and biochemical characteristics which distinguish them from other infectious agents. Three such characteristics which would appear to have special significance for this symposium are (i) their extraordinarily widespread occurrence in birds and mammals, (ii) their ability to produce latent or inapparent infections in their hosts, and (iii) their unique development cycle involving forms adapted for extracellular and intracellular survival. These three characteristics are interrelated, and we hope to demonstrate this in the following discussion.

HOST RANGE OF CHLAMYDIAE

No other group of pathogenic organisms is more widespread in nature than the chlamydiae. At a previous conference on Trachoma and Allied Diseases in 1966, Meyer gave a detailed and comprehensive review on the worldwide distribution of these organisms and the broad range of their parasitism (2). It is not necessary to detail all of his data, but a few findings will serve to illustrate the point.

Meyer stated that 130 species of birds have been found to be infected, and there is no doubt that this list can and will be extended as more birds are studied. Among the mammals found to be carriers of chlamydiae, and to be suffering from *Chlamydia*-induced diseases, are mice, cats, opossums, cattle, sheep, goats, hogs, dogs, northern fur seals, muskrats, hares, rats, and many other rodents. Chlamydiae have also been isolated from various arthropods associated with infected animals, but there is no evidence of spread by arthropod vectors. The extent of carriers in laboratory mice is emphasized by Meyer's statement that

TABLE 1. *Characteristics distinguishing Chlamydia species*

Species	Susceptibility to sulfonamide	Nature of microcolonies	Reaction to iodine	G + C[a] content of DNA (%)	Principal host
C. trachomatis	Susceptible	Rigid	Positive stain	~44	Humans
C. psittaci	Resistant	More diffuse	Not stained	~41	Nonprimate vertebrates

[a] Guanine plus cytosine.

"because no mice or hamsters free from latent PL [chlamydiae] infections could be located between 1940 and 1944, the search for specific human PL parasites in the respiratory tract was discontinued."

Although Grayston (3) has reported that the worldwide impact of trachoma has diminished significantly since World War II, *C. trachomatis* still remains one of the most common infectious agents in the world because of its tendency to reproduce long-term infectious processes. Infected individuals number in the hundreds of millions. It is conceivable that the occurrence of *C. trachomatis* in the genital tracts of humans may rival that of other venereal infections in frequency. The increasing recognition of chlamydiae in venereal infections indeed is one subject of this meeting.

LATENCY OF CHLAMYDIAL INFECTIONS

Another biological characteristic that is universal among chlamydiae is the production of latent, persistent, and inapparent infections in host animals. This can be illustrated by many reports, but only a few will be recorded here.

One well-documented example of psittacosis in man was reported by Meyer and Eddie (4). A young biochemist at the Hooper Foundation became acutely ill with psittacosis in April 1938. Upon recovery, he continued to shed organisms in his respiratory tract although his only symptom was an occasional period of cough with a thick tenacious sputum. During the next 8 years he served in the U.S. Army abroad, and in 1946 he was found to be a carrier of the organisms, which were isolated from sputum. After extensive penicillin therapy, his carrier state was terminated.

Persistence of organisms in persons with trachoma have been well documented, and Gale et al. reported that two Taiwan monkeys inoculated in 1960 with *C. trachomatis* developed chronic eye infections and the organisms were repeatedly recovered from the eyes over a 10-year period (5).

Hanna et al. (6) discussed latency in infections of Indian children in boarding school. Under excellent hygienic conditions, children originally showing active trachoma demonstrated dramatic clinical regression of the disease process, but they continued to carry chlamydiae in their eyes, whether they were treated with topical tetracycline or systemic sulfonamide.

Inapparent infections in flocks of domestic birds have often been revealed only upon outbreaks of disease in persons processing poultry for market, as in the case of turkey psittacosis (7).

In laboratory studies on persistence and latency, several model systems have been reported. Morgan and Bader (8) infected L cells in balanced salt solution with the 6BC strain of *C. psittaci* and found that these organisms lost all infectivity after 24 h. Such cells could be maintained for 3 days without cell death or production of infectious chlamydiae. If complete medium was added not later than day 3 postinfection, infectious organisms were present on day 4, and a typical growth curve could be plotted on days 4, 5, and 6.

In our laboratory, we succeeded in adapting the meningopneumonitis strain to growth in HeLa cells, and a persistent infection was established with about one-half the cells showing inclusions at any one time (9). Giemsa stains of infected cells showed evidence that the cells could possibly expel colonies from their cytoplasm without cell death. One bottle culture of infected HeLa cells was maintained for almost 1 year by twice-weekly addition of fresh medium. The cells maintained a healthy appearance and produced 10^8 to 10^9 infectious organisms weekly. The addition of penicillin almost completely depressed production of infectious organisms for 100 days, but, upon addition of penicillin-free medium, regular production of infectious elementary bodies (EB) commenced almost immediately.

More recently, Schoenholz (10) inoculated rabbit corneal cells in culture at 37°C with *C. psittaci* at near the limiting dilution. Such cultures were infectious for 1 to 2 days but were then noninfectious for eggs. When such cultures were incubated at 39°C, overt multiplication began. In animal studies, Schoenholz (11) reported that macrophages removed from mice previously injected with *C. psittaci* survived for 30 days or longer when explanted at 37°C and showed no evidence of *C. psittaci* growth. When such macrophage explants were incubated at 39.5°C, however, the chlamydiae resumed multiplication and the macrophage explant was destroyed within a few days. Lymphocytes from immune mice prevented macrophage destruction.

GROWTH CYCLE AND CELL ENVELOPES

As is true of essentially all microorganisms which require an intracellular environment, the adaptation of chlamydiae to growth in established cell lines has been a requirement for many quantitative studies on their morphology, differentiation, and biochemistry. Storz (12) has listed many of the cell lines that have been used for studies on the isolation, cultivation, and host-parasite interaction of chlamydiae, and these include cells from a wide variety of tissues from numerous vertebrates. Not all chlamydiae grow well in cell cultures, and production of large amounts of chlamydiae in this manner has usually been difficult.

One of the most successful systems was that developed by Higashi and associates (13, 14) in which the Cal-10 strain of *C. psittaci* was adapted to growth in suspended cultures of mouse fibroblasts (L cells). This very successful system has been modified for special purposes by Moulder (15), and our own studies to be reviewed briefly below have been completely dependent on this system. The organisms grow well on both monolayer and suspended cell cultures of L cells, and 10- to 100-mg amounts of highly purified organisms can be routinely prepared. This has made possible a variety of studies that we believe have relevance to all the chlamydiae and not only to *C. psittaci*.

In the process of survival and adaptation, the chlamydiae have developed a highly specialized and unique developmental cycle in which two morphological forms of the organism occur alternately (16). The first of these is commonly referred to as the EB. It is seen in electron micrographs as a spherical body about 0.3 μm in diameter with an electron-dense center. The EB has a resistant rigid cell envelope, it can survive under certain conditions for relatively long periods outside host cells, it is highly infectious, and it has a demonstrable hemagglutinin in its envelope which may be involved in the process of attachment and penetration of host cells. The EB envelope comprises about 15% of the dry weight of the organisms and resembles cell envelopes of gram-negative bacteria in a number of respects, including amino acid composition and the presence of a carbohydrate-lipopolysaccharide complex (17, 18) (Fig. 1).

FIG. 1. *Elementary body cell envelopes of C. psittaci shadow-cast with platinum-palladium alloy. Subunit layers in hexagonal array on the inner side of the cell wall and in flat cell wall fragments are clearly evident. Bar represents 250 nm. (Reprinted from J Bacteriol 104:1332–1337, 1970)*

The EB enter susceptible cells by a phagocytic process which creates a phagosome in which the microcolony develops. The phagosome membrane continues to enlarge during the entire growth cycle. This great expansion of the phagosome membrane occurs at a time when host cell protein synthesis has been essentially inhibited. Stokes (19, 20) has reported that no proteins synthesized by *C. psittaci* become associated with these membranes, but cycloheximide-resistant glycosylation of membrane fractions continues during the infection cycle, suggesting a chlamydia-mediated role in membrane synthesis. Within a short time after penetration, the EB begins to undergo changes in its envelope, as shown by its rapidly increasing sensitivity to sonic treatment. There is a dispersion of the central condensate to form a homogeneously arrayed cytoplasm in which strands of nucleic acid and many ribosomes can be seen. The mass of the organism enlarges and the second form of the organism, referred to as the initial or reticulate body (RB), is complete. The RB is a strictly intracellular resident. Its cell envelope is very fragile and highly permeable as compared with the EB. The envelope comprises only about 5% of the dry weight of the RB and contains no hemagglutinin (21, 22) (Fig. 2).

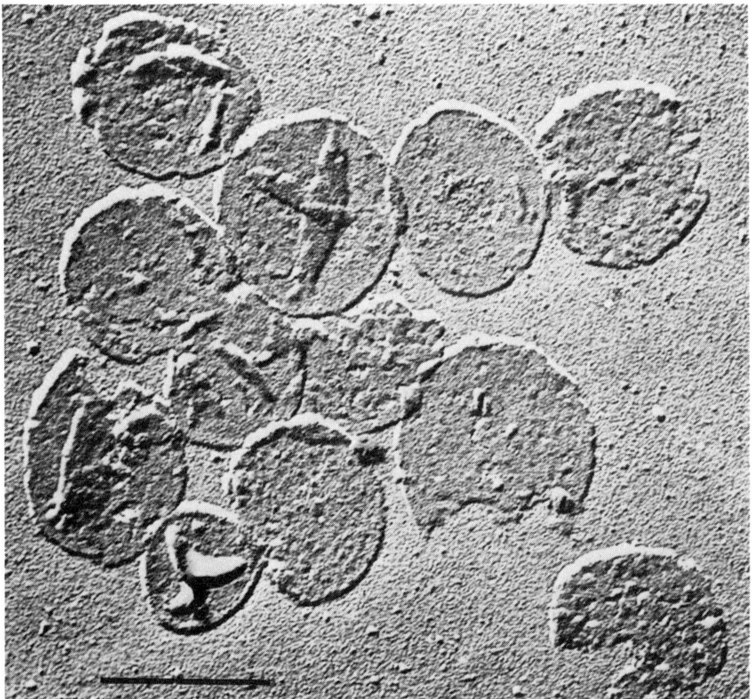

FIG. 2. *Reticulate body cell envelope of C. psittaci shadow-cast with platinum-palladium alloy. Overlapping edges indicate flexible fragile nature of envelopes. Bar represents 1 μm. (Reprinted from J Bacteriol 94: 1184–1188, 1967)*

The RB continue to grow in size, and 10 to 15 h after infection they begin to undergo binary fission. At 20 to 25 h postinfection, some of the RB begin to undergo a conversion process in which there is a central condensation of the cytoplasmic contents and a typical small EB is formed (Fig. 3). Some of the RB, however, continue to divide until the host cell cytoplasm is almost completely replaced by the colony of organisms. Little is known specifically of the process of release of organisms from the cells, although in highly adapted systems, such as *C. psittaci* in mouse L cells, the host cells surely die after 40 to 60 h of infection, with autolysis following. Todd and Storz (23) suggested that host cell lysosomal enzymes play a role in release of these organisms. Lysosomal enzymes remain localized within lysosomes until maturation of the organism, and they are then released into host cell cytoplasm and are always associated with host cell lysis.

The morphological changes that this organism undergoes during the growth cycle suggested that the RB may represent a naturally occurring protoplast, with the EB actually shedding its envelope to become an RB. It has, however, been clearly demonstrated that the cell envelope is continuous throughout the developmental cycle, but very significant changes in the envelope do occur during the conversion of EB to RB (24).

A. Tamura (Bacteriol. Proc., p. 90, 1971) has studied the kinetics of conversion of EB to RB in L cells by infection with EB heavily labeled with ^{32}P. Intact preparations of EB are highly resistant to sonic treatment, and their labeled DNA is

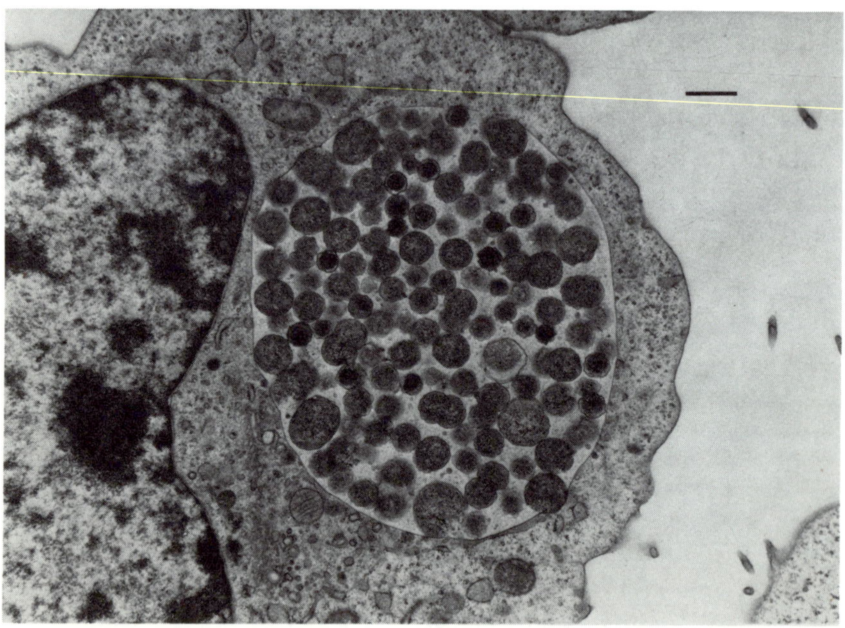

FIG. 3. *Inclusion of C. psittaci in L cell after 36-h infection. Phagosome membrane separating organisms from cell cytoplasm, newly formed elementary bodies, typical reticulate bodies, reticulate bodies undergoing binary fission, and organisms undergoing conversion to elementary bodies can be seen. Bar represents 1 μm.*

protected from subsequent deoxyribonuclease (DNase) treatment. Within 2 h after infection, however, the organisms begin to show increased sensitivity to sonic treatment, as evidenced by release of labeled DNA and its digestion by DNase. By 6 h, essentially all the DNA is released by sonic treatment, indicating that the loss of rigidity and stability of the EB envelope proceeds rapidly after entry. This process is initiated by the organism and not the host cell, as shown by the continued resistance of the envelopes of EB which had been irradiated with ultraviolet (UV) light prior to infection.

One of the most interesting changes is the disappearance of the monolayer of hexagonally arrayed subunits which line the inner aspect of EB envelopes. These subunits have been isolated from EB envelopes by hot formamide extraction and contain protein-carbohydrate complexes which are missing in RB envelopes, as determined by acrylamide gel analysis (25, 26). Soon after entry of EB into the host cell, this layer begins to break down. Apparently it disappears by dilution since fragments may be seen in RB up to 20 to 24 h. Synthesis of this layer is apparently necessary during the process of maturation of RB to form EB. Penicillin inhibits the formation of this subunit layer, and RB grown in penicillin-containing medium for 48 h or longer show no evidence of subunit formation.

Another very significant difference between EB and RB is their toxicity for mice and other animals. Most strains of chlamydiae will kill mice when 10^8 to 10^9 viable organisms are injected intravenously. We have previously shown that about 10^8 to 5×10^8 infectious EB of *C. psittaci* will kill a 10-g mouse in 24 h, even though the animals are treated with sufficient antibiotic to prevent infection. These toxic deaths can be prevented by passive or active immunization of test mice, or by in vitro neutralization of the EB prior to injection. EB envelope preparations or suspensions of heat-killed or UV-killed EB are not toxic, but will absorb antitoxin from antisera (27). There is thus a surface antigen on EB which may function in the toxic reaction and does specifically absorb antitoxin. The mechanism of the toxicity of live EB is unknown, but we believe it to be a result of EB-phagocytic cell, especially macrophage, interactions with perhaps a release of large amounts of lysosomal constituents from lysed phagocytes. At any rate, the live mature EB are harmful and destructive to certain cells involved in clearance of the organisms.

However, the RB appear to be completely nontoxic for white mice. Very large numbers of RB can be injected into mice intravenously without any noticeable effect, and suspensions of RB do not absorb antitoxin. Thus, the surface antigen which absorbs antitoxin and immunizes mice against toxic challenge appears to be completely absent during the RB phase of the growth cycle. This characteristic of RB is obviously related to the ability of chlamydiae to adapt to long-term residence in host cells without destruction of the cell. It also gives some support to the concept that the toxicity of EB is an expression of a mechanism designed to release mature EB from host cells at the end of the growth period.

The ability of chlamydiae to reside indefinitely within host cells as RB is also shown by experiments with penicillin-treated cultures (28, 29). Penicillin does not affect attachment to and penetration of susceptible cells by EB nor does it have any effect on the conversion of EB to RB. The RB continue to grow in mass until the host cell cytoplasm is filled with giant RB, and even after 40 to 48 h of growth the RB envelopes are essentially unchanged from 18-h RB envelopes. Penicillin does, however, prevent binary fission, inhibit the formation of the subunit layer, and block the conversion of RB to EB.

TABLE 2. *Characteristics of elementary bodies (EB) and reticulate bodies (RB)*

Characteristic	EB	RB
Morphology	Small, dense-centered	Large, flat, homogeneous
RNA:DNA	RNA:DNA = 1:1	RNA:DNA = 3:1
Density	Heavier	Lighter
Sonication	Relatively resistant	Sensitive
Effect of trypsin	Resistant	Sensitive, lysis
Infectivity	+	−
Toxicity for mice	+	−
Antitoxin absorption	+	−
Hemagglutinin	+	−
Permeability	Much reduced	+
Envelope subunit	+	−
Muramic acid in envelope	−	−

Numerous other studies have further illustrated the differences in EB and RB, and all are summarized in Table 2. None of these studies has shed a great deal of light on the mechanism of maintenance of latent or persistent infections by chlamydiae. What they do reflect, I think, is the adaptation of this group of organisms to maximal survival and maintenance of intracellular infection with a minimum of harm to host cells. When conditions develop in favor of the chlamydiae, maximal production of EB results, such as we see in nearly ideal models for organism production in L cells and chick embryos where the organisms reproduce rapidly and in a lethal manner. However, in many infected hosts, perhaps a great majority, the intracellular reproductive forms are relatively harmless and only compete with the host cell on a moderate basis for energy and nutrition.

In this discussion I have not answered many questions, but I hope I have posed a few to which I think attention should be directed. We particularly need greater insight into the biochemical processes that cause EB to begin their conversion to RB and the reverse conversion of RB to EB. The structure of the chlamydial envelope is undoubtedly of great importance in the pathogenesis of chlamydial disease and in development of resistance, yet we have a very limited understanding of the layer-by-layer antigenic and biochemical architecture of these envelopes. We are just beginning to understand the biological nature of the chlamydiae.

REFERENCES

1. Page LA: Order II. *Chlamydiales*, p 914–918. *In* RE Buchanan and NE Gibbons (ed), Bergey's manual of determinative bacteriology, 8th ed. The Williams & Wilkins Co., Baltimore, 1974
2. Meyer KF: The host spectrum of psittacosis-lymphogranuloma venereum (PL) agent. Am J Ophthalmol 63:1225/199–1246/220, 1967
3. Grayston JT, Wang SP: New knowledge of chlamydiae and the diseases they cause. J Infect Dis 132:87–105, 1975
4. Meyer KF, Eddie B: Human carrier of the psittacosis virus. J Infect Dis 88:109–125, 1951
5. Gale JL, Wang SP, Grayston JT: Chronic trachoma in two Taiwan monkeys ten years after infection, p 489–493. *In* RL Nichols (ed), Trachoma and related disorders caused by chlamydial agents. Excerpta Medica, Amsterdam, 1971

6. Hanna L, Dawson C, Briones O, et al: Latency in human infections with TRIC agents. J Immunol 101:43–50, 1968
7. Irons JV, Sullivan TD, Rowen J: Outbreaks of psittacosis from working with turkeys or chickens. Am J Public Health 41:931–937, 1951
8. Morgan HR, Bader JP: Latent viral infection of cells in tissue culture. IV. Latent infection of L cells with psittacosis virus. J Exp Med 106:39–44, 1957
9. Manire GP, Galasso GJ: Persistent infection of HeLa cells with meningopneumonitis virus. J Immunol 83:529–533, 1959
10. Schoenholz WK: Studies on Bedsonia latency. I. Induction of latency in rabbit cornea (SIRC) cell culture and its reversal by changes in culture conditions. Z Immunitaetsforsch 135:283–293, 1967
11. Schoenholz WK: Studies on Bedsonia latency. II. Effect of immune lymphocytes and of rabbit anti-lymphocyte globulin on infected macrophages exposed to increased incubation temperature *in vitro*. Z Immunitaetsforsch 139:359–371, 1970
12. Storz J: Chlamydia and chlamydia-induced diseases. Charles C Thomas, Publisher, Springfield, Ill., 1971, p 79–88
13. Higashi N, Notake K, Fukuda T: Growth characteristics of the meningopneumonitis virus in strain L cells. Annu Rep Inst Virus Res Kyoto Univ 2:23–56, 1959
14. Higashi N, Tamura A: A plaque assay for meningopneumonitis virus in monolayers of strain L cells. Virology 12:578–588, 1960
15. Moulder JW: A model for studying the biology of parasitism: *Chlamydia psittaci* and mouse fibroblasts (L cells). BioScience 18:875–881, 1969
16. Bedson SP, Bland JOW: Morphological studies of the psittacosis virus, with a description of the developmental cycle. Br J Exp Pathol 31:461–466, 1932
17. Manire GP, Tamura A: Preparation and chemical composition of the cell walls of mature infectious dense forms of meningopneumonitis organisms. J Bacteriol 94:1178–1183, 1967
18. Narita T, Manire GP: Protein-carbohydrate lipid complex isolated from cell envelopes of *Chlamydia psittaci* in alkaline buffer and ethylenediaminetetraacetate. J Bacteriol 125:308–316, 1976
19. Stokes GV: Formation and destruction of internal membranes in L cells infected with *Chlamydia psittaci*. Infect Immun 7:173–177, 1973
20. Stokes GV: Cycloheximide-resistant glycosylation in L cells infected with *Chlamydia psittaci*. Infect Immun 9:497–499, 1974
21. Tamura A, Manire GP: Preparation and chemical composition of the cell membranes of developmental reticulate forms of meningopneumonitis organisms. J Bacteriol 94:1184–1188, 1967
22. Tamura A, Manire GP: Hemagglutinin in cell walls of *Chlamydia psittaci*. J Bacteriol 118:144–148, 1974
23. Todd WJ, Storz J: Ultrastructural cytochemical evidence for the activation of lysosomes in cytocidal effect of *Chlamydia psittaci*. Infect Immun 12:638–646, 1975
24. Tamura A, Matsumoto A, Manire GP, et al: Electron microscopic observations on the structure of envelopes of elementary bodies and reticulate forms of *Chlamydia psittaci*. J Bacteriol 105:355–360, 1971
25. Matsumoto A, Manire GP: Electron microscopic observations on the fine structure of cell walls of *Chlamydia psittaci*. J Bacteriol 104:1332–1337, 1970
26. Tamura A, Tanaka A, Manire GP: Separation of polypeptides of *Chlamydia* and its cell walls by polyacrylamide gel electrophoresis. J Bacteriol 118:139–143, 1974
27. Christoffersen G, Manire GP: The toxicity of meningopneumonitis organisms (*C. psittaci*) at different stages of development. J Immunol 103:1085–1088, 1969
28. Matsumoto A, Manire GP: Electron microscopic observations of the effects of penicillin on the morphology of *Chlamydia psittaci*. J Bacteriol 101:278–285, 1970
29. Tamura A, Manire GP: Effect of penicillin on the multiplication of meningopneumonitis organisms. J Bacteriol 96:875–880, 1968

Antigens of *Chlamydia trachomatis*

CHO-CHOU KUO, HARLAN D. CALDWELL, SAN-PIN WANG, AND J. THOMAS GRAYSTON

Our knowledge about the antigenic structure of *Chlamydia trachomatis* is limited, although the organisms have been cultivated in vitro in embryonated chicken eggs since 1935 for lymphogranuloma venereum (LGV; 1, 2) and since 1957 for trachoma (3). Detailed immunological studies of *C. trachomatis* have been virtually prevented by the difficulty in freeing the organisms, which are obligate intracellular parasites, from host cell material. We have overcome this problem by using two culture systems, i.e., chick embryo yolk sac versus HeLa 229 cell culture or HeLa 229 versus L-cell culture (4–6). Another problem in antigenic analysis of *C. trachomatis* is to produce the mass of antigen required for biochemical methods of fractionation and analysis. This problem has been circumvented by use of microtechniques or immunochemical methods (4–7; H. D. Caldwell, C. C. Kuo, and S. P. Wang, Abstr. Annu. Meet. Am. Soc. Microbiol. 1976, p. 51, C151).

A summary description of our studies on antigenic analysis of *C. trachomatis* is presented in this report. It concerns the group-, delayed hypersensitivity-, species-, and type-specific antigens. The known properties of these antigens are summarized in Table 1.

I. GROUP ANTIGENS

All members of the genus *Chlamydia* share a common antigen demonstrable by the complement fixation (CF) test which is a heat-stable (100°C for 30 min) lipoprotein-carbohydrate complex (8, 9). This antigenicity is destroyed by periodate treatment (10). We have studied the antigens of chlamydiae by double immunodiffusion (4). The antiserum was obtained by immunizing rabbits with purified organisms grown in chick embryo yolk sacs (2×10^{10} to 4×10^{10} organisms/ml), and the test antigen was organisms grown in HeLa 229 cell culture (2×10^9 to 8×10^9 organisms/ml). A micro-immunodiffusion test was used (11). Two trachoma (B/TW-5/OT and A/G-17/OT) and two psittacosis (meningopneumonitis Cal-10 and mouse pneumonitis Nigg) strains were tested. Extensive sonic treatment was used to solubilize antigen. Three group antigens were demonstrated (Fig. 1). They were heat stable at 100°C for 30 min. Two were solubilized by the detergent sodium deoxycholate, and the remaining one was a cell wall antigen which contained carbohydrate. Lipid antigens prepared by the chloroform-methanol-water partition method of Folch gave one precipitation line which fused in a reaction of identity.

The chemical nature of the group CF antigen was elucidated by S. P. Dhir in our laboratory (12). A polysaccharide hapten antigen was obtained from the common lipid group antigen by alkali saponification. The lipid antigen was a polysaccharide ester of fatty acid. The group-reactive, water-soluble, polysaccharide hapten was shown to be common to the organisms of trachoma, LGV, and psittacosis. The antigen was an acidic polysaccharide of large molecular

TABLE 1. *Antigens of Chlamydia trachomatis*

Antigen	Test	No. of antigenic components recognized, or *C. trachomatis* immunotypes	Physicochemical characteristics	Antigenic determinant
Group	Double immunodiffusion	3	Heat stable, 100°C/30 min, lipid-carbohydrate	Acidic polysaccharide, 2-keto-3-deoxyoctanoic acid, mol wt 0.2 $\times 10^6$ to 2.0 $\times 10^6$
Delayed hypersensitivity (species)[a]	Skin test, lymphocyte blastogenesis		Heat stable, 100°C/30 min, acid-extractable	Protein
Species	Two-dimensional immunoelectrophoresis	18	15 are heat labile, 80°C/30 min	One immunodominant antigen is a protein, mol wt 1.55×10^5
Type	Micro-immunofluorescence (mouse toxicity prevention)	12 trachoma, 3 LGV[b]	Heat labile	Unknown

[a] Acid-extracted antigen.
[b] *C. trachomatis* immunotypes by micro-immunofluorescence.

weight (0.2 to 2 million). The acidic component was critical to antigenicity. This immunodominant group was shown to be a 2-keto-3-deoxyoctanoic acid which is similar to but not identical to that of *Salmonella* lipopolysaccharide.

II. DELAYED HYPERSENSITIVITY ANTIGEN

The presence of delayed hypersensitivity to chlamydial infection has been demonstrated by the Frei skin test (13). The skin test has been most successful with LGV (14). It has been used for psittacosis (15), but the results with trachoma were disappointing (16). Since the antigen used consists of whole organisms (inactivated by boiling or by formalin and phenol treatment), the reaction is group specific (8, 14). Species specificity has been demonstrated by use of acid-extracted antigen (14). We have used guinea pigs for study of delayed hypersensitivity (5). They were immunized with purified yolk sac-grown organisms incorporated in Freund complete adjuvant. Test antigens were prepared from organisms grown in HeLa 229 cell culture. Determination of hypersensitivity was by a skin test and in vitro peripheral blood lymphocyte blast transformation utilizing a microtechnique (5). The strains employed were trachoma (B/TW-5/OT and C/TW-3/OT), LGV (L_2/434/Bu), and psittacosis (meningopneumonitis Cal-10). The skin test allergen was found to be a heat-stable

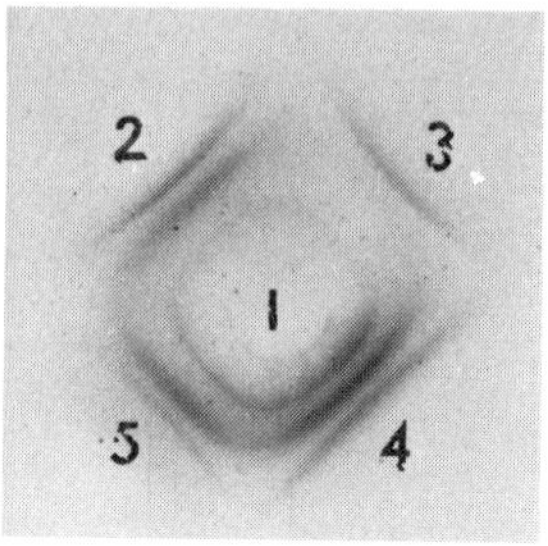

FIG. 1. *Demonstration of three heat-stable Chlamydia group antigens by double immunodiffusion. (1) Anti-trachoma-TW-5 antiserum; (2) antigen trachoma-TW-5; (3) antigen trachoma-G-17; (4) antigen psittacosis-mouse pneumonitis; (5) antigen psittacosis-meningopneumonitis.*

protein (100°C for 30 min) extractable with weak acid (0.05 N HCl at 56°C for 16 h). This antigen was shown to be species specific and able to differentiate between *C. trachomatis* and *C. psittaci*, but not between trachoma and LGV organisms.

III. SPECIES ANTIGEN

Species antigens are more difficult to study because of heat lability. The presence of species antigens has been demonstrated by the CF test after removing the group antibody by absorption of the serum with boiled antigen (14) and by double immunodiffusion (4, 17). By the double immunodiffusion test we were able to demonstrate the presence of seven antigenic components in trachoma and psittacosis (4). Three of them were shown to be heat-stable group antigens, and four were heat-labile species antigens.

Recently, we have been successful in studying species antigens by use of two-dimensional immunoelectrophoresis (6, 7). The organisms studied were trachoma (K/UW-31/Cx), LGV (L_2/434/Bu), and psittacosis (meningopneumonitis Cal-10). Mouse L-cell suspension culture was used for growth of 434 and meningopneumonitis to be used as the serological test antigens. The organisms were purified by one cycle of differential centrifugation ($500 \times g$ for 15 min and $15{,}000 \times g$ for 60 min) and then were pelleted through 30% Renografin (methyglucamine diatrizoate, 76% for injection, Squibb and Sons, New York, N.Y.) at $20{,}000 \times g$ for 1 h. They were solubilized with 1% (vol/vol) of the non-ionic detergent Triton X-100 (Calbiochem, San Diego, Calif.). The insoluble material was removed by centrifugation at $30{,}000 \times g$ for 30 min, and the supernatant fluid was used as test antigen. The immunogens for preparation of rabbit hyperimmune sera were the organisms of UW-31, 434, and meningopneumonitis grown in monolayer culture of HeLa 229 cells and purified by the same method. The rabbits were immunized with a total dose of 3×10^{11} organisms.

The method of two-dimensional immunoelectrophoresis has been described (6, 18). Briefly, glass mounting slides (4 by 4 cm) were covered with 3 ml of 0.5% agarose in pH 8.6 barbital buffer, ionic strength 0.05, and 0.5% Triton X-100 and 0.01% sodium azide. Five microliters (50 μg of protein) of solubilized antigen was subjected to electrophoresis in the first dimension at 2.7 V/cm

(5 mA/slide) for 1 h and 50 min. The second-dimension gel was 1.8 ml of agarose containing 0.2 to 0.4 ml of antiserum. Second-dimension electrophoresis was carried out at 2.7 V/cm for 6 h. After electrophoresis, the slides were washed, dried, and stained with Coomassie brilliant blue. Nineteen distinct components were observed with LGV-434 and 16 were observed with psittacosis-meningopneumonitis (Fig. 2). In the heterologous test between these two organisms, only one reciprocal cross-reacting antigen was observed, indicating that *C. trachomatis* and *C. psittaci* were antigenically heterogeneous. However, antigens within *C. trachomatis* were strikingly homogeneous, as shown when LGV-434 antigen was subjected to electrophoresis against anti-trachoma-UW-31 antiserum (Fig. 3). It was also shown that most of the species antigens were heat labile, since heating at 80°C for 30 min destroyed 16 of 19 LGV-434 antigens.

Because of the promising species specificity of Triton-solubilized antigens, the test was applied to human sera of patients with laboratory-proven *C. trachomatis* infection. It was found that one of the antigens of LGV-434

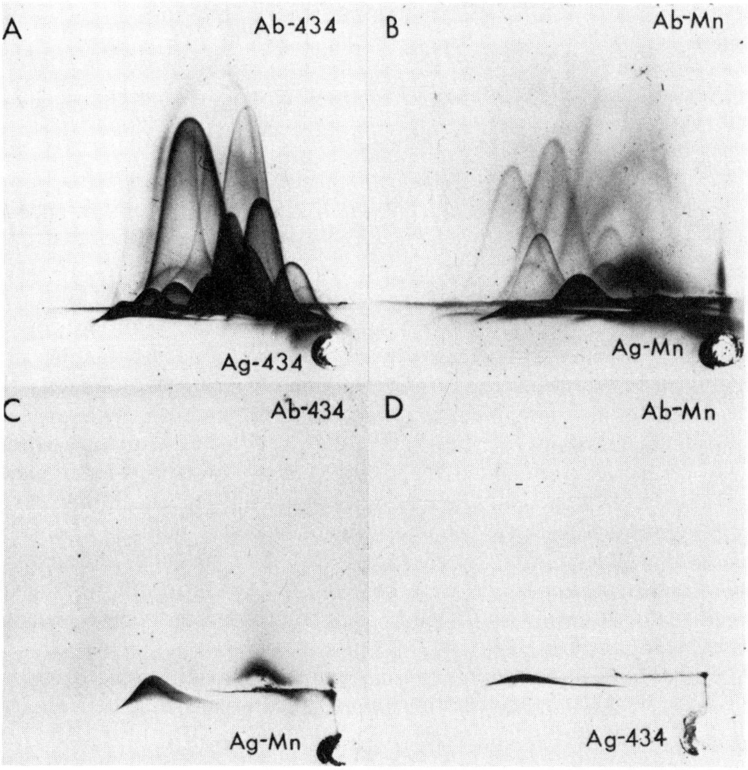

FIG. 2. *Serological cross-reactions of LGV and psittacosis by two-dimensional immunoelectrophoresis. (A) LGV-434 antigen versus anti-434; (B) psittacosis-meningopneumonitis (Mn) antigen versus anti-Mn; (C) Mn antigen versus anti-434; (D) 434 antigen versus anti-Mn.*

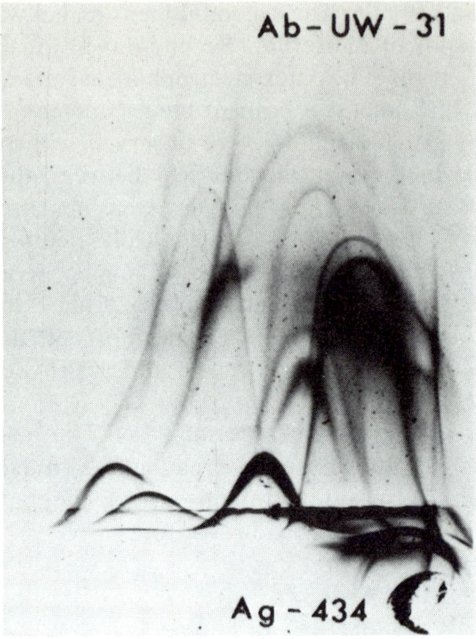

FIG. 3. *Serological cross-reactions between C. trachomatis strains LGV-434 and trachoma-UW-31 by two-dimensional immunoelectrophoresis (434 antigen versus anti-UW-31).*

was strongly and consistently precipitated by the patients' sera. Monospecific antiserum against this antigen was made in rabbits. The antigen was obtained by excising the antigen-antibody complex from two-dimensional gel made by reacting LGV-434 antigen and rabbit anti-LGV-434 antiserum. The antiserum obtained was monospecific in that a single precipitin line with electrophoretic mobility of 0.65 relative to bovine serum albumin was formed when it was allowed to react with LGV-434 antigen (Fig. 4). In heterologous tests, the monospecific antisera reacted in two-dimensional immunoelectrophoresis with the Triton-solubilized trachoma-LGV strains (B/TW/5/OT, K/UW-31/Cx, and L_2/434/Bu) but not with the mouse pneumonitis (Nigg) strain or the psittacosis strain meningopneumonitis. The trachoma-LGV specificity of this antigen was further shown by indirect immunofluorescence staining with monospecific antisera of chlamydial inclusions in infected HeLa 229 cells. The inclusions of all the 15 known immunotypes of trachoma and LGV were stained. Inclusions of *C. psittaci* strains meningopneumonitis, 6BC, and feline pneumonitis and the mouse pneumonitis strain were not stained.

Precipitins with specificity to this antigen were identified in 15 of 18 sera from patients with diagnosed *C. trachomatis* infections including LGV, trachoma, nongonococcal urethritis, and nongonococcal cervicitis. The test used was peak suppression in which the monospecific antiserum is added to the human serum in the second phase of electrophoresis and the profile of the precipitation line is compared with that from electrophoresis with the human serum alone (Fig. 5).

This antigen has now been purified by affinity chromatography performed with rabbit monospecific antiserum (Caldwell et al., Abstr. Annu. Meet. Am. Soc.

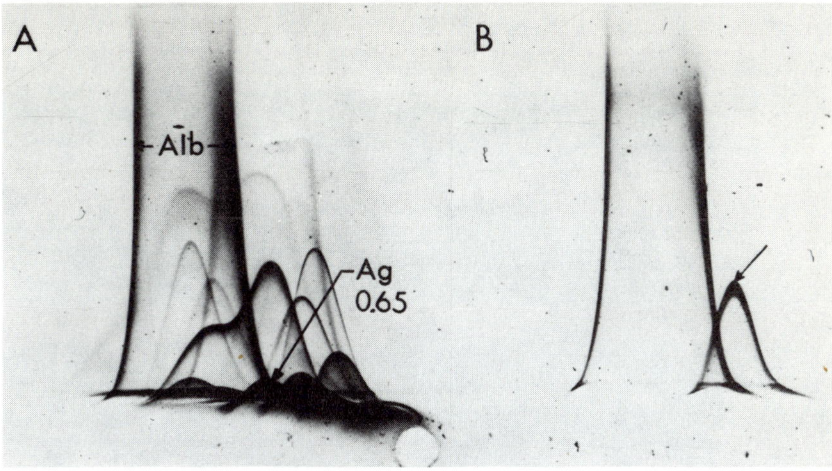

FIG. 4. *Two-dimensional immunoelectrophoresis profile of LGV-434 (A) and demonstration of the monospecificity of rabbit antiserum to LGV-434 antigen-0.65 (B). A: 5 μl (50 μg of protein) of LGV-434 antigen was subjected to electrophoresis in the first dimension (anode to the left) and to electrophoresis against 0.2 ml of rabbit antiserum to LGV-434 in the second dimension (anode at the top). Antigen-0.65 and bovine serum albumin (alb) are identified by the arrows. The electrophoretic mobility of antigen-0.65 was determined relative to bovine serum albumin. B: 5 μl (50 μg of protein) of LGV-434 antigen versus 0.2 ml of rabbit anti-antigen-0.65. Antigen-0.65 is identified by the arrow. The more anodal peak in the figure is the bovine albumin marker.*

Microbiol. 1976, p. 51, C151). The antigen was shown to be a heat-labile (65°C for 30 min) protein with an apparent molecular weight of 155,000. A simple, sensitive, and specific serological test for detection of antibody to this antigen in LGV patients by use of counterimmunoelectrophoresis has been developed; 43 of 45 (96%) known LGV cases tested were positive (Caldwell et al., Abstr. Annu. Meet. Am. Soc. Microbiol. 1976, p. 51, C151). However, the test was not sensitive enough to detect antibodies in other *C. trachomatis* infection; 50 cases of nongonococcal urethritis tested were negative. No false positive was found; 50 normal controls were negative by this test.

IV. TYPE-SPECIFIC ANTIGEN

Injection of a large dose of live LGV (19) or trachoma (20, 21) organisms intravenously into mice causes death within 24 h. The exact mechanism is not clear, but it is postulated that a toxic factor is present in the organisms (19). The "toxin" is associated with viable organisms and has not been separated from the organisms. It is heat labile. The death of mice can be prevented by prior immunization with a sublethal dose of live organisms or "toxoid" (formalinized organisms). Protection of mice from toxic death was utilized for the classification of *C. trachomatis* into different immunotypes by Bell and Theobald (22) and Wang and Grayston (21). By this method, our laboratory classified 125 trachoma agents into six immunological types, designated A through F (23).

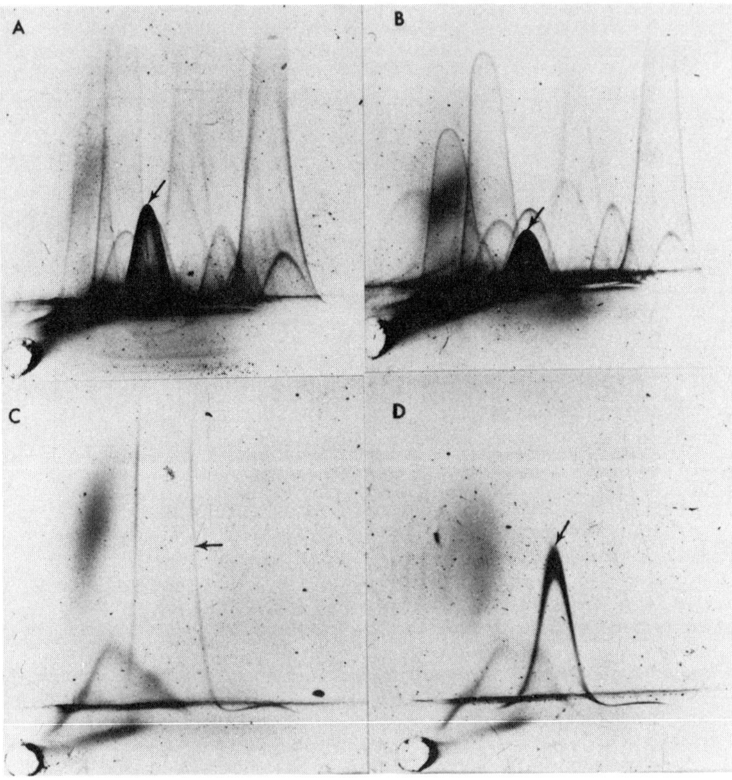

FIG. 5. *Identification of precipitins to LGV-434 antigen-0.65 in human sera from patients with C. trachomatis infections by peak suppression performed with monospecific rabbit antiserum to antigen-0.65. Antigen: 5 μl (50 μg of protein) of LGV-434. Antisera: A, 0.2 ml of serum from LGV patient; B, same as A plus 0.2 ml of anti-antigen-0.65; C, 0.2 ml of serum from ocular trachoma patient; D, same as C plus 0.2 ml of anti-antigen-0.65. Arrows indicated precipitate before and after suppression. First-dimension electrophoresis with the anode at the right; second-dimension electrophoresis with anode at the top.*

This cumbersome technique was later replaced by an indirect micro-immunofluorescence (micro-IF) test (24, 25). Immune mouse serum was used in the micro-IF test for typing. It was shown that the type-specific antibody is the antibody responsible for prevention of toxic death.

More than 500 *C. trachomatis* strains have been typed (26). The trachoma strains fall into 12 immunotypes (A through K plus Ba) and LGV strains fall into three types (L_1, L_2, and L_3). The micro-IF test also detects heterologous antibodies. The highest titers are obtained with homologous antigens, and the pattern of heterologous cross-reactions is characteristic for each immunotype. The serological cross-reaction pattern between immunotypes is presented in Table 2. Since the pattern of serological cross-reaction is sufficiently distinct for each immunotype, it is possible to designate the type of organisms by testing unknown antisera against a battery of prototype antigens in a one-way test (27).

TABLE 2. *Serological cross-reaction pattern of mouse antisera to 15 immunotypes of C. trachomatis*[a]

Mouse antiserum	Prototype antigens														
	C	J	A	H	I	K	L_3	L_2	L_1	D	E	Ba	B	G	F
C	4	3	2												
J	2	4	2												
A			4												
H				4											
I					4										
K						4	2								
L_3						4	4	1	1					1	
L_2						2	2	4	2						
L_1						2	2	3	4						
D							2	3	3	4	2	1		1	
E							2	4	4	4	4	2		1	
Ba							2	4	3	3	3	4	4	2	2
B											1	4	4		
G							1							4	3
F														2	4

[a] Homologous titer is given a score of 4. Each unit represents a twofold serum dilution.

One-way serological cross-reaction is observed between some immunotypes. As shown in Table 2, type C antiserum cross-reacted with type A antigen; however, no cross-reaction of type A antiserum with type C antigen was seen. This relationship has been called senior (type C)-junior (type A) relationship. A similar relationship is also observed with types ED and $L_{1,2,3}$.

Preliminary characterization of the type-specific antigen was done by S. P. Dhir in our laboratory (28). The antigen was solubilized by alkali treatment (pH 12.5, 37°C for 1 h) and was found to be present in the lipid extract from the chloroform-methanol-water partition of Folch. The antigen was not destroyed by trypsin (1%), Pronase (100 μg/ml), or periodate treatment. The results indicated that the specific antigen may be a lipoid hapten which is strongly bound to a protein. Since the micro-IF test detects surface antigen, the type-specific antigen is presumably located on the surface of the organisms.

ACKNOWLEDGMENT

This study was supported by U.S. Public Health Service research grant 5-R01-EY-00219 from the National Eye Institute.

REFERENCES

1. Miyagawa Y, Mitamura T, Yaoi H, et al: Studies on virus of lymphogranuloma inguinale Nicholas, Favre and Durand: cultivation of virus on chorioallantoic membrane of chick embryo. Jpn J Exp Med 13:733–738, 1935
2. Rake G, Jones HP: Studies on lymphogranuloma venereum. Development of the agent in the yolk sac of the chick embryo. J Exp Med 75:323–338, 1942
3. T'ang F, Chang H, Huang Y, et al: Studies on the etiology of trachoma with special reference to isolation of the virus in chick embryo. Chin Med J 75: 429–447, 1957

4. Kuo CC, Kenny GE, Wang SP: Trachoma and psittacosis antigens in agar gel double immunodiffusion, p 113–123. *In* RL Nichols (ed), Trachoma and related disorders caused by chlamydial agents. Excerpta Medica, Amsterdam, 1971
5. Kuo CC, Wang SP, Grayston JT: Studies on delayed hypersensitivity with trachoma organisms, p 158–176. *In* RL Nichols (ed), Trachoma and related disorders caused by chlamydial agents. Excerpta Medica, Amsterdam, 1971
6. Caldwell HD, Kuo CC, Kenny GE: Antigenic analysis of chlamydiae by two-dimensional immunoelectrophoresis. I. Antigenic heterogeneity between *C. trachomatis* and *C. psittaci*. J Immunol 115:963–968, 1975
7. Caldwell HD, Kuo CC, Kenny GE: Antigenic analysis of chlamydiae by two-dimensional immunoelectrophoresis. II. A trachoma-LGV-specific antigen. J Immunol 115:969–975, 1975
8. Rake G, Eaton MD, Shaffer MF: Similarities and possible relationships among viruses of psittacosis, meningopneumonitis and lymphogranuloma venereum. Proc Soc Exp Biol Med 48:528–531, 1941
9. Page LA: Revision of the family *Chlamydiaceae* Rake Rickettsiales: unification of the psittacosis-lymphogranuloma venereum-trachoma group of organisms in the genus *Chlamydia* Rake and Stearns, 1945. Int J Syst Bacteriol 16:223–252, 1966
10. Barwell CF: Some observation on the antigenic structure of psittacosis and lymphogranuloma venereum viruses. I. Preparation and use in complement-fixation tests of antisera from different sources. Br J Exp Pathol 33:258–267, 1952
11. Sharpless NS, LoGrippo GA: A standardized immunochemical method for quantitative determination of the immunoglobulins in serum. Henry Ford Hosp Bull 13: 55–73, 1965
12. Dhir SP, Hakomori S, Kenny GE, et al: Immunochemical studies on chlamydial group antigen (presence of a 2-keto-3-deoxycarbohydrate as immunodominant group). J Immunol 109:116–122, 1972
13. Frei W: Eine neue Hautreaktion bei "Lymphogranuloma inguinale." Klin Wochenschr 4:2148–2149, 1925
14. Bedson SP, Barwell CF, King EJ, et al: The laboratory diagnosis of lymphogranuloma venereum. J Clin Pathol 2:241–249, 1949
15. Terskih II: Early intradermal diagnosis of ornithosis. Acta Virol 1:211–215, 1957
16. Barwell CF, Dunlop EMC, Race JW: Results of complement fixation and intradermal tests for Bedsoniae in genital infection, diseases of the eye and Reiter's disease. Am J Ophthalmol 63:1527–1534, 1967
17. Collins AR, Barron AL: Demonstration of group- and species-specific antigens of chlamydial agents by gel diffusion. J Infect Dis 121:1–8, 1970
18. Thirkill CE, Kenny GE: Serological comparison of five arginine-utilizing *Mycoplasma* species by two-dimensional immunoelectrophoresis. Infect Immun 10:624–632, 1974
19. Rake G, Jones H: A toxic factor associated with the agent of lymphogranuloma venereum. Proc Soc Exp Biol Med 53:86–88, 1943
20. Bell SD, Jr, Snyder JC, Murray ES: Immunization of mice against toxic doses of homologous elementary bodies of trachoma. Science 130:626–627, 1959
21. Wang SP, Grayston JT: Classification of trachoma virus strains by protection of mice from toxic death. J Immunol 90:849–856, 1963
22. Bell SD, Jr, Theobald B: Differentiation of trachoma strains on the basis of immunization against toxic death of mice. Ann NY Acad Sci 98:337–346, 1962
23. Alexander ER, Wang SP, Grayston JT: Further classification of TRIC agents from ocular trachoma and other sources by the mouse toxicity prevention test. Am J Ophthalmol 63:1469–1478, 1967
24. Wang SP: A micro immunofluorescence method. Study of antibody response to TRIC organisms in mice, p 273–288. *In* RL Nichols (ed), Trachoma and related disorders caused by chlamydial agents. Excerpta Medica, Amsterdam, 1971

25. Wang SP, Grayston JT: Classification of TRIC and related strains with micro immunofluorescence, p 305–321. *In* RL Nichols (ed), Trachoma and related disorders caused by chlamydial agents. Excerpta Medica, Amsterdam, 1971
26. Grayston JT, Wang SP: New knowledge of chlamydiae and the disease they cause. J Infect Dis 132:87–105, 1975
27. Wang SP, Kuo CC, Grayston JT: A simplified method for immunological typing of trachoma-inclusion conjunctivitis–lymphogranuloma venereum organisms. Infect Immun 7:356–360, 1973
28. Dhir SP, Wang SP, Grayston JT: Type-specific antigens of trachoma organisms, p 133–141. *In* RL Nichols (ed), Trachoma and related disorders caused by chlamydial agents. Excerpta Medica, Amsterdam, 1971

Animal Models for the Study of Chlamydial Infections of the Eye and Genital Tract

S. DAROUGAR, M. A. MONNICKENDAM, H. EL-SHEIKH, J. D. TREHARNE, R. M. WOODLAND, AND B. R. JONES

Chlamydia trachomatis is responsible for a variety of ocular, oculogenital, and genital diseases in humans.

Hyperendemic trachoma, of eye-to-eye transmission, is the most common cause of blindness in the world, affecting 500 million people and causing blindness in some 2 million of them (1).

Paratrachoma, the endemic sexually transmitted TRIC (trachoma-inclusion conjunctivitis) agent infections of the eye, i.e., TRIC ophthalmia neonatorum, inclusion conjunctivitis, TRIC punctate keratoconjunctivitis, and trachoma (2), are not uncommon in the developed countries (3).

Nongonococcal genital infections including chlamydial genital infections, i.e., nonspecific urethritis, nonspecific cervicitis, and nonspecific proctitis, are the commonest sexually transmitted diseases, affecting several million people in the developed and developing countries (4–7).

However, little is known about the host-parasite relationship in chlamydial diseases of the eye, and even less is known about such relationships in chlamydial infections of the genital tract.

In recent years, studies have been carried out in human and animal models to elucidate the nature of the interactions between chlamydiae and their hosts. The purpose of this presentation is to discuss the present status of the use of animal models, and also to report the development of new animal models for the study of chlamydial infections of the eye and genital tract.

HUMAN VOLUNTEERS

Following the successful isolation of *C. trachomatis* in 1957 (8), several attempts were made to produce experimental infections in the eyes of humans. The result of these experimental studies demonstrated that *C. trachomatis* strains isolated from patients suffering from trachoma, inclusion conjunctivitis, and ophthalmia neonatorum are all pathogens for the eye and produce a follicular conjunctivitis similar to inclusion conjunctivitis. Pannus and scarring may develop in those patients who were infected previously. A partial and short-lived immunity may occur following a previous ocular infection or immunization with *C. trachomatis* vaccine (9–16).

In most countries however, ethical and other considerations greatly restrict the use of human volunteers for experimental studies.

NONHUMAN PRIMATES

Several species of nonhuman primates, i.e., baboons (*Papio* sp.), rhesus monkeys (*Macaca* sp.), Taiwan monkeys (*Macaca cyclopis*), and owl monkeys

TABLE 1. *Experimental inoculation of the eye of baboons with 0.2 ml of 20% (wt/vol) yolk sac suspension of Chlamydia*

No. of baboons	Sex	Chlamydial strain	Clinical	Microbiology	
				Scrapings (inclusions)	Isolation
1	F	TRIC/G/GB/IOL-238/R	Mild follicular conjunctivitis (10 weeks)	+ (10 days)	+ (7 weeks)
1	M	TRIC/G/GB/IOL-238/R	Mild follicular conjunctivitis (7 weeks)	+ (10 days)	+ (2 weeks)
1	F	TRIC/D/USA/Cal-8/ON	Mild follicular conjunctivitis (4 weeks)	ND[a]	+ (4 weeks)
1	M	TRIC/D/USA/Cal-8/ON	Mild follicular conjunctivitis (3 weeks)	ND	+ (2 weeks)
1	M	TRIC/B/IR/IOL-1597/OT	Moderate follicular conjunctivitis (6 weeks)	ND	+ (2 weeks)
3	M	LGV/ /GB/LH-B_2/Bu	—	—	—
3	M	PS/ /GB/IOL-83/J	—	—	—
1	M	*Chlamydia* sp./ /IR/IOL-207/OT	Mild follicular conjunctivitis (4 weeks)	ND	+ (3 months)

[a] Not done.

(*Aotus trivirgatus*), have been utilized as models for the study of chlamydial infections of the eye and genital tract, as an alternative to human volunteers (17–25).

In our laboratory, chlamydial agents were inoculated into the eye, genital tract, or joints of 12 male and female baboons (Tables 1–3).

In the eye, inoculation of *C. trachomatis* serotypes B (IOL-1597), D (Cal-8), and G (IOL-238), and an atypical iodine-negative chlamydial isolate (IOL-207) from the eye of a child with trachoma in Iran, all produced a mild to moderate follicular conjunctivitis for a period of 3 to 10 weeks, similar to inclusion conjunctivitis in man (Table 1). On the other hand, inoculation of a lymphogranuloma venereum (LGV) agent (LH-B_2) and a *C. psittaci* agent (IOL-83) isolated from the knee joint of a man with Reiter's disease failed to produce conjunctivitis (Table 1).

In the genital tract, intraurethral inoculation of *C. trachomatis* serotypes D (Cal-8), G (IOL-238), and LGV (LH-B_2) and of *C. psittaci* agent (IOL-83) produced a mild urethritis with discharge for a period of 1 to 2 weeks (Table 2). Chlamydiae were reisolated from the urethra of only those baboons inoculated with Cal-8 or IOL-238. Chlamydial inclusions were seen in scrapings from one animal inoculated with IOL-83 (Table 2). Intraurethral inoculation of IOL-207 produced

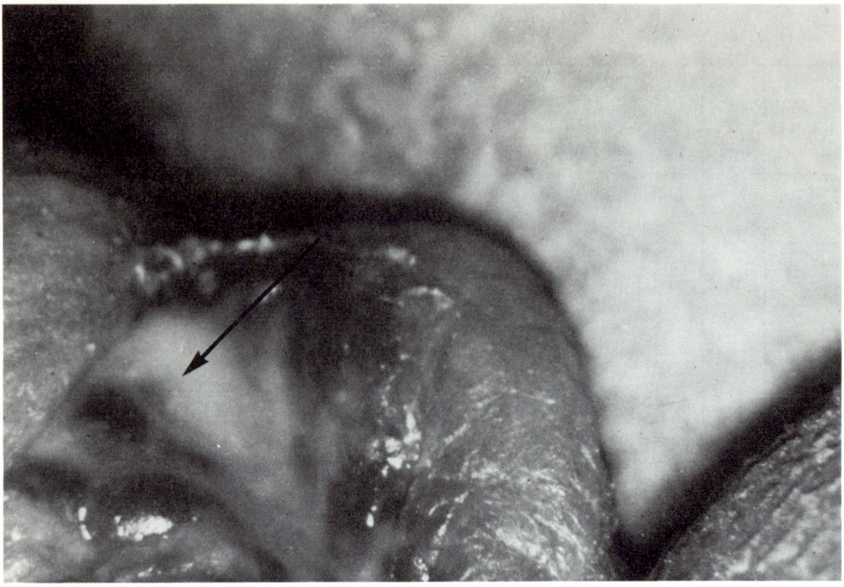

FIG. 1. *Urethral discharge in baboon after intraurethral inoculation with Chlamydia.*

a moderately severe urethritis with severe discharge (Fig. 1), and chlamydiae were reisolated from the urethra for up to 3 months after inoculation (Table 2).

Chlamydial cervicitis has been produced in baboons (17). In our hands, inoculation of the genital tract of a female baboon with *C. trachomatis* serotype G (IOL-238) and of another female baboon with serotype D (Cal-8) failed to produce a demonstrable cervicitis, and chlamydiae were not reisolated from these animals.

TABLE 2. *Experimental inoculation of the urethra of male baboons with 0.5 ml of 20% (wt/vol) yolk sac suspension of Chlamydia*

No. of baboons	Chlamydial strain	Clinical			Microbiology		
		Discharge	Meatitis	Inguinal adenopathy	Polymorphs	Scrapings (inclusions)	Isolation
1	TRIC/G/GB/IOL-238/R	+ (2 weeks)	+	+	++	−	+ (2 weeks)
1	TRIC/D/USA/Cal-8/ON	+ (2 weeks)	−	−	++	−	+ (1 week)
3	LGV/ /GB/LH-B_2/Bu	+ (1/3) (2 weeks)	−	+	−	−	−
3	PS/ /GB/IOL-83/J	+ (2/3) (1 week)	+ (2/3)	+ (2/3)	−	+ (1/3) (2 weeks)	−
1	*Chlamydia* sp./ /IR/IOL-207/OT	+++	+++	+	+++	ND[a]	+ (3 months)

[a] Not done.

TABLE 3. *Experimental inoculation of the knee joints of baboons with 0.5 ml of 20% (wt/vol) yolk sac suspension of Chlamydia*

No. of baboons	Chlamydial strain	Clinical		Microbiology	
		Swellings	Synovial fluid aspiration	Scrapings (inclusions)	Isolation
2 (M, F)	TRIC/G/GB/IOL-238/R	+ (4 weeks)	+ (7 weeks)	ND[a]	−
3 (M)	LGV/ /GB/LH-B_2/Bu	+ (2/3) (3 weeks)	+ (2/3) (3 weeks)	+ (1/3) (day 14)	+ (2/3) (days 21 and 28)
3 (M)	PS/ /GB/IOL-83/J	−	+ (1/3) (1 week)	−	−
1 (M)	*Chlamydia* sp./ /IR/IOL-207/OT	+ (2 weeks)	+ (3 weeks)	ND	+ (3 months, from synovial membrane)

[a] Not done.

The knee joints of nine baboons were inoculated with *C. trachomatis* serotype G (IOL-238), an LGV agent (LH-B_2), a *C. psittaci* isolate (IOL-83), or an atypical chlamydial agent (IOL-207). Mild to moderate arthritis, with swelling and increased temperature of joints, and synovial fluids, developed in all baboons except those inoculated with IOL-83 (Table 3). Chlamydial agents were reisolated from the synovial fluids of two baboons inoculated with the LGV agent (LH-B_2) and from the synovial fluid and synovial membrane of another baboon inoculated with IOL-207 (Table 3). This result demonstrates that *Chlamydia* may be pathogenic for synovial membranes.

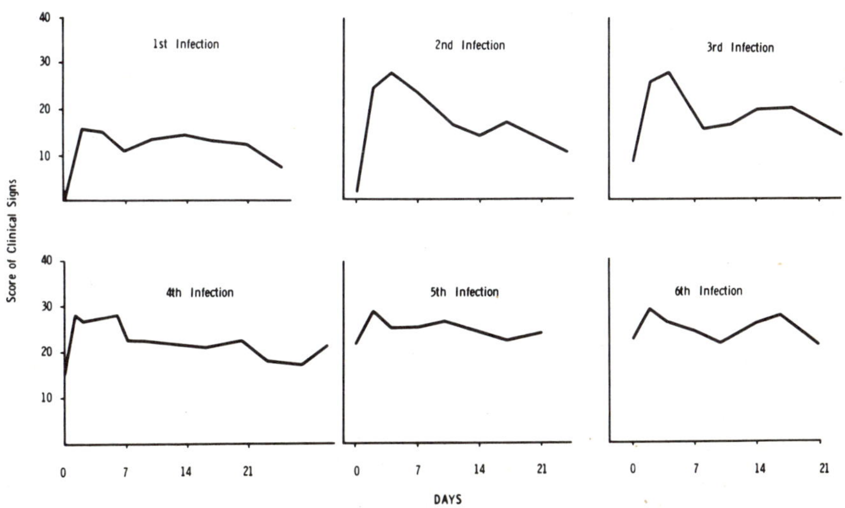

FIG. 2. *Duration and severity of clinical signs in guinea pigs repeatedly infected with GPIC agent.*

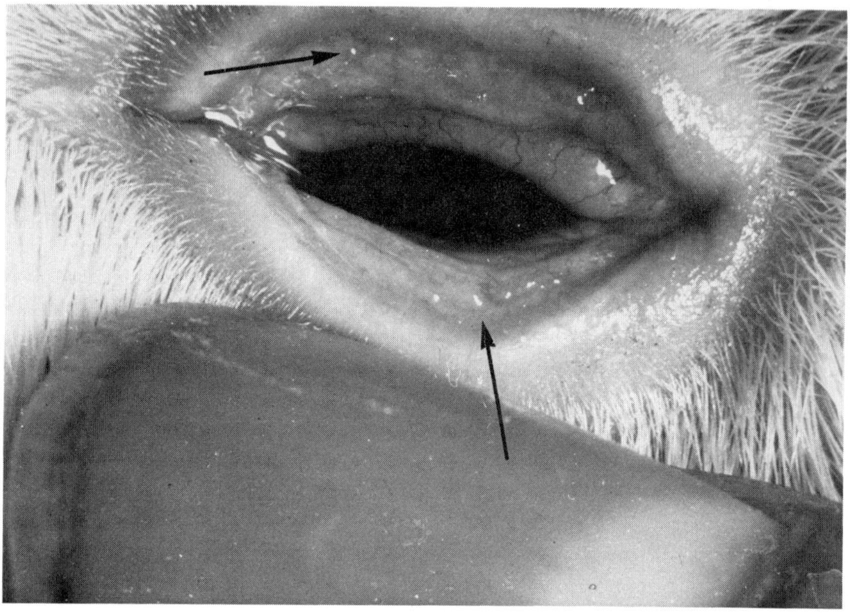

FIG. 3. *Follicular reaction in upper and lower lid conjunctiva of a guinea pig repeatedly infected with GPIC agent.*

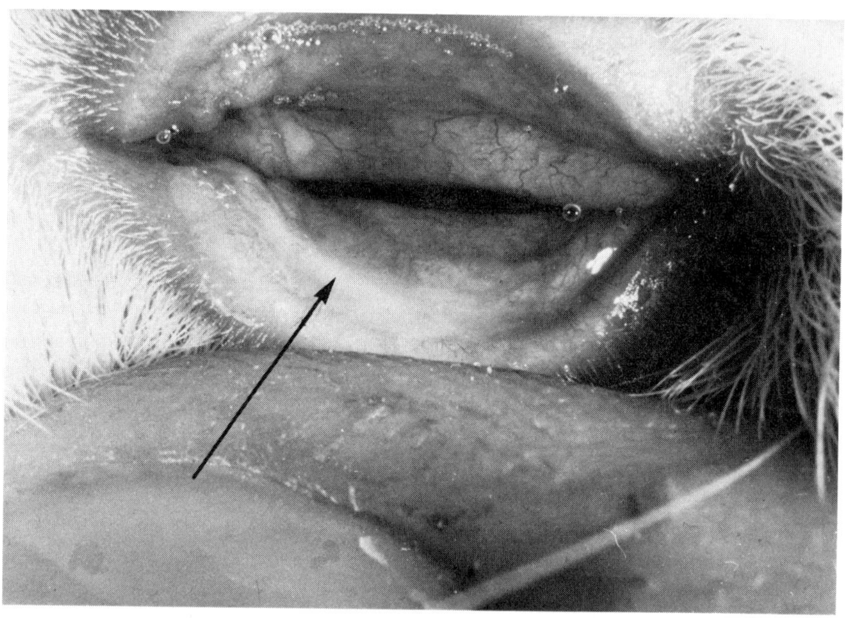

FIG. 4. *Severe diffuse scarring in lower lid conjunctiva of a guinea pig repeatedly infected with GPIC agent.*

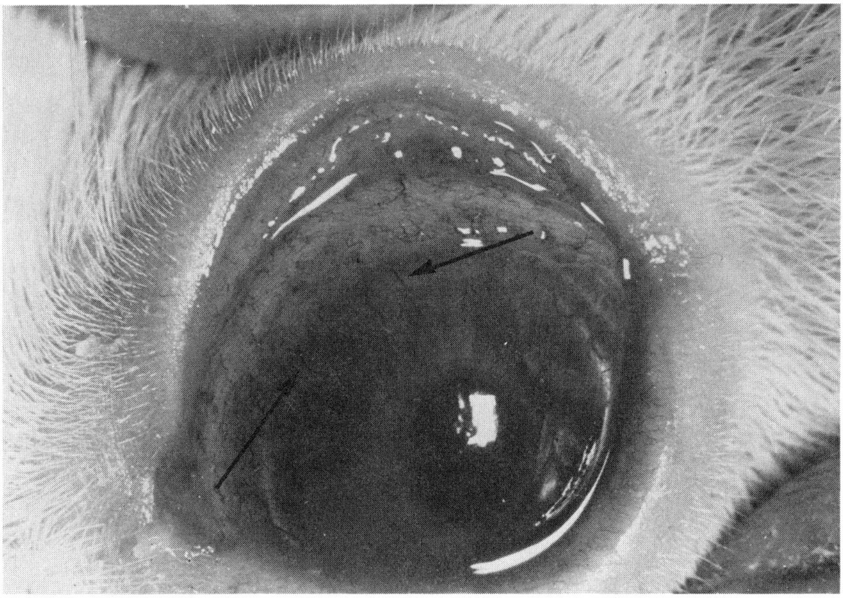

FIG. 5. *Corneal vascularization (pannus) in a guinea pig repeatedly infected with GPIC agent.*

The results of our work and other studies (17–25) indicate that, although nonhuman primates are not natural hosts to *C. trachomatis*, experimental chlamydial infections of the eye and genital tract in them are closely similar to those of humans. They can, therefore, be used as models for the study of chlamydial infections of the eye and genital tract.

SMALL LABORATORY ANIMALS

Since large-scale studies on nonhuman primates are very expensive and require adequate facilities and space, small laboratory animals with naturally occurring chlamydial infections offer particular attraction as models for the study of chlamydial diseases.

GUINEA PIGS

Guinea pig inclusion conjunctivitis (GPIC) is a naturally occurring disease caused by the GPIC agent, a member of *C. psittaci*. The eye infection is common among herds of guinea pigs and may cause a mild to severe but short-lived conjunctivitis of 30 to 40 days. The inflammation spontaneously resolves without damage to the cornea or conjunctiva (26, 27).

Experimental inoculation of the eye of guinea pigs with GPIC agent generally also produces an acute but short-lived and benign conjunctivitis that resolves spontaneously (26–28).

Experimental intraurethral inoculation of male and intravaginal inoculation of female guinea pigs with GPIC agent produces genital infection. Evidence of

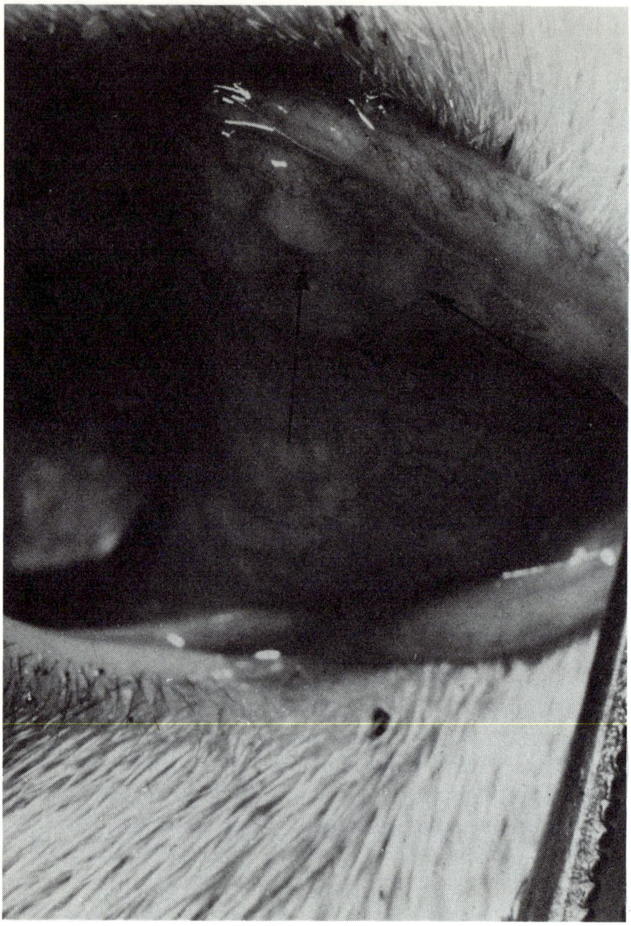

FIG. 6. *Follicular reaction in the nictitating membrane of a cat infected with FKC agent.*

infection in these animals was obtained by detection of GPIC agent inclusions in cells and antibodies in serum. Ocular infection could be detected in offspring of infected mothers as early as 3 days after birth. Venereal transmission of the infection was shown to occur between males and females (29, 30).

Earlier work with this animal model indicated that, during ocular infection, guinea pigs develop antichlamydial antibodies in the serum and eye secretions, as well as cellular hypersensitivity detectable by skin tests and by macrophage migration inhibition tests (31–35). These cellular and secretory immune mechanisms appear to have an important role in the resistance to reinfection.

In our laboratory, a programmed sequence of reinfection of guinea pig eyes with one drop of yolk sac suspension of GPIC agent, containing 10^6 EID_{50}, instilled approximately once a month, produced a chronic but more severe disease lasting for several months (Fig. 2). The disease was associated with follicles and scarring of the conjunctiva (Fig. 3 and 4) and neovascularization of the cornea (Fig. 5), resembling trachoma in humans.

This guinea pig model is being used in extensive clinical, microbiological, and histological studies in order to define the interactions between the GPIC agent and the animal host, with a view to determining their immunopathological basis. The results of these studies may help to elucidate the pathways to blindness in hyperendemic trachoma.

CATS

Chlamydial keratoconjunctivitis in cats caused by the feline keratoconjunctivitis (FKC) agent, a member of *C. psittaci*, is a common entity among domestic cats as well as those in catteries. In domestic cats, the infection is sporadic and presents as an acute mucopurulent conjunctivitis. In catteries, the eye infection may be endemic, and a large percentage of cats may have chronic conjunctivitis for periods of 1 year or more (36–39).

In the United Kingdom, the cat agent was first isolated in 1971 from the eyes of young kittens (4 to 6 weeks old) in the cattery at the Institute of Ophthalmology. Subsequently, several other isolates were obtained from the eyes of cats in another cattery in London and from domestic cats.

In cats with naturally occurring chlamydial infections, clinical examination with a Haag-Streit slit lamp showed the presence of a mixed papillary and follicular conjunctivitis, with occasional development of scarring of the conjunctiva and pannus formation.

Experimental inoculation of the eyes of cats with the FKC agent produced a moderate to severe mucopurulent conjunctivitis, with diffuse infiltration and papillary hypertrophy, for a period of 2 to 3 weeks. After this initial stage, the animals developed a chronic conjunctivitis with papillary and follicular reactions. The follicles were present in the upper and lower lid conjunctivae and in the inner surface of the nictitating membrane (Fig. 6). In some cases, the chronic course of the disease led to the development of scarring of the conjunctiva and pannus (Fig. 7 and 8). These clinical features are strikingly similar to human trachoma.

Studies in catteries in London have indicated that genital chlamydial infections in cats may occur naturally. FKC agent was isolated from the genital tract of a queen. Experimentally, we have been able to produce a mild to moderate urethritis in males by means of intraurethral inoculation of FKC agent. The urethritis continued for 4 weeks and was associated with discharge and marked inflammation in the anterior part of the urethra. The agent was reisolated for up to 2 weeks after the inoculation. In the female cats, the intravaginal inoculation of the agent produced a moderate vaginitis with discharge, continuing for 4 weeks. Chlamydiae were reisolated for up to 2 weeks after the inoculation.

Serotyping of the FKC isolates obtained in London with a micro-immunofluorescence test (40, 41) showed close antigenic similarity between these isolates and some of the isolates obtained from the eyes of cats in the United States (isolates kindly provided by J. Schachter, San Francisco, Calif.).

We have used the cat chlamydial ocular infection as a model for the study of the effect of concomitant bacterial and chlamydial infections in the eye. The eyes were

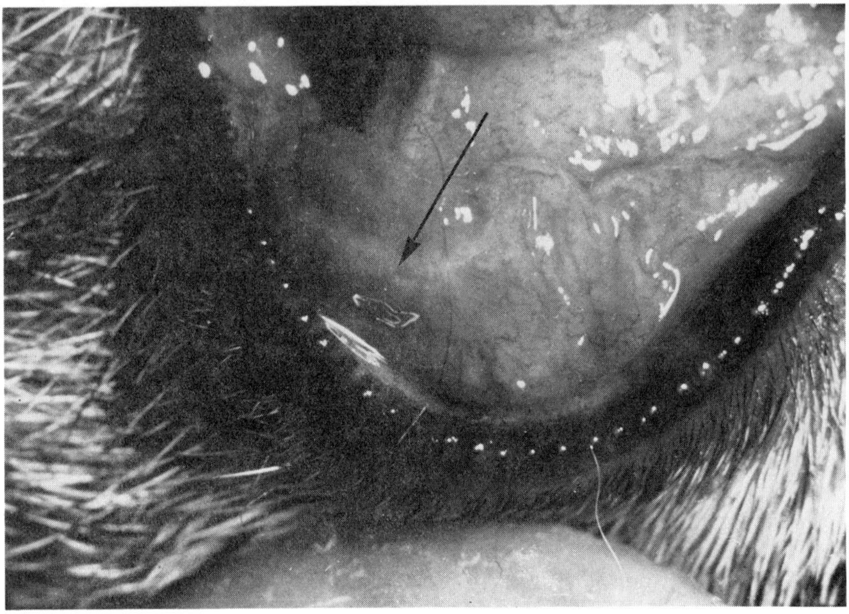

FIG. 7. *Linear scar in lower lid conjunctiva of a cat infected with FKC agent.*

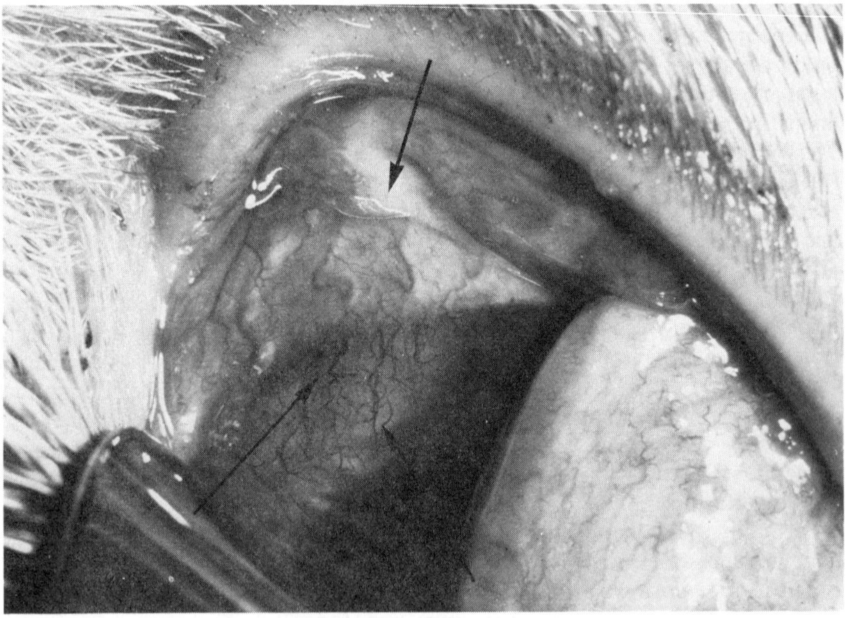

FIG. 8. *Pannus (lower arrow) and diffuse scarring of upper lid conjunctiva (upper arrow) of a cat infected with FKC agent.*

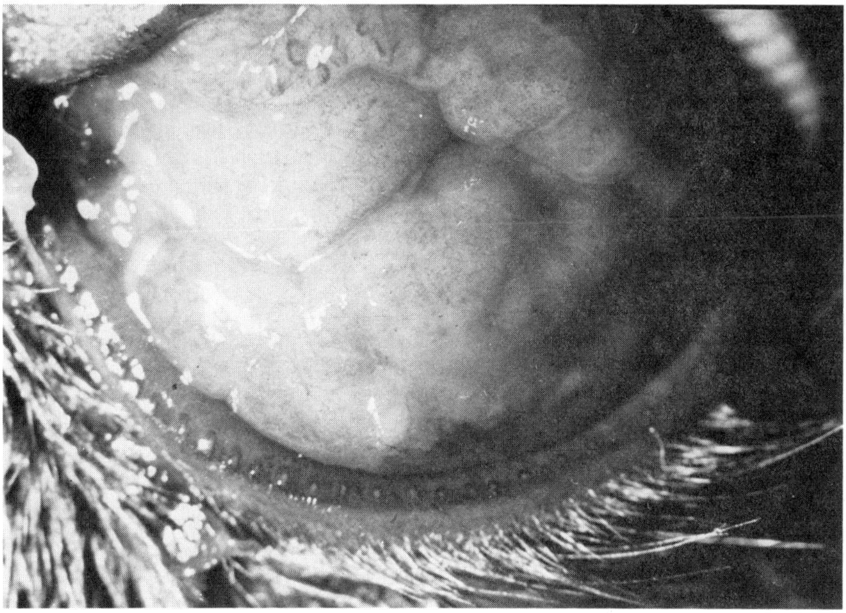

FIG. 9. *Severe papillary reaction and edema of upper and lower lid conjunctiva of a cat following combined experimental inoculation of streptococci and FKC agent.*

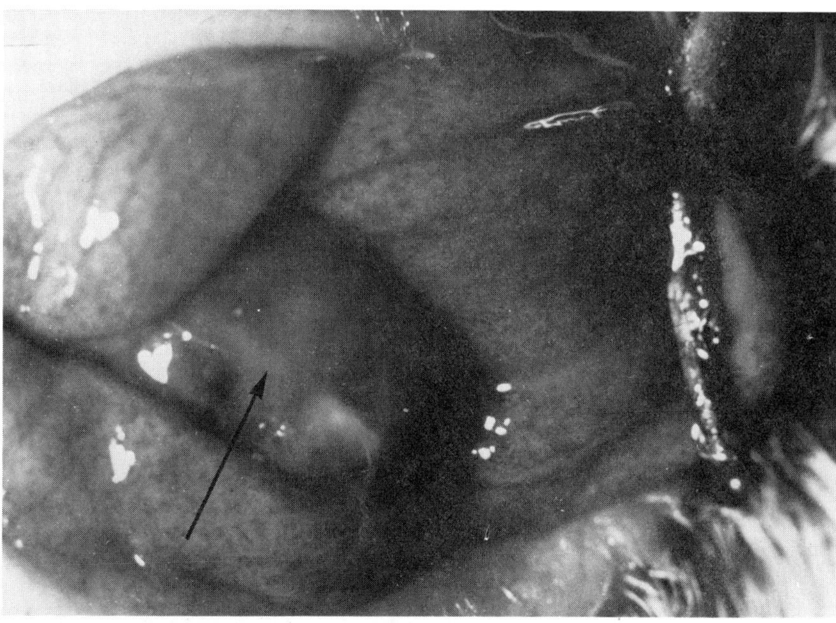

FIG. 10. *Corneal ulceration and edema in a cat following combined experimental inoculation of streptococci and FKC agent.*

inoculated with an isolate of *Streptococcus* sp. initially isolated from the eye of a naturally infected cat. A mild, transient conjunctivitis with slight conjunctival infiltration and hyperemia developed. The combined inoculation of this bacterial isolate and an FKC agent produced a far more severe keratoconjunctivitis with corneal ulceration and permanent damage to the eye in half the cats (Fig. 9 and 10), in contrast to the milder disease produced by either *Chlamydia* or bacteria alone.

The development of chronic keratoconjunctivitis in cats and guinea pigs, which we have shown to be analogous to human trachoma, and, in addition, chlamydial infections of the genital tract in cats constitute important experimental animal models for the study of chlamydial infections of the eye and genital tract in humans.

ACKNOWLEDGMENTS

We are indebted to J. Kinnison and A. Tilbury for technical assistance, to T. Tarrant for the graphics, and to the Audio-Visual Department of the Institute of Ophthalmology for help in preparation of photographs.

We are grateful to May and Baker Ltd. for providing facilities, space, and animals for the study of cat keratoconjunctivitis.

This work was supported by grants from the Medical Research Council and an anonymous donor, for the study of chlamydial infections.

REFERENCES

1. Tarizzo M (ed): Field methods for the control of trachoma, p 9, World Health Organization, Geneva, 1973
2. Jones BR: Ocular syndromes of TRIC virus infection and their possible genital significance. Br J Vener Dis 40:3–18, 1964
3. Jones BR: The prevention of blindness from trachoma. Trans Ophthalmol Soc UK 95:16–33, 1975
4. Catterall RD: The situation of gonococcal and non-gonococcal infections in the United Kingdom, p 5–13. *In* D Danielsson (ed), Genital infections and their complications. Almquist and Wiksell, Stockholm, 1975
5. Juhlin L: The situation of gonococcal and non-gonococcal infections in Sweden and other Scandinavian countries, p 25–27. *In* D Danielsson (ed), Genital infections and their complications. Almquist and Wiksell, Stockholm, 1975
6. Nichol CS: The situation in the United Kingdom, p 13–25. *In* RD Catterall (ed), Sexually transmitted diseases. Academic Press, London, 1976
7. Weisner PJ, Holmes KK: Current view of the epidemiology of sexually transmitted diseases in the United States, p 15–24. *In* D Danielsson (ed), Genital infections and their complications. Almquist and Wiksell, Stockholm, 1975
8. T'ang FF, Chang WL, Huang YT, et al: Studies on the etiology of trachoma with special reference to isolation of the virus in chick embryo. Chin Med J 75:429–447, 1957
9. Grayston JT, Woolridge RL, Wang S-P: Trachoma vaccine studies in Taiwan. Ann NY Acad Sci 98:352–366, 1962
10. Jones BR: Human inoculations with trachoma virus. Rev Int Trach 37:600–608, 1960
11. Jones BR, Collier LH: Inoculation of man with inclusion blenorrhoea virus. Ann NY Acad Sci 98:212–226, 1962
12. Mitsui Y, Konishi K, Nishimura A, et al: Experiments in human volunteers with trachoma and inclusion conjunctivitis agent. Br J Ophthalmol 46:651–659, 1962
13. Snyder JC, Bell SD Jr, Murray ES, et al: Attempts to immunize a volunteer (with formalin inactivated virus) against experimental trachoma induced by Saudi Arabian strain 2. Ann NY Acad Sci 98:368–375, 1962

14. Tarizzo M, Nataf R, Nabli B: Experimental inoculation of thirteen volunteers with agent isolated from inclusion conjunctivitis. Am J Ophthalmol 63:1120–1128, 1967
15. Thygeson P: The limbus and cornea in experimental and natural trachoma and inclusion conjunctivitis. Ann NY Acad Sci 98:201–209, 1962
16. Thygeson P, Hanna L, Dawson C, et al: Inoculation of human volunteers with egg-grown inclusion conjunctivitis virus. Am J Ophthalmol 53:786–795, 1962
17. Alexander ER, Chiang WT: Infection of pregnant monkeys and their offspring with TRIC agents. Am J Ophthalmol 63:1145–1153, 1967
18. Collier LH: Experimental infection of baboons with inclusion blenorrhoea and trachoma. Ann NY Acad Sci 98:188–196, 1962
19. Darougar S, Kinnison JR, Jones BR: Chlamydial isolates from the rectum in association with chlamydial infections of the eye or genital tract. I. Laboratory aspects, p 501–506. *In* RL Nichols (ed), Trachoma and related disorders caused by chlamydial agents. Excerpta Medica, Amsterdam, 1971
20. Digiacomo RF, Gale JL, Wang S-P, et al: Chlamydial infection of the male baboon urethra. Br J Vener Dis 51:310–313, 1975
21. Murray ES, Fraser CEO, Peters JH, et al: The owl monkey as an experimental primate model for conjunctival trachoma infection, p 386–395. *In* RL Nichols (ed), Trachoma and related disorders caused by chlamydial agents. Excerpta Medica, Amsterdam, 1971
22. Thygeson P, Dawson C, Hanna L, et al: Observations on experimental trachoma in monkeys produced by strains of virus propagated in yolk-sac. Am J Ophthalmol 50:907–917, 1960
23. Wang S-P, Grayston JT: Trachoma in the Taiwan monkey (*Macaca cyclopis*). Ann NY Acad Sci 98:177–187, 1962
24. Wang, S-P, Grayston JT: Pannus with experimental trachoma and inclusion conjunctivitis agent infection of Taiwan monkey. Am J Ophthalmol 63:1133–1145, 1967
25. Wang S-P, Grayston JT, Alexander ER: Trachoma vaccine studies in monkeys. Am J Ophthalmol 63:1615–1630, 1967
26. Kazdan JJ, Schachter J, Okumoto M: Inclusion conjunctivitis in the guinea pig. Am J Ophthalmol 64:116–124, 1967
27. Murray ES: Guinea pig inclusion conjunctivitis virus. I. Isolation and identification as a member of Psittacosis-Lymphogranuloma-Trachoma group. J Infect Dis 114:1–12, 1964
28. Murray ES, Charbonnet LT: Experimental conjunctival infection of guinea pigs with the guinea pig inclusion conjunctivitis organism, p 369–376. *In* RL Nichols (ed), Trachoma and related disorders caused by chlamydial agents. Excerpta Medica, Amsterdam, 1971
29. Mount DT, Bigazzi PE, Barron AL: Infection of genital tract and transmission of ocular infections to newborns by the agent of guinea pig inclusion conjunctivitis. Infect Immun 5:921–926, 1972
30. Mount DT, Bigazzi PE, Barron AL: Experimental genital infection of male guinea pigs with the agent of guinea pig inclusion conjunctivitis and transmission to females. Infect Immun 8:925–930, 1973
31. Modabber F, Bear SE, Cerny J: The effect of cyclophosphamide on the recovery from a local chlamydial infection. Guinea pig inclusion conjunctivitis (GPIC). Immunology 30:929–933, 1976
32. Murray ES, Charbonnet LT, MacDonald AB: Immunity to chlamydial infections of the eye. I. The role of circulatory and secretory antibodies in resistance to reinfection with guinea pig inclusion conjunctivitis. J Immunol 110:1518–1525, 1973
33. Murray ES, Radcliffe FT: Immunologic studies in guinea pigs with guinea pig inclusion conjunctivitis (GPIC) Bedsonia, p 1263–1269. *In* RL Nichols (ed), Trachoma and related disorders caused by chlamydial agents. Excerpta Medica, Amsterdam, 1971

34. Watson RR, MacDonald AB, Murray ES, et al: Immunity to chlamydial infections of the eye. III. Presence and duration of delayed hypersensitivity to guinea pig inclusion conjunctivitis. J Immunol 111:618–623, 1973
35. Watson RR, Mull JD, MacDonald AB, et al: Immunity to chlamydial infections of the eye. II. Studies of passively transferred serum antibody in resistance to infection with guinea pig inclusion conjunctivitis. Infect Immun 7:597–599, 1973
36. Cello RM: Ocular infections in animals with PLT (Bedsonia) group agents. Am J Ophthalmol 63:1270–1273, 1967
37. Ostler HB, Schachter J, Dawson CR: Acute follicular conjunctivitis of epizootic origin—feline pneumonitis. Arch Ophthalmol 82:587–591, 1969
38. Storz J: Chlamydia and chlamydial-induced diseases, p 119. Charles C Thomas, Publisher, Springfield, Ill., 1971
39. Yerasimides TG: Isolation of a new strain of feline pneumonitis virus from a domestic cat. J Infect Dis 106:290–296, 1960
40. Treharne JD, Katzenelson E, Davey SJ, et al: Comparison of serotyping subgroup A chlamydial isolates by a one-way cross-reaction and by a two-way cross-reaction immunofluorescence test, p 289–304. *In* RL Nichols (ed), Trachoma and related disorders caused by chlamydial agents. Excerpta Medica, Amsterdam, 1971
41. Wang, S-P: A Micro-immunofluorescence method study of antibody response to TRIC organisms in mice, p 273–288. *In* RL Nichols (ed), Trachoma and related disorders caused by chlamydial agents. Excerpta Medica, Amsterdam, 1971

Review of Clinical, Epidemiological, and Immunological Studies of Guinea Pig Inclusion Conjunctivitis Infection in Guinea Pigs

EDWARD S. MURRAY

During an experiment in 1960 designed to determine whether vitamin-deficient diets would render guinea pigs susceptible to infection by trachoma organisms, many inclusion bodies were found in the Giemsa-stained cells of uninoculated guinea pigs in normal as well as in vitamin-deficient animals (M. Behar and E. Murray, unpublished data). In 1964, Murray (1) published studies on the characteristics of the chlamydial agent causing this infection, which occurs as a natural eye infection of guinea pigs. Subsequently, a number of other investigations have added to our knowledge of this guinea pig infection. Since many of the studies produced overlapping data, it seems appropriate to discuss the disease under general headings pointing out the findings of the various investigators in each area.

NATURAL GPIC INFECTION OF GUINEA PIGS

Epidemiologically, Murray (1) found that guinea pig inclusion conjunctivitis (GPIC) was endemic in four out of five of the herds of commercial guinea pig suppliers in Boston, Mass. In these herds the disease appeared in guinea pigs examined toward the end of the first or beginning of the second month of life. Most infections were clinically mild, consisting of moderate edema and inflammation of the conjunctiva; most infections escaped notice until the organism was demonstrated in conjunctival scrapings. Kazdan et al. (2) also found the disease endemic in commercial herds in the area surrounding San Francisco, Calif., but noted infections in newborn guinea pigs as early as 2 weeks after birth; they also noted the mild and self-limited character of the natural infection, which caused no corneal damage.

EXPERIMENTAL INFECTION OF NORMAL GUINEA PIGS WITH GPIC ORGANISMS

Normal guinea pigs inoculated in the eye with GPIC organisms developed clinical and microbiological evidence of conjunctival infection between 1 and 5 days after inoculation, depending on the size of the infecting dose (1). Guinea pigs of all ages were susceptible, with little or no clinical or microbiological difference in infections of newborn or 1-year-old guinea pigs (1).

Kazdan et al. (2) studied the cytological and other effects of experimental eye infection. They described an acute keratoconjunctivitis with follicular hypertrophy, chemosis, and superficial epithelial keratitis; a self-limited mild micropannus developed. No scarring occurred, and all pathological signs disappeared by the end of 4 weeks. The major cytological findings included an intense polymorphonuclear exudate accompanied by typical chlamydial inclusions.

Treharne (3) made a detailed study of experimental infections produced with large doses (10^6 ID_{50}), and his findings confirmed those of earlier investigators. In addition, he plotted out the duration of the disease using the large infecting dose and compared yolk sac isolation, tissue culture, Giemsa stain, and immunofluorescence in detecting chlamydiae in the end stages of acute GPIC. Culture by use of irradiated McCoy cells was the most sensitive method for isolating GPIC agents. Using this method, Treharne detected the organisms in the eye scrapings up to 30 days after inoculation. Giemsa stain and immunofluorescence techniques were approximately equivalent but somewhat less senstitive.

CHARACTERISTICS OF THE GPIC CHLAMYDIA

Growth and Glycogen Content of the Inclusion Matrix (1, 4)

The GPIC organisms grow luxuriantly in the yolk sacs of embryonated eggs, reaching 10^8 to 10^{10} organisms per ml. Yolk sac pools of organisms are an excellent source for producing experimental infections. Inclusions in the conjunctivae as well as in irradiated McCoy cells do not stain with iodine; hence, the matrix of organisms appears to have no glycogen (1, 4).

Susceptibility to Antibiotics

The growth of GPIC organisms is inhibited by tetracycline (1) and is inhibited very slightly by D-cycloserine (4). It is unaffected by sulfadiazine and streptomycin (1).

Group B Character of the GPIC Organism

Gordon et al. (4) added to the characteristics noted by Murray (1), demonstrating that the GPIC agent is a typical group B *Chlamydia* with (i) absence of glycogen in the matrix, (ii) a relatively loose and irregular inclusion, (iii) nonsusceptibility to sulfadiazine, and (iv) almost complete resistance to the inhibitory action of D-cycloserine.

Toxic Effect in Mice (1)

Massive doses of the GPIC agent kill mice within 24 to 48 h. Mice which survive a moderate infective dose are immune to toxic challenge 4 to 6 weeks later. Guinea pigs also succumb to massive intracerebral GPIC doses in 20 h; this is apparently a toxic phenomenon.

ATTEMPTS TO INDUCE GPIC IN OTHER ANIMALS (1)

Attempts to infect the conjunctiva of mice, hamsters, and rabbits were unsuccessful. On one occasion during four attempts, very minimal and brief conjunctival infection was noted in a single mongolian gerbil.

PATHOGENICITY OF OTHER CHLAMYDIAL STRAINS FOR THE GUINEA PIG CONJUNCTIVA (1)

Attempts were made to infect the conjunctivae of guinea pigs with other chlamydiae. Two guinea pigs received in each eye 0.04 ml of a 50% yolk sac suspension of the following chlamydial strains: meningopneumonitis, lymphogranuloma venereum, ovine abortion, psittacosis (6BC strain), and feline pneumonitis. None of these strains produced any macroscopic signs of conjunctivitis or any microscopic evidence of infection over a period of 29 days after inoculation. Furthermore, when the 10 guinea pigs inoculated with these chlamydiae were challenged with GPIC 38 days after their original infection, all 10 animals developed conjunctivitis, with 12 to 30% of the conjunctival cells showing inclusion bodies in the cytoplasm.

Sixteen attempts were made to infect guinea pig eyes with various strains of trachoma: two attempts with LT-1, one with Gambia-1, two with TE-55, six with SA-2, four with Hong Kong, and one with Egypt-2. (According to the official recommendation for nomenclature of trachoma strains [Gear et al., Am J Trop Med Hyg 12:440, 1963], these strains would be listed, respectively, as: TRIC/GB/MRC-7/OT, TRIC/D/WAG/MRC-1/OT, TE-55, TRIC/B/SAU/Har-2/OT, TRIC/C/HK/HAR-29/OT, and TRIC/A/ET/HAR-13/OT.) All animals received 0.04 ml of a 50% suspension of infected yolk sac in each eye (approximately 10^7 EID_{50}), and the conjunctivae were scraped every 3 or 4 days from day 0 to day 14. No inclusion bodies were observed at any time.

Several chlamydial strains from exotic sources (the opossum B and "Storz" strains) did produce mild conjunctivitis in guinea pigs, but only when massive doses of these agents were used as inocula (5). The opossum B strain was isolated in Colombia in 1945 by Roca-Garcia (6). Storz (7) isolated a chlamydial agent from organs of young guinea pigs that died shortly after birth. However, this fatal disease in guinea pigs occurred at a time when sheep housed in juxtaposition to the guinea pig herd were dying from some type of chlamydial infection (Storz, personal communication).

GPIC AS A GUINEA PIG VACCINE MODEL

Murray and Radcliffe (5) explored the possibility of using the guinea pig as a model for testing chlamydial strains as vaccines. Formalin-killed preparations of the GPIC, psittacosis, opossum, and "Storz" strains were given intraperitoneally and intramuscularly as single doses or with booster inoculations. All failed to produce any significant immunity to subsequent live GPIC conjunctival challenge infection. In the one experiment in which suggestive immunity was produced, the apparent immunity was later shown to be due almost certainly to cross-contamination with live GPIC agent during the experiment.

GPIC AS A SEXUALLY TRANSMITTED DISEASE OF GUINEA PIGS

Mount et al. (8, 9) demonstrated that GPIC is a sexually transmissible disease experimentally. They infected pregnant female guinea pigs intravaginally with GPIC and made cytological and histopathological studies of the resulting vaginal infections. They also studied the newborn guinea pigs of vaginally infected mothers and found that these newborns uniformly developed conjunctival GPIC infections between 3 and 5 days after birth. Barron's group also demonstrated (9) that male guinea pigs could be infected intraurethrally with the GPIC agent. They studied the cytological and histopathological lesions in the male urethra and demonstrated that male guinea pigs infected intraurethrally almost certainly transmit GPIC infection to females during mating.

These experimental sexual transmission studies strongly suggest that the natural GPIC infection of guinea pigs is a sexually transmitted disease analogous to inclusion conjunctivitis of the human newborn. However, the organisms causing GPIC in guinea pigs and inclusion conjunctivitis in humans are quite different, being, respectively, group B and group A chlamydiae.

NATURAL AND VACCINE-INDUCED IMMUNITY

Murray et al. (10, 11) studied the role of circulatory and secretory antibodies in resistance to reinfection with the GPIC *Chlamydia*. Different types of immunological stimuli were given to two separate groups of guinea pigs. One group received active conjunctival infection with live GPIC; the other was given formalin-killed GPIC organisms intraperitoneally. Neutralizing complement-dependent serum antibodies were produced by both stimuli. However, immunity to conjunctival rechallenge infection developed only in those guinea pigs originally infected conjunctivally with live organisms. The latter animals were also the only ones to develop specific GPIC secretory immunoglobulin A (S-IgA) antibodies in eye secretions. These studies cast doubt on the importance of specific serum antibodies in resistance to GPIC conjunctival infection. Studies by Watson and colleagues (12) cast further doubt on the value of humoral antibodies; circulatory antibodies which were passively transferred from immune animals with a previous infection to normal animals were ineffective in protecting the conjunctivae from challenge infection.

Watson and colleagues (R. R. Watson, E. S. Murray, and S. E. Thompson, Fed. Proc. 32:1029, 1973) demonstrated that pregnant guinea pigs infected with GPIC transferred GPIC circulating antibodies in utero to their unborn young. Nevertheless, such newborn guinea pigs were fully susceptible to infectious GPIC challenge of the eye within the first week after birth. The mothers, on the other hand, were immune to reinfection with the same challenge infection given their newborns.

Several groups studied the microbiological and S-IgA responses to GPIC infections of the guinea pig eye, vagina, and male urethra and compared them with the resultant immunity to live GPIC challenge of the same or distant secretory sites (13; R. R. Watson and E. S. Murray, Fed. Proc. 33:763, 1974; E. S. Murray et al., submitted for publication). These studies produced four major findings:

1. Immunity produced at any one secretory site was complete against a moderate challenge infection at the same secretory site between 4 and 8 weeks after a primary infection. However, immunity was relative, since it was complete only when challenge doses of 100 to 3,000 ID_{50} were used and was only partial against challenge doses of 10^6 ID_{50} or higher. Moreover, even at the lower challenge doses (100 to 3,000 ID_{50}), immunity waned and became partial 3 to 4 months after primary infection.

2. Infectious challenges of 1,000 to 3,000 ID_{50} were administered at secretory sites different from the primary secretory site of infection. A primary GPIC infection of the eye produced strong immunity to vaginal or male urethral challenge infections 4 to 6 weeks later. A primary vaginal infection produced marked but less solid immunity to a subsequent eye infection challenge 4 weeks later, whereas a primary urethral infection produced slight partial immunity to an eye challenge infection 6 weeks later.

3. There was little or no evidence that serum antibodies bore any significant relationship to immunity demonstrated in secretory site infections.

4. The height of the S-IgA titer in tears and vaginal fluids did not appear to correlate well with the degree of immunity demonstrated by resistance to challenge infections at the same site or sites distant from the primary infection.

PRELIMINARY ATTEMPTS TO MEASURE THE DEVELOPMENT OF CELLULAR IMMUNITY IN GPIC INFECTION

Watson et al. (14) induced long-lasting immune responses in guinea pigs infected with GPIC, as measured by macrophage inhibition assays and skin sensitivity. Four to six days after conjunctival infection with GPIC, positive skin tests appeared and developed to a maximal response between days 11 and 83. They declined in intensity thereafter but were still observable at days 350 and 850 postinfection. Positive macrophage inhibition assay responses correlated well with positive skin test results.

Modabber et al. (15) used cyclophosphamide to produce marked cellular depletion in the B-cell areas of the spleen and lymph nodes of guinea pigs during the GPIC infection. The cyclophosphamide produced marked delay (10 to 14 days) in the appearance of serum and S-IgA antibodies in eye secretions. However, this was not accompanied by any significant alteration in the character of the clinical and microbiological acute infection and recovery up to day 18; complete recovery nevertheless was delayed. This study suggests that T cells play a major role in immunity to GPIC, with B cells being of secondary importance; these results correlate well with the findings of Murray et al. (submitted for publication).

ACKNOWLEDGMENTS

This work was supported by Public Health Service grant 9-ROI-EY-00812 from the National Eye Institute and grants AI 00177 (training grant), 5-TOI-AI-0221, and AI-06251 from the National Institute of Allergy and Infectious Diseases, and by a grant from the Arabian American Oil Co., Dhahran, Saudi Arabia.

We are deeply indebted to the following colleagues who, through advice and/or technical assistance, made this work possible: L. Sullivan, S. Torrey, M. P. O'Leary, K. J. Welch, and A. Grimshaw.

REFERENCES

1. Murray ES: Guinea pig inclusion conjunctivitis. I. Isolation and identification as a member of the psittacosis-lymphogranuloma-trachoma group. J Infect Dis 114:1–12, 1964
2. Kazdan JJ, Schachter J, Okumoto M: Inclusion conjunctivitis in the guinea pig. Am J Ophthalmol 64:116–124, 1967

3. Treharne JD: A comparison of microbiological techniques for the detection of GPIC agent in the conjunctiva of infected guinea pigs, p 435–445. *In* RL Nichols (ed), Trachoma and related disorders caused by chlamydial agents. Excerpta Medica, Amsterdam, 1971
4. Gordon FB, Weiss E, Quan AL, et al: Observations of guinea pig inclusion conjunctivitis agent. J Infect Dis 116:203–207, 1966
5. Murray ES, Radcliffe FT: Immunologic studies in guinea pigs with guinea pig inclusion conjunctivitis (GPIC) Bedsonia. Am J Ophthalmol 63:1263/237-1269/243, 1967
6. Roca-Garcia M: Viruses of the lymphogranuloma-psittacosis group isolated from opossum in Colombia: opossum virus B. Nature (London) 213:502, 1967
7. Storz J: Ueber eine naturliche Infektion eines Meerschweinschenbestandes mit einem Erreger aus der Psittakose-Lymphogranuloma Gruppe. Zentralbl Bakteriol Parasitenkd Infektionskr Hyg Abt 1 Orig 193:432, 1964
8. Mount DT, Bigazzi PE, Barron AL: Infection of genital tract and transmission of ocular infection to newborns by the agent of guinea pig inclusion conjunctivitis. Infect Immun 5:921–926, 1972
9. Mount DT, Bigazzi PE, Barron AL: Experimental genital infection of male guinea pigs with the agent of guinea pig inclusion conjunctivitis and transmission to females. Infect Immun 8:925–930, 1973
10. Murray ES, Charbonnet LT: Experimental conjunctival infection of guinea pigs with the guinea pig inclusion conjunctivitis organism, p 369–376. *In* RL Nichols (ed), Trachoma and related disorders caused by chlamydial agents. Excerpta Medica, Amsterdam, 1971
11. Murray ES, Charbonnet LT, MacDonald AB: Immunity to chlamydial infections of the eye. I. The role of circulatory and secretory antibodies in resistance to reinfection with guinea pig inclusion conjunctivitis. J Immunol 110:1518–1525, 1973
12. Watson RR, Mull JD, MacDonald AB, et al: Immunity to chlamydial infections of the eye. II. Studies of passively transferred serum antibody in resistance to infection with guinea pig inclusion conjunctivitis. Infect Immun 7:597–599, 1973
13. Howard LV, O'Leary MP, Nichols RL: Animal model studies of genital chlamydial infections: immunity to reinfection with guinea pig inclusion conjunctivitis agent in the urethra and eye of male guinea pigs. Br J Vener Dis 52:261–265, 1976
14. Watson RR, MacDonald AB, Murray ES, et al: Immunity to chlamydial infections of the eye. III. Presence and duration of delayed hypersensitivity to guinea pig inclusion conjunctivitis. J Immunol 111:618–623, 1973
15. Modabber F, Bear SE, Cerny J: The effect of cyclophosphamide on the recovery from a local chlamydial infection—guinea-pig inclusion conjunctivitis (GPIC). Immunology 30:929–933, 1976

Experimental Nonhuman Primate Urethral Infection with *Chlamydia trachomatis* and *Ureaplasma* (T-Mycoplasma)

JAMES L. GALE, RONALD F. DiGIACOMO, MARK D. KIVIAT,
SAN-PIN WANG, AND WILLIAM R. BOWIE

INTRODUCTION

Genital infection with *Chlamydia trachomatis* has been reported in male (1, 2) and female (3–5) nonhuman primates. Because of the quantitative importance of nongonococcal urethritis relative to gonorrhea (6), we undertook further studies of nonhuman primate *C. trachomatis* genital infection with organisms considered candidate etiological agents in this disease. Our successful urethral infection of baboons (*Papio cynocephalus*) with production of follicular lesions has been reported (7). In this paper, we present further details in two additional species, together with preliminary results of our efforts to infect nonhuman primates intraurethrally with *Ureaplasma urealyticum* (T-mycoplasma).

METHODS

Monkeys. Mature male monkeys were used, including two baboons (*Papio cynocephalus*), four pigtailed macaques (*Macaca nemestrina*), and two capuchins (*Cebus albifrons*). The four pigtailed macaques, weighing 8 to 10 kg, previously had been inoculated intraurethrally with *Neisseria gonorrheae* without being colonized (unpublished data). Two animals were castrate and two were intact. The baboons, weighing about 30 kg, had each been inoculated intraurethrally with *C. trachomatis* several times 20 months prior to the present experiment, as described previously (7). The *Cebus* monkeys, weighing about 2.5 kg, had also previously been experimentally inoculated intraurethrally with *N. gonorrheae*, without colonization.

Organisms. Inocula consisted of 5-iodo-2′-deoxyuridine (IUdR)-treated McCoy cell samples of *C. trachomatis* strain E/TW-187/C or *U. urealyticum* broth cultures, approximately 18 h old. *C. trachomatis* inocula generally titered between 10^4 and 10^6 inclusion-forming units (IFU) in McCoy cells. The *Ureaplasma* inoculum contained 1 to 5 infecting units per 0.5 ml, as determined by serial broth dilutions.

Urethral inoculation. Organism suspensions were inoculated (0.25 ml) through a #3½ FR catheter inserted 3 to 5 cm into the urethra of monkeys sedated with ketamine hydrochloride (Vetalar; Parke-Davis & Co.). The glans was subsequently compressed for approximately 5 min. Each urethra had been immediately previously traumatized by the three swabs obtained for attempted preinoculation isolation.

Isolations. For attempted *C. trachomatis* isolation, calcium alginate aluminum swabs were placed in sucrose-phosphate buffer following insertion into the primate urethra. This broth was subsequently inoculated into IUdR-treated McCoy cells

(8). For attempted *Ureaplasma* isolations, swabs were placed in urea broth and swabbed directly on T-strain agar. Subsequent anaerobic incubation and identification of *Ureaplasma* were carried out in the laboratories of King K. Holmes (9).

Antibody determinations. Serum antibody to *C. trachomatis* was measured by an indirect micro-immunofluorescence (micro-IF) technique (10). Fluorescein isothiocyanate-conjugated goat antiserum to human immunoglobulin M (IgM), IgG, and IgA (obtained from Hyland Laboratories, Los Angeles, Calif.) was used for detection of antibodies to 12 chlamydial immunotypes pooled into eight test antigens (11).

Endoscopy. Urethral endoscopy was carried out in sedated baboons and pigtailed macaques immediately prior to urethral inoculation or as indicated under Results, by use of a pediatric endoscope.

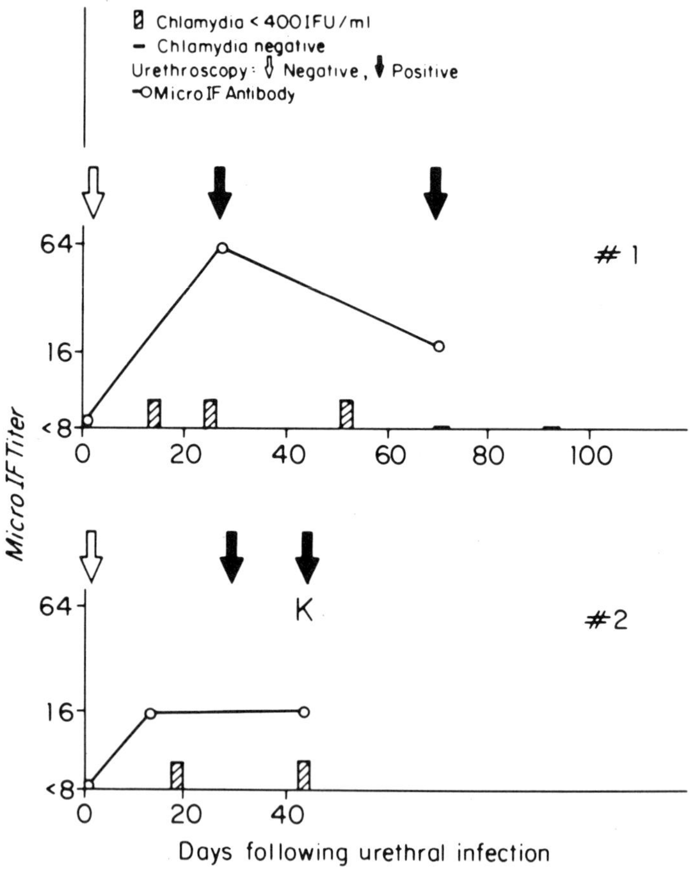

FIG. 1. *Clinical and serological response and microbial colonization following urethral inoculation of two Macaca nemestrina primates with* 10^5 *IFU of Chlamydia trachomatis organisms per ml.*

Pathological examination. En bloc resection of the genitourinary tract of two pigtailed macaques was performed. Tissue blocks were fixed in 10% formalin, and sections were stained with hematoxylin and eosin.

RESULTS

In the first experiment, the four pigtailed macaques were inoculated intraurethrally with 4×10^5 IFU of the TW-187 strain. Two animals (one intact and one castrate) were negative for urethral chlamydial organisms at the first examination 2 weeks after inoculation. They remained culturally negative up to 45 days after inoculation. Both demonstrated a micro-IF antibody titer of 1:32 to type E antigen about 30 days after inoculation. Urethroscopy did not show any urethral lesions.

The other two macaques were shedding chlamydial organisms from the urethra at 14 days postinoculation, although in very low titers (Fig. 1). The intact male (#1)

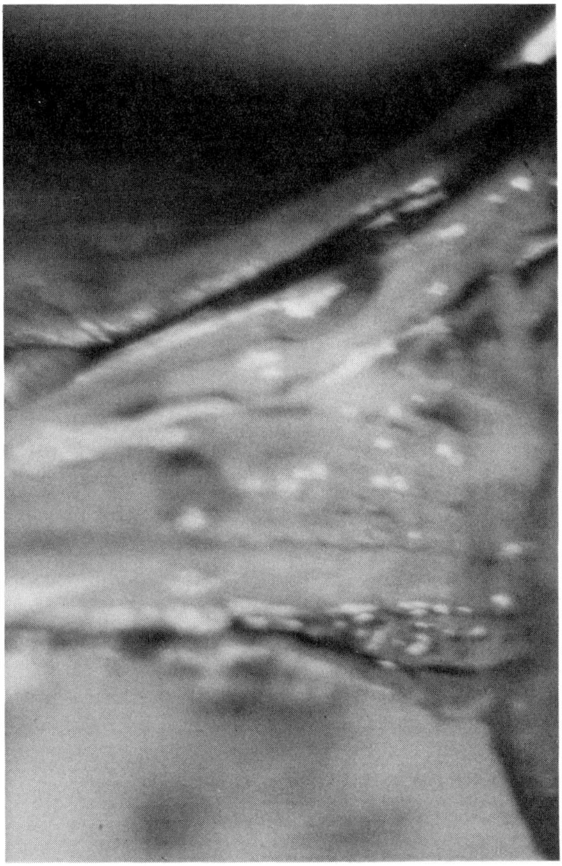

FIG. 2. *Urethral follicles of Macaca nemestrina #2 (see Fig. 1) at autopsy.*

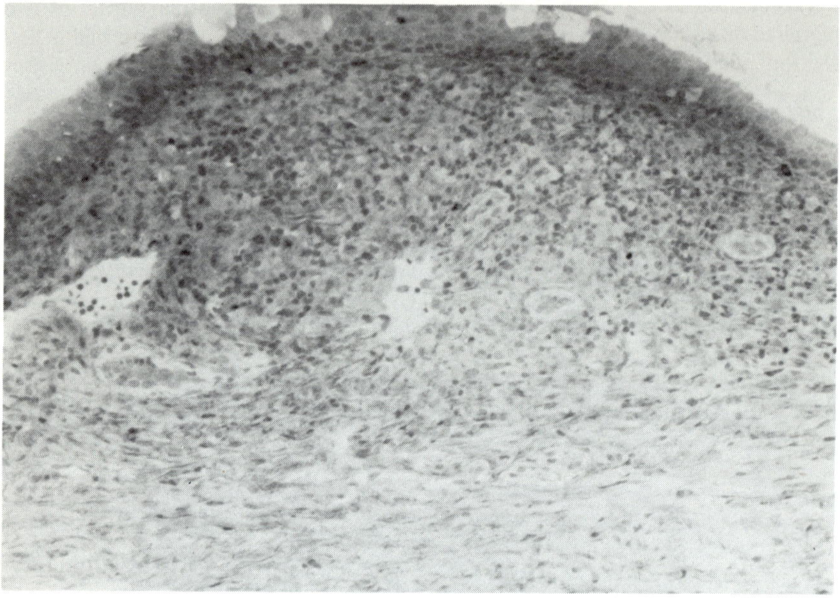

FIG. 3. *Urethral follicle showing subepithelial aggregation of mononuclear cells, predominantly macrophages. (Thick section, toluidine blue, ×50.)*

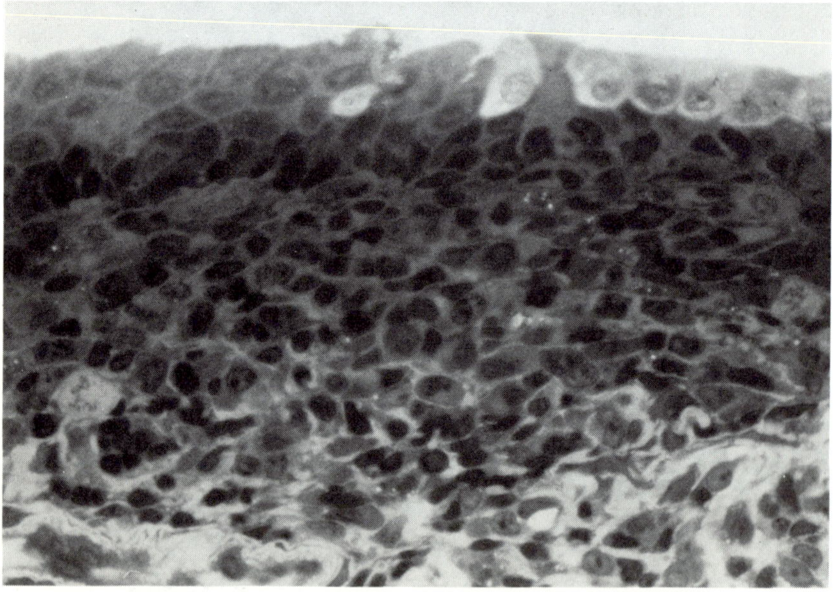

FIG. 4. *Urethral follicle showing subepithelial infiltration of macrophages with scattered lymphocytes. Lighter-staining epithelial cells contain inclusions suggestive of Chlamydia trachomatis infection. (Thick section, toluidine blue, ×200.)*

demonstrated an antibody titer of 1:64 and urethral follicles 30 days after inoculation. Follicles seen by urethroscopy were few in number and confined to the fossa navicularis of the penile urethra. This monkey continued to shed chlamydiae in low titer for about 55 days, and the urethral follicles were still present 10 weeks after inoculation, although a urethral exudate was never noted. The castrated male (#2) exhibited a moderate number of urethral follicles of much larger size in both the fossa navicularis and bulbous portion of the urethra. This monkey continued to shed chlamydiae for 40 days after inoculation, at which time it was killed. The micro-IF titer at that time was 1:16.

At necropsy, genital tissue specimens were obtained for *C. trachomatis* culture and for light and electron microscopy. The urethra, exposed along its entire length, exhibited numerous follicles in the fossa navicularis (Fig. 2) and the proximal and distal bulbous urethra. No exudate was present. The lesions were firm, pale yellow, and well delineated; they varied in size from 1 to 4 mm. The membranous and

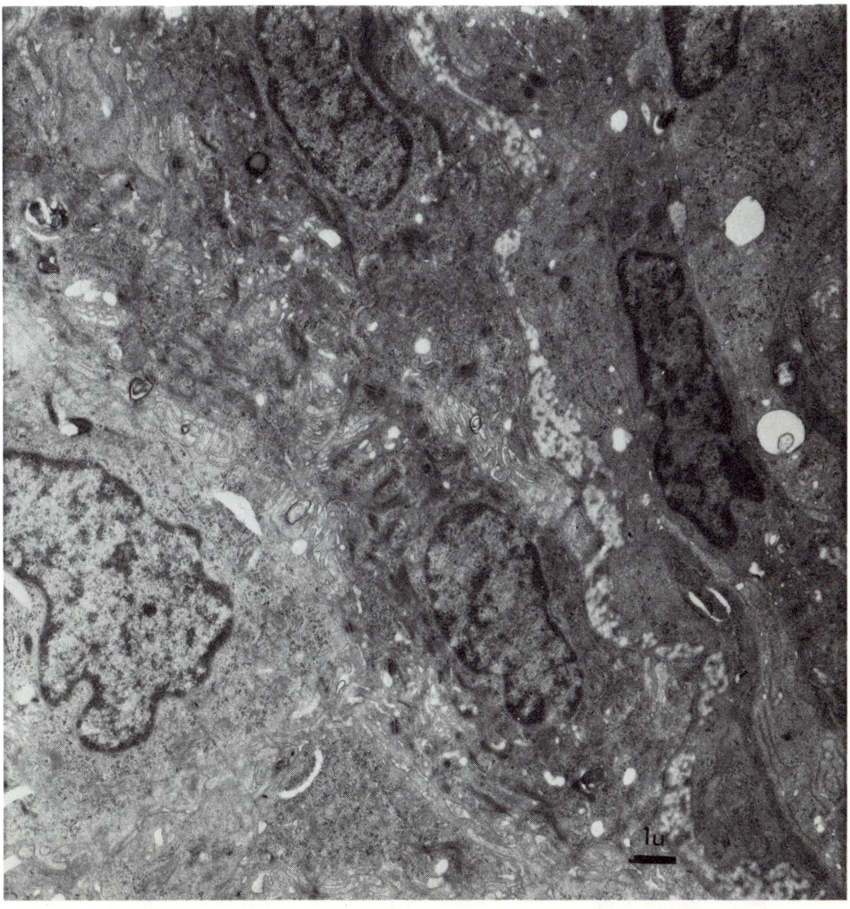

FIG. 5. *Urethral follicle showing dense nature of lesion (× 12,000).*

prostatic urethra were unaffected. Lesions were not observed in the bladder. The testes were absent, and the seminal vesicles and the prostate were atrophic.

Swabs taken at various sites in the urethra were all positive for chlamydiae, yielding from 10 to 40 IFU per ml of secretion. Tissue specimens were homogenized and cultured. Urethral tissue yielded 70 IFU per ml of suspension. The prostate, seminal vesicles, vas deferens, urinary bladder, ureters, kidneys, and inguinal lymph node tissue specimens were negative. Urine obtained by bladder aspiration and conjunctival swabs were also negative. All specimens were passed at least once in further McCoy cultures before being called positive or negative.

Light microscopy showed an intact urethral epithelium overlying a granulomatous lesion consisting of a dense collection of macrophages and plasma cells (Fig. 3 and 4). Few polymorphonuclear leukocytes were seen. No obvious edema or scar tissue was present. Electron micrographs confirmed the dense nature of the lesion (Fig. 5). Structures resembling *C. trachomatis* inclusions containing reticulate bodies were identified infrequently in macrophages (Fig. 6). A structure resembling

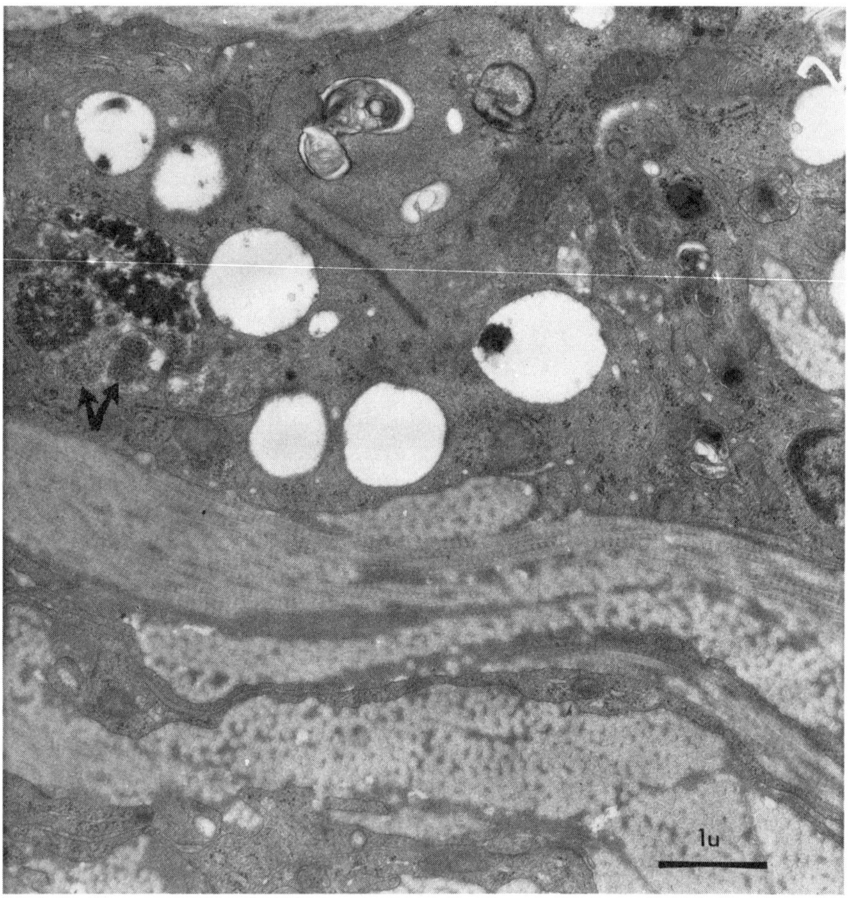

FIG. 6. *Urethral follicle with macrophage-containing structure resembling Chlamydia trachomatis reticulate bodies (× 24,400).*

a previously published electron micrograph of what was identified as a degenerating *C. trachomatis* inclusion in cell cultures (12) was also seen in the vacuole of a macrophage (Fig. 7).

The second experiment consisted of a urethral inoculation of *Ureaplasma* into two baboons, two pigtailed macaques, and two capuchins. One pigtailed macaque shed over 10^5 organisms per ml of urine in the urine sampled 1 week after inoculation with *Ureaplasma*. Urethral swabs have been consistently positive on T-strain agar and urea broth through the 4th week, although subsequent quantitative urine cultures have not been obtained.

Endoscopy 21 and 42 days after *Ureaplasma* urethral inoculation showed follicular lesions in the region of the bulbous urethra, circumferentially. These lesions were essentially indistinguishable from those described previously in animals with *C. trachomatis* infections. No other monkeys have shed *Ureaplasma*, and no other baboon or pigtailed macaque has had any positive endoscopic findings.

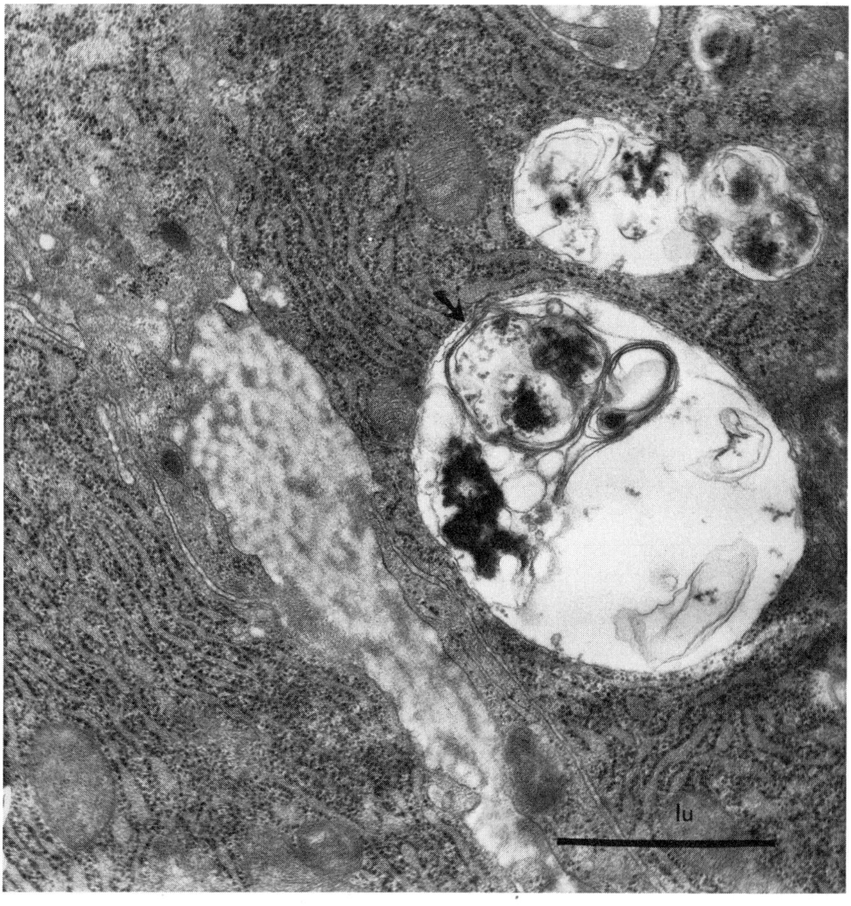

FIG. 7. *Urethral follicle with macrophage vacuole containing structure resembling degenerating Chlamydia trachomatis inclusion (×36,000).*

DISCUSSION

These preliminary results suggest that the pigtailed macaque, like the baboon, is a suitable nonhuman primate for more definitive studies with genital tract microorganisms having a possible etiological role in human nongonococcal urethritis. Although based on small numbers of monkeys, it appears that the macaque shed *C. trachomatis* to a lesser degree and for a shorter period than the baboon. The latter has been shown to shed organisms from the urethra for about 90 days, and in greater quantities (7). However, availability of monkeys may be the deciding factor in selection of an adequate simian model.

The finding of follicles in the macaque urethra after chlamydial infection confirms earlier studies in the baboon (7). Grossly, these appeared similar to lesions previously reported. Although appropriate control monkeys were not used in the experiments reported here, we have never seen follicles prior to inoculation of any organism, even when the monkey has undergone endoscopy for prior attempted urethral colonization experiments or in the absence of urethral shedding of organisms.

The finding of urethral follicles, similar to those in the *C. trachomatis*-infected baboons and macaques, after colonization of the macaque urethra with *U. urealyticum* was unexpected. This particular monkey had been previously inoculated with *C. trachomatis* and shed 10^2 organisms per ml of urethral discharge for 4 weeks after urethral inoculation. Two cultures for *C. trachomatis* 10 and 12 weeks after inoculation were negative. Two cultures for *C. trachomatis* during shedding of *Ureaplasma* were also negative. Preinoculation endoscopy revealed no urethral lesions. These findings clearly need further documentation.

The histology of the urethral follicles was similar to that described in the human after experimental infection (13). The cellular response consisted principally of macrophages, with few polymorphonuclear leukocytes noted. The pathogenesis of this lesion seems worthy of further study.

REFERENCES

1. Darougar S, Kinnison JR, Jones BR: Chlamydial isolates from the rectum in association with chlamydial infection of the eye or genital tract. I. Laboratory aspects, p 501–506. *In* RL Nichols (ed), Trachoma and related disorders caused by chlamydial agents. Excerpta Medica, Amsterdam, 1971
2. Smith DE, James PG, Schachter J, et al: Experimental bedsonial arthritis. Arthritis Rheum 16:21–29, 1973
3. Thygeson P, Mengert WF: The virus of inclusion conjunctivitis. Arch Ophthalmol 15:377–410, 1936
4. Alexander ER, Chiang WT: Infection of pregnant monkeys and their offspring with TRIC agents. Am J Ophthalmol 63:1145–1153, 1967
5. Pukhner AF, Kozlova VI: [Virological study of chronic cervicitis and urethritis of obscure etiology]. Akush Ginekol 41:47–52, 1965
6. King A: Failure to control venereal disease. Br Med J 1:451–457, 1970
7. DiGiacomo RF, Gale JL, Wang SP, et al: Chlamydial infection of the male baboon urethra. Br J Vener Dis 51:310–313, 1975
8. Wentworth BB, and Alexander ER: Isolation of *Chlamydia trachomatis* by use of 5-iodo-2-deoxyuridine-treated cells. Appl Microbiol 27:912–916, 1974
9. Holmes KK, Handsfield HH, Wang SP, et al: Etiology of nongonococcal urethritis. N Engl J Med 292:1199–1205, 1975

10. Wang SP: A micro-immunofluorescence method. Study of antibody response to TRIC organisms in mice, p 273–288. *In* RL Nichols (ed), Trachoma and related disorders caused by chlamydial agents. Excerpta Medica, Amsterdam, 1971
11. Wang SP, Grayston JT: Local and systemic antibody response to trachoma eye infection in monkeys, p 217–232. *In* RL Nichols (ed), Trachoma and related disorders caused by chlamydial agents. Excerpta Medica, Amsterdam, 1971
12. Friis RR: Interaction of L cells and *Chlamydia psittici*: entry of the parasite and host response to its development. J Bacteriol 110:706–721, 1972
13. Yukihiko M, Kitamuro T, Fujimoto M: Adaptation of TRIC agents to tissue culture and characteristics of tissue culture adopted variants. Am J Ophthalmol 63:1191–1205, 1967

Susceptibility of Chlamydiae to Chemotherapeutic Agents

JOHN D. TREHARNE, JILL DAY, C. K. YEO, BARRIE R. JONES, AND S. SQUIRES

The susceptibility of *Chlamydia trachomatis* agents to certain antimicrobial drugs has been confirmed in laboratory studies (1–8) and in clinical trials (9–13). Until recently, treatment trials in the field have relied on clinical evaluation of drug effects, and most laboratory studies have been limited to antimicrobial sensitivity tests against *C. trachomatis* grown in the yolk sac of the developing chick embryo.

The problems of drug binding and other factors, such as thermal inactivation of antimicrobial compounds during the long incubation periods required in the egg culture system (5 to 13 days), have yielded equivocal results (14). The cell culture system described here probably parallels more closely the in vivo situation, particularly as applied to the topical treatment of ocular infection due to *C. trachomatis* (15).

The intentions of this study were as follows:

1. To design a simple, rapid, and reproducible cell culture method which could be used to assay the susceptibility of *C. trachomatis* isolates to a range of antimicrobial compounds.
2. To compare the susceptibility of a range of *C. trachomatis* serotypes to several antimicrobial drugs.
3. To determine the most effective antichlamydial compounds in vitro.
4. To test a number of antimicrobial compounds with little or no antichlamydial activity, which could be useful in combating extraneous bacterial or fungal contaminants present in clinical specimens collected for isolation of *C. trachomatis* in cell culture.
5. To test the sensitivity of TRIC (trachoma-inclusion conjunctivitis) agents which had been reisolated from patients treated with antichlamydial drugs.
6. To use the guinea pig inclusion conjunctivitis (GPIC) animal model (16) as an in vivo system for testing the most effective antichlamydial drugs found from the in vitro studies.

MATERIALS AND METHODS

Chlamydial agents. The following *C. trachomatis* serotypes were used to study susceptibility to chloramphenicol, rifampin, and spiramycin.

Serotype A: A/MRC-17/OT, A/HAR-1/OT, A/IOL-574/OT
Serotype B: B/TW-5/OT, B/HAR-2/OT, B/IOL-1065/OT, B/IOL-1573/OT
Serotype C: C/TW-3/OT, C/IOL-223/OT
Serotype D: D/UW-3/GCx, D/MRC-1/OT, D/IOL-1895/GU
Serotype E: E/UW-5/GCx, E/DK-20/ON, E/IOL-268/OT
Serotype F: F/UW-6/GCx, F/MRC-301/GU
Serotype G: G/IOL-238/R, G/Cal-392/OC, G/IOL-1417/GU, G/IOL-215/GU
Serotype H: H/UW-4/GCx, H/IOL-1459/GU
Serotype K: K/UW-31/GCx, K/IOL-1419/GU

In subsequent studies, the *C. trachomatis* isolate D/UW-3/GCx was the only serotype used to study the antichlamydial activity of the complete range of antimicrobial compounds.

Stock pools of all serotypes were grown in irradiated McCoy cell cultures. They were divided into 1.0-ml samples and stored frozen at −70°C until required. Each sample was thawed and diluted with Eagle MEM (minimal essential medium) before use, so that the inoculum would produce 100 to 500 inclusions per ml.

The GPIC strain A-10 was used in all in vivo experiments, each animal being inoculated with approximately 1,000 ELD_{50} by droplet instillation into the conjunctival sac.

Cell culture. The method of preparation and inoculation of irradiated McCoy cell monolayers has been described previously (17). The main exception was that no antibiotics were added to the growth medium.

To determine the approximate minimal inhibitory concentration (MIC) of any particular compound, 96-hole flat-bottomed microtiter plates (Flow Laboratories Ltd.) were used in which cells were grown on 5.5-mm-diameter cover slips placed in each well. For more accurate determination of MICs, cells were grown on conventional 13-mm-diameter cover slips placed in individual plastic tubes.

Antimicrobial compounds. All antimicrobial solutions were freshly prepared, made up either from sterile powder or from solutions which were passed through a sterilizing membrane filter.

Doubling dilutions of antimicrobial solutions were made up in Eagle complete medium with additional glutamine (CM) but excluding any antibiotic other than the one to be tested.

Each well of alternate rows of microtiter plates previously seeded with irradiated McCoy cells was inoculated with 0.2 ml of the appropriate *C. trachomatis* serotype diluted in CM. Plates were then sealed and centrifuged for 60 min at 2,000 × *g* at 35°C. They were then incubated at 35°C for a further 2 h before removal of medium from all wells. This medium was then replaced with 0.2 ml of CM containing the appropriate concentration of antimicrobial compound. Five wells were used for each concentration of antimicrobial compound tested. Antimicrobial concentrations under test were also added to McCoy cells which had not been inoculated with *C. trachomatis*; these served as controls for cellular toxicity.

Microtiter plates were incubated for a further 48 h at 35°C, after which cover slips were removed, stained with Giemsa stain, and examined for the presence of inclusions by use of both bright-field and dark-field microscopy at a magnification of ×400.

The MIC was regarded as the lowest antimicrobial concentration which completely inhibited the formation of either initial body or elementary body inclusions in the cell culture monolayer.

Essentially the same technique was used with the tubes containing 13-mm-diameter cover slips. Five tubes were used for each antimicrobial dilution tested, and the tubes were centrifuged, incubated, stained, and examined by the same methods as described for the microtiter plates.

In addition to determining the MIC of a particular compound, the minimal lethal concentration (MLC) was also determined for several selected antimicrobial agents. The MLC was defined as the lowest concentration of drug which completely inhibited further multiplication of *C. trachomatis* after removal of the drug. Those infected cell monolayers treated with drug concentrations which had inhibited the formation of *C. trachomatis* inclusions were rinsed three times with fresh growth medium. The cells were then harvested, homogenized, and passed into a further series of uninfected, irradiated McCoy cell monolayers in which the antimicrobial compound had been omitted from the growth medium. The MLC was determined after two such passages in the absence of the antimicrobial compound had been completed and no further *C. trachomatis* inclusions had appeared.

Guinea pigs. Duncan-Hartley guinea pigs weighing 300 to 400 g were used in all experiments. All animals were inoculated with the A-10 strain of GPIC agent, and conjunctival scrapings for the detection of chlamydial inclusions were collected and examined as described previously (16). In all infected guinea pigs undergoing treatment, 1% ointment of the relevant antimicrobial drug was instilled into the conjunctival sac.

RESULTS

Relative antimicrobial susceptibility of a range of *C. trachomatis* serotypes. Twenty-five *C. trachomatis* isolates representing nine different micro-immunofluorescence (micro-IF) serotypes were examined for their susceptibility to chloramphenicol, spiramycin, and rifampin (Fig. 1). Against all serotypes, the MIC of chloramphenicol was 1.0 μg/ml and that of spiramycin was 0.5 to 1.0 μg/ml. There was a slightly wider range of sensitivity to rifampin, a few isolates of A, C, and E serotypes showing an occasional increase from the MIC of 0.05 to 0.1 μg/ml. In general, these results failed to demonstrate a great variation in the antimicrobial susceptibility of different serotypes, and it was therefore decided to use only isolate UW-3 (serotype D) in all subsequent cell culture work.

MICs of antimicrobial agents for *C. trachomatis*. Twenty antimicrobial compounds were tested in the cell culture system to determine their MICs for *C. trachomatis* (serotype D).

Rifampin showed the greatest degree of antichlamydial activity, with an MIC of 0.5 μg/ml (Table 1). The three tetracyclines, oxytetracycline, doxycycline, and minocycline, all exhibited the same degree of inhibition, with an MIC of 0.2 μg/ml. The closely related macrolide antibiotics erythromycin and spiramycin, and also sulfamethoxazole, had MICs of 0.5 μg/ml. Chloramphenicol and the penicillins were less inhibitory, but it was interesting to note that the antiviral drug adenine arabinoside 5′-monophosphate (Ara-AMP) did show some antichlamydial activity.

Although rifampin clearly showed the greatest antichlamydial activity as determined by the MIC, the MLC of this drug was nearly 20 times greater than its MIC. The MLCs of the two tetracyclines and spiramycin, however, were the same as their MICs, and these drugs showed greater antichlamydial activity than rifampin when determined by this method.

Antimicrobials useful in primary isolation of *Chlamydia*. Table 2 lists some antimicrobial compounds which may be of use in combating bacterial or fungal contamination of clinical specimens for the isolation of *Chlamydia* in cell culture systems. The concentrations tabulated are those which are noninhibitory to *C. trachomatis* as well as being nontoxic to irradiated McCoy cells.

Screening for resistant chlamydial strains during clinical treatment trials. The MIC of rifampin for chlamydial agents isolated from the eyes of four patients before

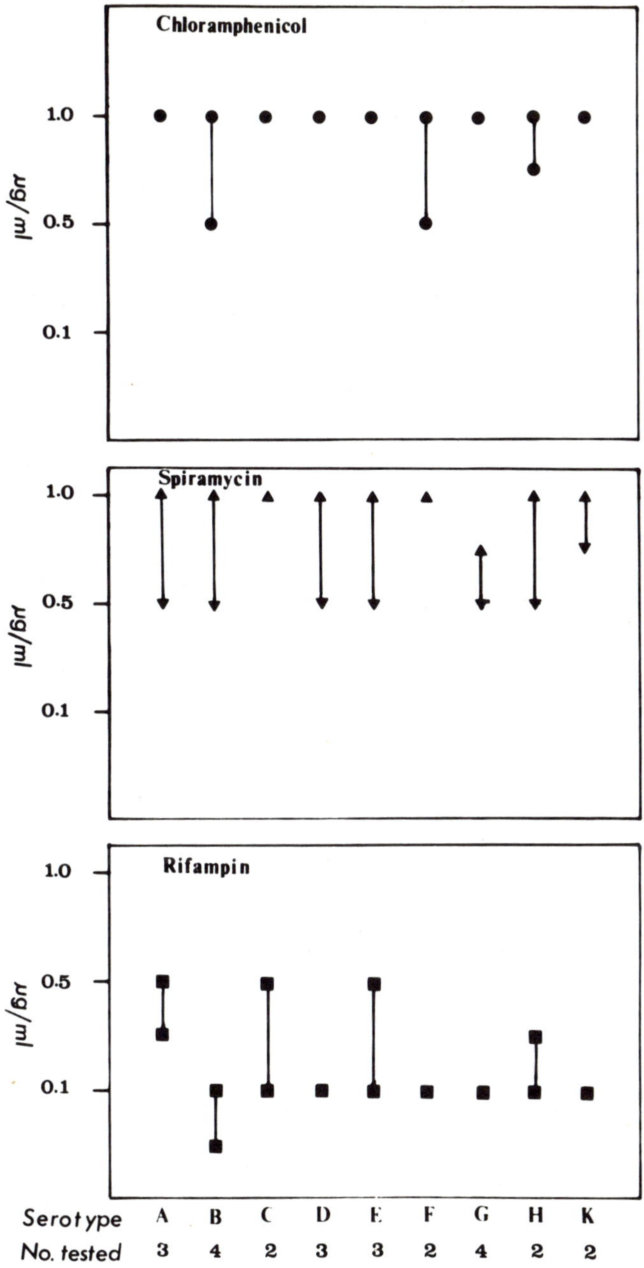

FIG. 1. *Minimal inhibitory concentrations of chloramphenicol, spiramycin, and rifampin for 25 C. trachomatis isolates.*

TABLE 1. *Minimal inhibitory and minimal lethal concentrations of antimicrobial agents for C. trachomatis (TRIC serotype D)*

Drug	MIC (μg/ml)	MLC (μg/ml)
Rifampin	0.05–0.1	1.0
Oxytetracycline	0.2–0.5	0.5
Doxycycline (Vibramycin)	0.2	0.2
Minocycline (Minocin)	0.2	—
Spiramycin	0.5	0.5
Erythromycin	0.5	—
Sulfamethoxazole	0.5	—
Chloramphenicol	1.0	5.0
Benzylpenicillin	1.0	—
Septrin (trimethoprim and sulfamethoxazole)	1.0	—
Ara-AMP	5.0	—
Trimethoprim	20.0	
Novobiocin	>50	
Kanamycin	>50	
Soframycin	>200	
Gentamicin	>500	
Colistin	>500[a]	
Vancomycin	>1,000	
5-Fluorocytosine	>50	
Nystatin	>500[a]	

[a] Units per milliliter.

treatment and after topical rifampin treatment for 4 to 6 weeks was determined in the cell culture system. None of the isolates showed any increase in resistance to rifampin as determined by their MICs (0.05 to 0.1 μg/ml).

Antimicrobial activity of topically administered drugs in the GPIC animal model. Figure 2A shows the results of an experiment in which guinea pigs infected with the GPIC agent were given topical 1% eye ointment, beginning 2 h after the GPIC inoculum had been instilled into the conjunctiva. Treatment either two times a day or five times a day was continued for 5 days. Conjunctival inclusions could not be detected in animals which were actively undergoing treatment, but when this period was ended, animals in both groups treated with rifampin quickly developed signs of

TABLE 2. *Concentrations of compounds useful for combating bacterial or fungal contamination of clinical specimens*

Drug	Maximal concn
Vancomycin	500 μg/ml
Gentamicin	100 μg/ml
Kanamycin	50 μg/ml
Colistin	100 U/ml
Soframycin	20 μg/ml
Nystatin (Mycostatin)	125 U/ml
5-Fluorocytosine	20 μg/ml

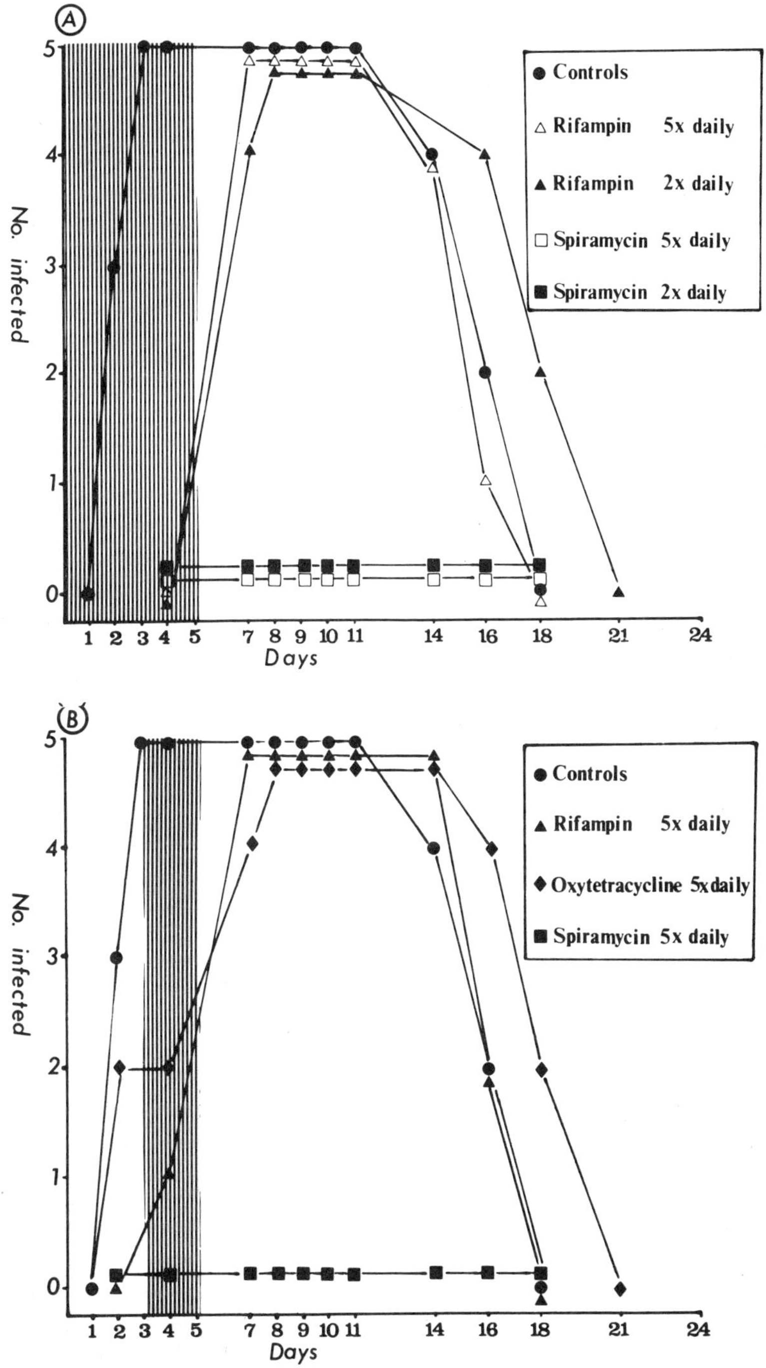

FIG. 2. *Antichlamydial activity of topically administered drugs in infected guinea pigs by demonstration of GPIC inclusions. The area filled with vertical lines indicates the period of treatment.*

clinical infection and chlamydial inclusions were detected in the conjunctival epithelial cells of all animals. The two groups of guinea pigs treated with 1% spiramycin failed to show clinical signs of infection, and inclusions could not be seen in their conjunctival cells even after the 5-day treatment period had ended.

In a second experiment (Fig. 2B), guinea pigs were inoculated with GPIC agent and topical treatment was begun 3 days after the initiation of the infection and lasted for a period of 2 days. There was some evidence of a prolongation of the incubation period of infection in animals treated with oxytetracycline and rifampin, but when the period of treatment was ended, all animals in both these drug-treated groups developed clinical and microbiological signs of infection with the GPIC agent. Those animals treated with spiramycin, however, showed no chlamydial inclusions and only minimal clinical signs of infection before treatment was begun at day 3. However, the conjunctival inflammatory picture which developed suggested that they were in fact infected with the GPIC agent.

DISCUSSION

It has been shown by many investigators that TRIC agents are sensitive to a wide variety of antimicrobial drugs (1, 3, 4, 8, 18). Most laboratory studies have only tested drug susceptibilities of a single or at best a few strains of *C. trachomatis* and have relied on the relatively complex and insensitive chick embryo yolk sac culture method for determining drug inhibitory effects.

With the development of the micro-IF test (Treharne et al., p. 249, this volume), *C. trachomatis* isolates can be divided into serotypes which correspond with the clinical and epidemiological type of chlamydial infection implicated (11). Thus, it is important to determine whether differing serotypes show variation in their sensitivity to antichlamydial drugs. Results obtained by use of the chick embryo culture technique (6) showed some chlamydial strains to be slightly more resistant than others to particular antimicrobial drugs. In our studies using three different antibiotics, we were unable to demonstrate any substantial differences in the drug susceptibility of 25 *C. trachomatis* isolates representing 9 different serotypes. From these results, it would seem that any of the epidemiological classes of disease caused by *C. trachomatis* could be treated effectively with any active antichlamydial drug.

In our in vitro cell culture studies, rifampin was the most potent antichlamydial drug in terms of its MIC. The MLC of rifampin was, however, higher than that of oxytetracycline, doxycycline, or spiramycin. These data concerning rifampin are not in agreement with the results of others who suggested that the antichlamydial action of rifampin is irreversible (1, 2). Initial results obtained with a combination of five parts sulfamethoxazole to one part trimethoprim suggested a potentiating effect of these two drugs used together. Subsequent experiments, however, failed repeatedly to demonstrate any potentiating action of this drug combination (18).

Cell culture techniques for the primary isolation of *C. trachomatis* from clinical specimens, although very sensitive, are susceptible to contamination by other extraneous microorganisms such as bacteria or fungi, thus rendering such tests invalid. Bacterial contamination is a particular problem when attempting to isolate *C. trachomatis* from the genital tract or rectum. The most commonly used antibiotics incorporated into cell culture media to combat such contamination have been streptomycin and vancomycin (3, 7), both of which have a limited antibacterial spectrum. From our work, additional or alternative antibiotics to help reduce or eradicate bacterial contaminants would be soframycin, gentamicin, and colistin. These drugs when used at very high concentrations do have some inhibitory effect on *C. trachomatis*, but used in the irradiated McCoy cell system at the concentrations suggested they resulted in no antichlamydial activity. Fungal contamination of clinical specimens can also be a problem in tropical or semitropical areas of high humidity, and, in addition to nystatin, the antifungal drug 5-fluorocytosine was found to be useful in controlling such contaminants.

In clinical treatment trials of follicular conjunctivitis due to TRIC agent infection, patients were treated topically with 1% rifampin ointment for 3 to 6 weeks. Several patients in this trial who initially converted from *C. trachomatis* isolation positive to isolation negative later relapsed and once again became *C. trachomatis* isolation positive. To determine whether these patients had developed rifampin-resistant *C. trachomatis* strains, the MICs of rifampin for pre- and post-treatment isolates were tested. No apparent change in the MICs could be detected, and it was thus assumed that the findings in these patients were probably due to faulty self-medication. However, it is possible that accidental ocular reinfection could have occurred, as some 80% of women and 50% of men presenting in London with ocular TRIC agent infection have concurrent genital infections with these organisms (Darougar et al., in preparation). In areas where trachoma is hyperendemic and patients are subject to a high pressure of reinfection, topical drug treatment of this type is particularly prone to failure, unless the whole epidemiological group is treated simultaneously.

In the limited number of in vivo trials carried out using the GPIC model of ocular chlamydial infection, we have demonstrated that, with a particular regimen of administration, certain topically applied drugs (rifampin and oxytetracycline) failed to inhibit or eradicate the GPIC agent completely from the conjunctiva. It should be noted that this animal model is a *C. psittaci* infection although it does resemble, in several striking ways, human ocular infections due to *C. trachomatis*; thus, our results with this model may help to explain the apparent relapses in the topical treatment trials of rifampin in TRIC agent infection of humans. In baboons (2) it was observed that topical rifampin treatment of established *C. trachomatis* infection rapidly improved the clinical appearance of the infected eye; however, no microbiological evidence of TRIC agent eradication was examined. Spiramycin, a macrolide antibiotic closely resembling erythromycin, gave encouraging results in our GPIC animal model. In the treatment regimens tested, the drug, when applied topically, appeared to eradicate the GPIC agent completely from the infected conjunctivae, with no relapses occurring after treatment was ended.

In conclusion, we recommend the cell culture system described as a simple and effective way of screening potential antichlamydial compounds and for testing apparently resistant TRIC agent isolates. Combination of this in vitro system with an in vivo system such as the GPIC animal model may give meaningful laboratory parameters for the estimation of antichlamydial drug activity.

ACKNOWLEDGMENTS

These studies were supported by grants from Lepetit Pharmaceuticals Ltd., the Department of Health and Social Security, the Wellcome Trust, and the World Health Organization.

REFERENCES

1. Becker Y, Zakay-Rones Z: Rifampin—a new antitrachoma drug. Nature (London) 222:851–853, 1969
2. Becker Y, Asher Y, Himmel N, et al: Rifampicin inhibition of trachoma agent in vivo. Nature (London) 224:33–34, 1969
3. Gordon FB, Quan AL: Drug susceptibilities of the psittacosis and trachoma agents. Ann NY Acad Sci 98:261–270, 1962
4. Nabli B: The effect of rifampicin on the growth of TRIC agents in embryonated eggs, p 34–39. *In* RL Nichols (ed), Trachoma and related disorders caused by chlamydial agents. Excerpta Medica, Amsterdam, 1971
5. Ridgway GL, Owen JM, Oriel JD: A method for testing the antibiotic susceptibility of Chlamydia trachomatis in a cell culture system. J Antimicrob Chemother 2:71–76, 1976
6. Shiao LC, Wang SP, Grayston JT: Sensitivity and resistance of TRIC agents to penicillin tetracycline and sulpha drugs. Am J Ophthalmol 63:1558–1568, 1967
7. Sowa J, Race MW: Sensitivity of trachoma agent to streptomycin and related antibiotics. J Hyg 69:709–716, 1971
8. Tarizzo ML, Nabli B: The effect of antibiotics on the growth of TRIC agents in embryonated eggs. Am J Ophthalmol 63:1550–1557, 1967
9. Dawson CR, Ostler HB, Hanna L, et al: Tetracyclines in the treatment of chronic trachoma in American Indians. J Infect Dis 124:255–263, 1971
10. Hoshiwara I, Ostler HB, Hanna L, et al: Doxycycline treatment of chronic trachoma. J Am Med Assoc 224:220–223, 1973
11. Jones BR: The prevention of blindness from trachoma. Trans Ophthalmol Soc UK 95:16–33, 1975
12. Ostler HB, Hassa L, Hoshiwara I, et al: A comparison of tetracycline and doxycycline in chronic trachoma of American Indians, p 540–544. *In* RL Nichols (ed), Trachoma and related disorders caused by chlamydial agents. Excerpta Medica, Amsterdam, 1971
13. Tarizzo ML, Nataf R: The treatment of trachoma. Rev Int Trach 46:9–48, 1970
14. Ghione M, Brivio R, Sanfilippo A, et al: Research on factors influencing experimental chemotherapy tests in trachoma. Am J Ophthalmol 63:1573–1577, 1967
15. Jawetz E: Chemotherapy of chlamydial infections. Adv Pharmacol Chemother 7:253–282, 1969
16. Treharne JD: A comparison of microbiological techniques for the detection of GP-IC agent in the conjunctiva of infected guinea pigs, p 435–444. *In* RL Nichols (ed), Trachoma and related disorders caused by chlamydial agents. Excerpta Medica, Amsterdam, 1971
17. Gordon FB, Harper IA, Quan AL, et al: Detection of Chlamydia (Bedsonia) in certain infections of man. 1. Laboratory procedures: comparison of yolk sac and cell culture for detection and isolation. J Infect Dis 120:451–462, 1969
18. Reeve P, Taverne J, Bushby SRM: Inhibition by pyrimidine analogues of the folic acid by trachoma agents. J Hyg 66:295–306, 1968

Antibiotic Susceptibility of *Chlamydia trachomatis* in Cell Culture

E. RUSSELL ALEXANDER, PATRICIA SKAHEN, AND KING K. HOLMES

INTRODUCTION

This discussion will describe a method of estimating antibiotic susceptibility, comment upon our results with various aminocylitol antibiotics against *Chlamydia trachomatis*, and note differences in the activity of lincomycin and clindamycin against this organism.

MATERIALS AND METHODS

Our method utilized McCoy cell cultures pretreated with 5-iodo-2′-deoxyuridine. The medium employed was that described by Wentworth and Alexander (1) except that cells used in these studies were grown in antibiotic-free medium for at least two passages. Twofold dilutions of antibiotic were added to the overlay medium. In the initial experiments, inocula containing 10^1 to 10^5 inclusion-forming units (IFU)/ml were evaluated. Concentrations of 10^4 and above were cytotoxic. Concentrations below 10^4 were all satisfactory. For the remainder of the experiments, inocula of approximately 3.5×10^3 IFU/ml were used. Cover slips were stained with iodine after 72 h of incubation, inclusions were counted, and the total number of inclusions per cover slip was calculated. The IC_{50} was that concentration of antibiotic which reduced the number of inclusions to less than 50% of the control; the minimal inhibitory concentration (MIC) was that concentration which completely prevented formation of any inclusions.

RESULTS

For the initial evaluation of aminocyclitol susceptibility, four strains of *C. trachomatis* were tested. These included strains U-047, U-211, and U-043-2 from urethritis cases, immunotyped by S.-P. Wang as D, G, and B, respectively. Each of these three isolates had been originally collected in transport medium containing 50 μg of streptomycin plus 25 U of nystatin per ml and had been passed in tissue culture in medium containing 10 μg of gentamicin per ml. The fourth strain, #434, was provided by J. Schachter. This strain was isolated from the lymph node of a patient with lymphogranuloma venereum (LGV) and was immunotyped as LGV II. Not unexpectedly, the four strains of *C. trachomatis* were resistant to streptomycin and gentamicin at concentrations of $>1{,}024$ μg/ml. However, some susceptibility to spectinomycin was demonstrable. The IC_{50} was 64 μg/ml and the MIC was 256 μg/ml for all four strains.

To determine whether more resistant strains might emerge after spectinomycin treatment, we studied 10 strains of *C. trachomatis* recovered from patients previously treated with spectinomycin for their gonorrhea (Table 1). Four were cervical isolates and six were from men with postgonococcal urethritis (PGU).

TABLE 1. *Resistance of 10 genital chlamydial strains to spectinomycin*

Strain	Site	Spectinomycin (μg/ml)	
		IC_{50}	MIC
U-119-4	Urethral	32	256
U-174-1	Urethral	32	256
U-130-2	Urethral	32	256
U-198-2	Urethral	32	256
U-263-2	Urethral	32	128
U-274-2	Urethral	32	256
RX-023-1	Cervical	32	128
RX-115-1	Cervical	32	256
RX-040-1	Cervical	64	256
RX-054-1	Cervical	64	256
Mean values		37	223

All 10 strains were resistant to streptomycin and gentamicin at concentrations of >1,024 μg/ml, but there was no evidence of increased resistance to spectinomycin. The IC_{50} for these 10 strains was consistently 32 or 64 μg/ml, and the MIC was 126 to 256 μg/ml. The clinical significance of the modest activity exhibited by spectinomycin against *C. trachomatis* is uncertain, but it is of some interest that Karney et al. (2) had reported that the incidence of PGU was slightly, but significantly, lower in men given 4 g of spectinomycin for their gonorrhea than in men given 2 g. This could be related to the marginal activity of spectinomycin against *C. trachomatis* in vitro, or, as discussed elsewhere in this volume (Bowie et al., p. 19), against *U. urealyticum* in vivo. The serum levels achieved after a 4-g intramuscular dose only transiently exceeded the IC_{50} level for *C. trachomatis* and approached the MIC level which we found.

We were not able to circumvent completely the obvious problem of exposure of clinical isolates to streptomycin or gentamicin before susceptibility testing because of contamination problems. However, after repeated unsuccessful attempts, we were able to recover four strains from transport medium containing only 5 μg of streptomycin per ml. These were isolated in antibiotic-free medium and were found to have spectinomycin susceptibilities similar to those of strains isolated in the usual manner (Table 2); these four strains were all resistant to streptomycin and gentamicin at concentrations of >1,024 μg/ml.

In addition, we examined the susceptibility of *C. trachomatis* to erythromycin and lincomycin. Most cases of nongonococcal urethritis (NGU) respond to

TABLE 2. *Resistance to spectinomycin of four genital chlamydial strains isolated without antibiotic*

Strain	IC_{50} (μg/ml)	MIC (μg/ml)
WM 009	64	128
V73-84	64	256
V74-15	64	256
RX-027	32[a]	128

[a] Low input (approximately 10 inclusion-forming units).

erythromycin treatment but not to lincomycin. This has been used to support the etiological role of *Ureaplasma urealyticum* in NGU, since this organism is also susceptible to erythromycin but resistant to lincomycin. We employed the same four strains of *C. trachomatis* that were first tested against the aminocyclitols. All four were completely inhibited by 1.0 μg of erythromycin per ml, but the MIC of lincomycin for these strains was 512 μg/ml. The IC_{50} of erythromycin was <0.06 μg/ml, and that of lincomycin was 64 μg/ml. Thus, the susceptibility of *C. trachomatis* to erythromycin and lincomycin is consistent with the observed response of NGU to these agents. Of interest is the fact that all four strains were highly susceptible to clindamycin, the MIC being 1 μg/ml and the IC_{50} being 0.25 or 0.125 μg/ml. This is of interest for two reasons: it might have some therapeutic potential, and it is also unusual to see such disparity between clindamycin and lincomycin.

Finally, we show the consistency of response of 16 genital chlamydial strains to spectinomycin and lincomycin (Table 3).

DISCUSSION

Earliest studies of antibiotic susceptibility of *C. trachomatis* strains were conducted in embryonated eggs (3, 4). The first reports of antibiotic susceptibility measurements in cell culture (5, 6) utilized the "fast" variants of *C. trachomatis*, because they were the only strains which could grow in the cell culture systems then in use. As these strains have subsequently been shown to be immunologically identical to LGV strains, their relevance to genital and ocular trachoma strains is in question (7). Ridgeway et al. have reported a method of susceptibility testing (8) which is essentially similar to the method reported here. Their emphasis appeared to be on screening a variety of antibiotics against a few standard strains, rather than testing the larger number of strains which we have examined. Nevertheless, although we appear to differ on the specific measurement of MIC, the relative susceptibilities appeared to be very similar.

TABLE 3. *Resistance of 16 genital chlamydial strains to spectinomycin and lincomycin*

Strain	Site	Spectinomycin (μg/ml)		Lincomycin (μg/ml)	
		IC_{50}	MIC	IC_{50}	MIC
SU-784-3	Urethral	128	256	32	512
SU-210-3	Urethral	64	256	32	512
SU-959-2	Urethral	64	256	32	512
SU-758-3	Urethral	64	256	32	512
SU-091-3	Urethral	64	256	64	512
SU-188	Urethral	128	256	32	256
SU-1617	Urethral	32	256	32	256
U-050-1	Urethral	128	256	32	256
U-315-1	Urethral	64	256	32	512
V74-44	Cervical	64	512	32	512
S-84-3	Cervical	64	256	32	512
CXT-1140	Cervical	32	256	64	256
S-17-1	Cervical	64	256	64	256
S-81-9	Cervical	64	256	32	128
V73-99	Cervical	64	256	32	512
S-97-9	Cervical	64	256	64	256
Mean values		67	267	38	362

We believe that we have demonstrated a simple method of antibiotic susceptibility testing that will yield reproducible, useful data. It is of interest that we did not demonstrate any increased level of resistance in cultures taken later in disease, but we have not pursued that further.

ACKNOWLEDGMENTS

This research was supported by Public Health Service research grant AI-12192 from the National Institute of Allergy and Infectious Diseases and by a grant from The Upjohn Co.

REFERENCES

1. Wentworth BB, Alexander ER: Isolation of *Chamydia trachomatis* by use of 5-iodo-2-deoxyuridine-treated cells. Appl Microbiol 27:912–916, 1974
2. Karney WM, Pederson AHB, Nelson M, et al: Spectinomycin versus tetracycline for the treatment of gonorrhea. N Engl J Med 296:889–894, 1977
3. Shiao LC, Wang SP, Grayson JT: Sensitivity and resistance of TRIC agents to penicillin, tetracycline and sulfa drugs. Am J Ophthalmol 63:1558–1568, 1967
4. Jawetz E, Hanna L: Trachoma viruses isolated in the United States. 3. Strain specific effects of tetracycline and penicillin. Proc Soc Exp Biol Med 105:320–323, 1960
5. Reeve P, Traverne J: Inhibition by pyrimidine analogues of the synthesis of folic acid by trachoma agents. J Hyg 66:295–306, 1968
6. Barker LF: Determinations of antibiotic susceptibility of rickettsiae and chlamydiae in BS-C-1 cell cultures. Antimicrob Agents Chemother 1968, p 425–434, 1969
7. Wang SP, Grayston JT: Studies on the identity of fast egg-killing chlamydial strains, p 322–326. *In* RL Nichols (ed), Trachoma and related disorders caused by chlamydial agents. Excerpta Medica, Amsterdam, 1971
8. Ridgeway GL, Owen JM, Oriel JD: A method for testing the antibiotic susceptibility of *Chlamydia trachomatis* in a cell culture system. J Antimicrob Chemother 2:71–76, 1976

Effect of Penicillin on *Chlamydia trachomatis* In Vitro and In Vivo

DEREK HOBSON, ELISABETH REES, F. WILLIAM A. JOHNSON, AND I. ANNE TAIT

Concurrent genital infections due to *Chlamydia trachomatis* and *Neisseria gonorrhoeae* are common in women (1, 2). The diagnosis of gonococcal infection can frequently be made immediately by smear and can be confirmed rapidly by bacteriological culture. Hence, many of these patients may be treated immediately by administration of a single large dose of penicillin in the clinic before it is realized that *C. trachomatis* is also present in the genital tract.

Clinical results in men with double infections have suggested that, although single-dose penicillin treatment is effective in curing gonorrhea, it may fail to eradicate concurrent chlamydial urethritis. For example, Richmond et al. found that postgonococcal urethritis occurred more often in men in whom both *N. gonorrhoeae* and *C. trachomatis* had been present before treatment (3), and Holmes et al. isolated *C. trachomatis* from 9 of 20 men who developed postgonococcal urethritis 1 to 6 weeks after treatment of gonorrhea with penicillin (4). Oriel (p. 38, this volume) has shown that even a single dose of 2 g of ampicillin given together with 1 g of probenicid failed to cure *C. trachomatis* infection associated with gonorrhea in 14 of 15 men, although the latter was cured in all cases. It is thus of practical importance to consider the clinical results of using a single dose of penicillin to treat women with concurrent gonococcal and chlamydial infections of the cervix, as well as the effect of this antibiotic on *Chlamydia* in vitro. This report gives preliminary data on both these aspects.

A series of McCoy monolayers was simultaneously infected with *C. trachomatis* SAL strain at a dilution calculated to contain 10,000 inclusion-forming units (IFU) per culture after 48 h of incubation at 37°C. No antibiotics were used other than benzylpenicillin, which was added at the time of infection. Details of our procedure are described elsewhere (5).

The effect of incubation throughout the whole growth period with increasing concentrations of penicillin from 0.1 to 1.0 U/ml was progressive in the sense that normal inclusions, containing Giemsa-stained elementary bodies (EB) which fluoresced brightly under dark-ground microscopy, decreased in numbers and could not be found at penicillin concentrations greater than 1.0 U/ml. However, this decline in the number of normal inclusions was accompanied by a rising number of small abnormal inclusions, which did not fluoresce under dark-ground illumination and which contained only a few large swollen particles staining a reddish color with Giemsa, instead of the normal packed mass of deep blue-stained EB. It was difficult to count these abortive inclusions accurately, but in cultures incubated with 0.5 to 1.0 U of penicillin per ml, where their numbers were highest, they represented over 50% of the number of IFU of *C. trachomatis* originally added to the culture. Even in infected cultures incubated with penicillin concentrations from 5 to 100 U/ml, abnormal inclusions were found, but they became progressively fewer and their contents became increasingly sparse and abnormal, with amorphous red-staining debris replacing the particulate forms. However, even at 100 U of penicillin per ml, the number of abortive inclusions represented approximately 10% of the number of IFU of *C. trachomatis* originally inoculated.

It seems probable that the effect of penicillin on chlamydiae is exerted only at the stage where the progeny growing within the McCoy cell are beginning to synthesize their compact cell wall and become definitive EB. Thus, the abnormal swollen bodies seen within the abortive inclusions in penicillin-treated McCoy cells probably represent initial bodies, the further maturation of which is arrested (Manire, p. 167, this volume). It might thus be expected that normal maturation, with the development of infective EB, would be resumed when the penicillin is removed. So far, we have failed to confirm that reversion can occur in vitro. However, in the very different intracellular environment of chlamydiae growing in columnar epithelium of the genital tract in vivo, it seems probable that similar abnormal inclusions may develop under the effect of penicillin, but that maturation to EB may recommence after the last traces of the antibiotic have disappeared from the system. It is thus of interest to look at the effect of single-dose therapy with penicillin in patients with both gonococcal and chlamydial genital infection.

In our preliminary studies we followed the course of events in 22 women with proven gonococcal infection who were treated at once for this infection with procaine penicillin (1.2 megaunits) plus amoxycillin (500 mg) before it was suspected that a concurrent infection with *C. trachomatis* might be present. In all these patients, penicillin eradicated the gonococcal infection.

In 12 of these women, *C. trachomatis* was in fact isolated from pretreatment swabs, but only after penicillin had already been given for the gonococcal infection. In five of them, *C. trachomatis* was not reisolated after penicillin adminstration during subsequent periods of observation of 22 days to 9 months. The number of post-treatment swabs was five in one case, four in two cases, and three and two in one case each. In the other seven of these women with proven gonococcal plus chlamydial infection, *C. trachomatis* was reisolated in swabs taken 5 to 28 days after penicillin treatment. In one of these patients, although *C. trachomatis* was not isolated 7 days after penicillin administration, it was recovered from later swabs. In the other six patients the first post-treatment swab was positive, and in one case the inclusion count was significantly higher than in the pretreatment swab (18,000 and 900 per cover slip, respectively).

In a further eight women there had been no pretreatment investigation for chlamydiae, but there was persistent pus in the endocervical canal after the gonococcal infection had been eradicated by penicillin. *C. trachomatis* was isolated from all eight women from swabs taken 15 to 34 days after penicillin administration. In five of these cases, *C. trachomatis* remained present in second confirmatory swabs taken from 28 to 130 days after penicillin administration, before the beginning of oxytetracycline therapy for their chlamydial infection.

In the remaining two patients in this series, *C. trachomatis* was isolated 20 and 27 days, respectively, after penicillin administration, although they had had no signs or symptoms of cervical infection. This is commonly the case in women where gonococci or chlamydiae, or both, are found in the cervix (Rees et al., p. 67, this volume).

These preliminary results suggest that many women who present themselves at STD (sexually transmitted disease) clinics, and who are in reality infective reservoirs of both gonococci and chlamydiae, may be recognized and treated only for gonococcal infection. The chlamydial infection may persist undiminished for weeks or months after the gonococcal infection has been cured; in some but not all cases, the clinical signs may persist, and there would seem to be a distinct risk that many of these women, after treatment for gonococcal infection, could return to their community as persistent but unsuspected disseminators of chlamydial STD.

Because of the recent emergence of β-lactamase-producing strains of *N. gonorrhoeae* (6), the single-dose regimen of penicillin for the routine treatment of gonorrhea may in the near future need to be replaced by some other form of antibacterial therapy. It is to be hoped that any new antibiotic or new regimen devised for this purpose will take account of the frequency with which gonorrhea is accompanied by infection due to *C. trachomatis*, the eradication of which cannot necessarily be guaranteed by treatment with antibiotics active against *Neisseria*.

ACKNOWLEDGMENT

These studies were supported by a project grant from the Medical Research Council of Great Britain.

REFERENCES

1. Hilton AL, Richmond SJ, Milne JD, et al: Chlamydia A in the female genital tract. Br J Vener Dis 50:1–10, 1974
2. Oriel JD, Powis PA, Reeve P, et al: Chlamydial infections of the cervix. Br J Vener Dis 50:11–16, 1974
3. Richmond SJ, Hilton AL, Clarke SKR: Role of chlamydia sub-group A in non-gonococcal and post-gonococcal urethritis. Br J Vener Dis 48:437–444, 1972
4. Holmes KK, Handsfield HH, Wang SP, et al: Etiology of nongonococcal urethritis. N Engl J Med 292:1199–1205, 1975
5. Johnson FWA, Hobson D: The effect of penicillin on genital strains of *Chlamydia trachomatis* in tissue culture. J Antimicrob Chemother 3:49–56, 1977
6. Percival A, Rowlands J, Corkill JE, et al: Penicillinase-producing gonococci in Liverpool. Lancet ii:1379–1382, 1976

Chlamydia trachomatis and Postgonococcal Genital Infection

J. D. ORIEL

Postgonococcal urethritis (PGU) occurs in a proportion of men after treatment for gonococcal urethritis; their urethral discharge returns and an excess of polymorphonuclear leukocytes is seen on microscopy of urethral material, but cultures for *Neisseria gonorrhoeae* are negative. It has now become apparent that many cases of PGU are due to *Chlamydia trachomatis* infections which have not been eradicated by the antimicrobial therapy used for the patients' gonorrhea.

Throughout the world physicians rely on single-dose therapy for the treatment of gonorrhea as this has clear epidemiological advantages, and many drugs have been effectively used in this way. However, since between 20 and 30% of men with gonorrhea have a simultaneous infection with *C. trachomatis*, the effect of these single-dose treatments on this organism needs to be considered.

Gentamicin has no significant effect against *C. trachomatis*, but it is active against *N. gonorrhoeae*. I cured 44 men of gonorrhea with a single injection of gentamicin (240 mg), but this treatment had no effect against associated *C. trachomatis* (Table 1). Eleven men who gave isolates before treatment still did so afterwards; four men who were isolation negative before treatment became positive later, perhaps because the passage of time increased the number of available organisms. All 15 *Chlamydia*-positive men developed PGU. Of the whole group of 44 patients, 26 (59%) developed PGU, and 11 of these gave isolates of *C. trachomatis* pre- or post-treatment.

Ampicillin has weak in vitro activity against *C. trachomatis*, similar to that of penicillin. I successfully treated 46 men with gonorrhea with ampicillin (2 g) plus probenecid (1 g). The lack of effect against associated *C. trachomatis* is shown in Table 2. Of the 15 patients whose cultures yielded *C. trachomatis*, 12 developed PGU. Of the whole group of 46 patients, 17 (37%) developed PGU, and 12 of these gave *Chlamydia* isolates either before or after therapy.

Spectinomycin has in vitro activity against *C. trachomatis* similar to that of penicillin and ampicillin, and a single dose similarly fails to have a significant

TABLE 1. *Effect of gentamicin (240 mg) on C. trachomatis isolation and postgonococcal urethritis (PGU) in 44 men successfully treated for gonococcal urethritis*

C. trachomatis isolation		No. of patients	
Pretreatment	Post-treatment	Total	With PGU
Positive	Positive	11	11
Positive	Negative	0	
Negative	Positive	4	4
Negative	Negative	29	11

TABLE 2. *Effect of ampicillin (2 g) plus probenecid (1 g) on C. trachomatis isolation and postgonococcal urethritis (PGU) in 46 men successfully treated for gonococcal urethritis*

C. trachomatis isolation		No. of patients	
Pretreatment	Post-treatment	Total	With PGU
Positive	Positive	11	10
Positive	Negative	1	1
Negative	Positive	3	1
Negative	Negative	31	5

effect on the microorganism. I treated 54 men with gonorrhea with injected spectinomycin (2 g) with uniform eradication of *N. gonorrhoeae*. However, associated *C. trachomatis* was unaffected, and all of 16 isolation-positive men developed PGU. Of the whole group of 54 patients, 24 (46%) developed PGU, and 16 of these yielded *C. trachomatis* before or after treatment (Table 3).

Between 35 and 60% of women with gonorrhea have cervical infections due to *C. trachomatis*. As in men, treatment with ampicillin (2 g) plus probenecid (1 g) or with spectinomycin (2 g) rarely eliminates *C. trachomatis* from the genital tract in women (Table 4). I have observed chlamydial infection persisting after the treatment of gonorrhea for periods of time up to 12 weeks.

The above data support the observations of Gale et al., Alexander et al., and Hobson et al. (p. 205, 223, and 227, this volume). It is plain that several antimicrobial agents used for the single-dose treatment of gonorrhea fail to eradicate an associated chlamydial infection. Persistence of *C. trachomatis* in men will probably cause PGU which will be diagnosed either on routine follow-up or because it causes symptoms. In either event, treatment with a tetracycline can be given. But women with persisting chlamydial infection are unlikely to be diagnosed (unless isolation of *C. trachomatis* is attempted); women do not develop PGU, and in them much chlamydial infection is asymptomatic. It is quite possible for a woman successfully treated for gonorrhea to be discharged as cured despite the persistence of an undiagnosed chlamydial infection.

The existence of synchronous infections by *N. gonorrhoeae* and *C. trachomatis* thus presents problems for the clinician. It is known that patients treated for gonorrhea with tetracyclines show a decreased incidence of PGU, although no

TABLE 3. *Effect of spectinomycin (2 g) on C. trachomatis isolation and postgonococcal urethritis (PGU) in 54 men successfully treated for gonococcal urethritis*

C. trachomatis isolation		No. of patients	
Pretreatment	Post-treatment	Total	With PGU
Positive	Positive	10	10
Positive	Negative	0	
Negative	Positive	6	6
Negative	Negative	38	8

TABLE 4. *C. trachomatis isolations from the cervix in women with gonorrhea before and after treatment with ampicillin (2 g) plus probenecid (1 g) or with spectinomycin (2 g)*

C. trachomatis isolation		No. of patients	
Pretreatment	Post-treatment	Treated with ampicillin + probenecid	Treated with spectinomycin
Positive	Positive	18	14
Positive	Negative	5	4
Negative	Positive	2	4
Negative	Negative	19	25

prospective studies on the effects of such treatment on *C. trachomatis* have yet been performed. When treatment schedules for gonorrhea are revised, or new ones are introduced, it may be thought desirable to assess the action of these against *C. trachomatis* by appropriate in vitro and clinical studies.

Chairman's Report and Discussion

HOWARD M. JENKIN

In this session, an introduction to the biology and some biochemistry of *Chlamydia* agents was presented by P. Manire; antigens of the *C. trachomatis* group were discussed by C. C. Kuo; the use of cats, guinea pigs, and primates as animal models for the study of chlamydial infections was described by S. Darougar, E. Murray, and J. Gale; and antibiotic susceptibility of various chlamydial agents was discussed by J. Treharne, E. R. Alexander, D. Hobson, and J. D. Oriel.

Dr. Manire discussed the ubiquitous nature of chlamydial organisms, their tendency to produce latent infections, and their unique growth cycle. The complex developmental cycle of *Chlamydia* is an expression of this organism's adaptability and is associated with changes in the cell envelope. The mechanism of conversion of the reticulated body to the elementary body and the nature of the toxic component in the cell envelope still require further investigation. The structure of the chlamydial envelope appears to be of importance in the pathogenesis of chlamydial infection and in development of host resistance, but there is limited understanding of the antigenic and biochemical architecture of the cell envelopes. By utilizing two cell culture systems to cultivate *C. trachomatis* agents, Dr. Kuo and his co-workers were able to bypass problems of contamination by host-cell antigens generally encountered when using micro- and immunochemical techniques. A separation of three antigenic groups was accomplished by use of a double immunodiffusion test. The antigens were heat-stable lipid-carbohydrate complexes. Dr. Kuo described the separation of 18 distinct species-specific antigenic components of *C. trachomatis* agents by rocket (two-dimensional) immunoelectrophoresis, most components proving to be heat labile. One immunodominant antigen from this separation technique was successfully used for serological diagnosis of lymphogranuloma venereum (LGV). A type-specific antigen was demonstrated by Dr. Kuo by the micro-immunofluorescence test in which 12 trachoma and 3 LGV immunotypes could be characterized. A heat-stable, species-specific protein antigen causing delayed hypersensitivity was studied in guinea pigs.

Dr. Murray described a guinea pig model for studying the infectious nature of the guinea pig conjunctivitis agent, associated with sexual transmission of the organism in the eye, vagina, and male urethra. Immunity in this disease does not appear to bear any relationship to humoral antibodies and an uncertain relationship to secretory antibodies.

Dr. Darougar presented pictorial evidence of acute and chronic infections with the agent of guinea pig inclusion conjunctivitis in guinea pigs. These pictures gave evidence of trachoma-like infection of the guinea pig eye, with follicular formation and scarring of the conjunctiva. Dr. Darougar also demonstrated that the cat can be used as a model to study cat keratoconjunctivitis (subgroup B *Chlamydia*) agent. This organism produced a trachoma-like conjunctivitis in nature, as well as when experimentally inoculated into the genital tract of cats, where it produced an active disease of the urethra and cervix.

Dr. Gale described the use of male monkeys (*Macaca nemistrina*) as models for studying type E *C. trachomatis* in the genital tract; follicular lesions of the distal urethra could be produced, and the organism was recovered from several urethral sites. *Ureaplasma urealyticum* used under the same conditions produced lesions in the urethra, and the organism was continually shed for 4 weeks. One monkey inoculated intraurethrally with *Ureaplasma* did not shed this organism, even after subsequent serial inoculation of *Ureaplasma*. The ureaplasma did establish an infection and continued to be shed. However, the urethral lesions at postmortem resembled those seen in experimental chlamydial urethritis.

Dr. Alexander presented a method for estimating antibiotic susceptibility of *C. trachomatis*. Emphasis was placed on the screening of lincomycin and clindamycin against a number of *C. trachomatis* strains by using McCoy cells pretreated with iododeoxyuridine. The minimal inhibitory concentration, used as an end point, represented the amount of antibiotic that completely prevented formation of inclusions. As had been reported earlier, streptomycin and gentamicin did not affect chlamydial growth, although some susceptibility to spectinomycin was observed. Erythromycin was effective in treating nonspecific urethritis infections associated with *C. trachomatis*, but lincomycin was not effective. In vitro, the same general observations were made. Four *C. trachomatis* strains were susceptible to clindamycin in vitro. This was unusual, since lincomycin, which is in the same family of aminocyclitols, was ineffective. *U. urealyticum* has the same pattern of response as *C. trachomatis*. Spectinomycin was as ineffective an inhibitor as lincomycin against 16 genital tract isolates of *C. trachomatis*. Oriel and Hobson presented clinical and laboratory data associated with antibiotic susceptibility of *C. trachomatis*; the failure of short courses of penicillin to eradicate chlamydial infections coincident with gonorrhea was discussed in relation to the limited in vitro activity of this antibiotic on *C. trachomatis* in McCoy cells.

To put this session in perspective briefly, new animal laboratory models are now available to study nonspecific urethritis infections related to *C. trachomatis* strains, and the antibiotic armament has been further defined by using cell culture sensitivity tests which can be related to the treatment of clinical disease. However, the problems of understanding the mechanisms of immune response to chlamydial agents, and characterization of their long-term "protective" antigens, remain to be resolved, particularly the nature of the toxic antigen(s). Mechanisms of the disease process related to differences in pathogenicity of chlamydial agents deserve further exploration.

Session V

LABORATORY DIAGNOSIS OF *CHLAMYDIA TRACHOMATIS* INFECTIONS

Serodiagnosis of *Chlamydia trachomatis* Infection with the Micro-Immunofluorescence Test

SAN-PIN WANG, J. THOMAS GRAYSTON, CHO-CHOU KUO,
E. RUSSELL ALEXANDER, AND KING K. HOLMES

Knowledge concerning *Chlamydia trachomatis* infection has increased rapidly in recent years (1). In addition to the long-standing problem of trachoma eye disease in the remaining endemic areas of the world, there is increasing recognition of genital trachoma infections around the world. These organisms are one of the major causes of urethritis and possibly of cervicitis (2). Laboratory techniques for diagnosis of *C. trachomatis* infection are of increasing interest. Efforts have been made to improve isolation of the organisms (3–8), direct demonstration of the organisms in infected tissues (9), and detection of antibodies in sera or eye secretions (10–15). Serology is the concern of this paper.

An effective serological test should be sensitive, specific, and relatively simple to perform. We believe that our revised micro-immunofluorescence (micro-IF) test meets these criteria. The method was devised initially (16, 17) to replace the cumbersome mouse toxicity prevention test (18) for immunotyping of strains (19–23). The technique has been applied to detect human antibodies to *C. trachomatis* infection (11–15). More recently, its application has been demonstrated in rickettsial infections (24).

Through studies with the micro-IF test in mice (17), primates (22, 25), and humans (14, 15), we found that the type-specific patterns of antibody response were essentially similar with different hosts. In this communication, we review the immunotypes of *C. trachomatis*, knowledge of which is important for interpretation of the test, and we present results of simplification of the technique for more practical application in serodiagnosis.

MATERIALS AND METHODS

(Micro-IF) Test

Details of the technique of the micro-IF test have been presented (17). Representative *C. trachomatis* (trachoma-lymphogranuloma venereum [LGV]) strains used for test were A/G-17/OT, B/TW-5/OT, C/TW-3/OT, D/UW-3/Cx, E/UW-5/Cx, F/UW-6/Cx, G/UW-57/Cx, H/UW-43/Cx, I/UW-12/Ur, J/UW-36/Cx, K/UW-31/Cx, L_1/440/Bu, L_2/434/Bu, and L_3/404/Bu. All strains except UW-6 were grown in HeLa 229 cells. Details of growth in HeLa 229 cell culture are presented separately by Kuo et al. (p. 328, this volume). Test antigens were prepared by mixing a homogenate of highly infected HeLa 229 cells with 5% of normal yolk sac suspension. UW-6 antigen was prepared in embryonated eggs as previously described. Pooled antigens were made for C/J, E/D, G/F, K/L_3, and L_1/L_2 in which the test antigens of corresponding individual strains were mixed in an equal proportion.

Commercial fluorescein conjugates were used for indirect fluorescent-antibody staining. Goat antisera to human IgM (μ-chain specific), to human IgG (γ-chain specific), and to human immunoglobulins (IgM, IgA, IgG), all labeled with fluorescein isothiocyanate, were obtained in lyophilized form from Hyland Laboratories, Los Angeles, Calif. Specificity of anti-human IgM and of anti-human IgG conjugates was reconfirmed by immunodiffusion. Rhodamine-conjugated bovine albumin used as counterstain was obtained in lyophilized form from Microbiological Associates, Inc., Bethesda, Md.

Immunotyping of Trachoma-LGV Strains

Details of immunotyping by micro-IF test have been reported (18). Many recent isolates identified in X-irradiated or 5-iodo-2′-deoxyuridine (IUdR)-treated McCoy cell culture (5) were further passed to HeLa 229 cells prior to mouse immunization. This was necessary to provide adequate antigen and to avoid immediate death of mice (apparently from reaction to infected McCoy cells). Many strains were typed in the simplified method (23), which often does not separate E from D or G from F.

Collection of Tear Specimens

Whatman no. 41 filter paper was used for collection tears. The paper strip was 5 by 20 mm^2 in size. It was autoclaved and dried in a sterilized container. The paper strip was inserted in the lower conjunctival sac and allowed to saturate with eye secretions. The strip was then eluted in a screw-capped vial containing 0.2 ml of 0.01 M phosphate-buffered saline (pH 7.0). The vial was kept at 4°C overnight and was frozen until tested. The eluate of the filter-paper strip has been shown to be approximately a 1:10 dilution as determined by the average amount (vol/wt) of tear collected.

RESULTS

Immunotypes of *C. trachomatis*

Using the micro-IF test, we have described 15 separate immunotypes of trachoma-LGV organisms (1, 19–22). They are lettered A–L, with L_1, L_2, L_3 being the three recognized LGV types. We have typed 563 strains isolated from 514 patients from around the world, some provided by other laboratories. The immunotype of these strains has been determined either by the two-way (19–22) or by the one-way simplified (23) typing test. Table 1 shows the distribution of strains according to immunotype and origin. Repeat isolates or paired isolates from the eye and the genital tract were obtained from 52 persons and were of the same immunotype, except in two cases where different types were isolated at different times.

Endemic-area ocular trachoma with its eye-to-eye transmission has been associated by isolation with types A, B, Ba, C, and D. Of these, only types A and C have not been isolated from the genital tract. However, types J and "CJ," which are closely related to C, have been isolated from the genital tract but not from endemic ocular trachoma. In non-trachoma-endemic countries, types isolated from the eye were similar to those isolated from the genital tract.

TABLE 1. *Distribution of 563 trachoma-LGV strains (derived from 514 patients) according to immunotype and origin*

Origin	Trachoma types										LGV types		
	A	B	Ba	C	D/E	F/G	H	I	J[a]	K	L_1	L_2	L_3
Ocular													
Trachoma-endemic area[b] ..	10	34	13	104	4[c]								
Trachoma-nonendemic area .		2	4	2	40	13	1	4	2	1			
Genital		17			134	91	23	16	10	16		4	1
LGV-bubo											3	11	3

[a] Two "CJ" strains are included.
[b] American Indian reservations are included.
[c] All are type D.

Complicated immunological interrelationships exist among the 15 immunotypes. Determination of immunotype relies on mouse antiserum prepared by intravenous immunization with whole organisms and collected at an early stage of antibody response.

In the simplified one-way typing test (23), the type of each strain is determined by the pattern of reaction produced by its antiserum to homologous and heterologous prototype strain antigens. In a test utilizing unknown antiserum only, it is usually not possible to separate several of the closely related types, including C/J, K/L_3, L_1/L_2, E/D, B/Ba, and G/F. For complete typing, it is necessary to utilize also the unknown strain as antigens against prototype antisera.

Serological Responses in Humans

The use of the micro-IF test with human serum or tear secretion is a one-way test similar to the simplified typing test in which the unknown antibody is tested against prototype antigens. There cannot be the assurance with human material of the specificity provided by 10-day monovalent mouse antiserum. Nevertheless, the patterns of response to the prototype antigens which have been observed are related to those which have been demonstrated in the mouse antiserum and frequently allow type-specific diagnosis. In order to increase the practicality of the test with minimal loss of immunotype specificity and sensitivity, some of the 15 types have been pooled or dropped from routine use. Those dropped from use were Ba and the three LGV antigens; C and J, E and D, and G and F antigens were pooled. The reactions to Ba antigen do not differ from those to B antigen. The LGV antigens have similarly not provided any specific diagnostic advantage. Diagnosis of LGV from serum has depended on very high-titer reactions to several antigens without higher-titer antibody to the LGV antigens. Identical reactions to C and J, E and D, and G and F are not always seen, but it is often not possible to differentiate between these pairs of antigens from serum reactions alone. Therefore, both antigens are used but they are pooled.

Figure 1 presents a schematic expression of the reaction patterns expected in human sera after infection with different immunotypes.

A type-specific diagnosis from antibody may depend upon the pattern of reaction and not only on a monotypic response. For example, with a senior infecting antigen such as type C—even in young children with first infection—antibody reaction will frequently be seen with type K and often A, H, and I. Infections with type H or I do not show antibody reaction with other types unless there has been previous infection with other types. On the basis of anti-

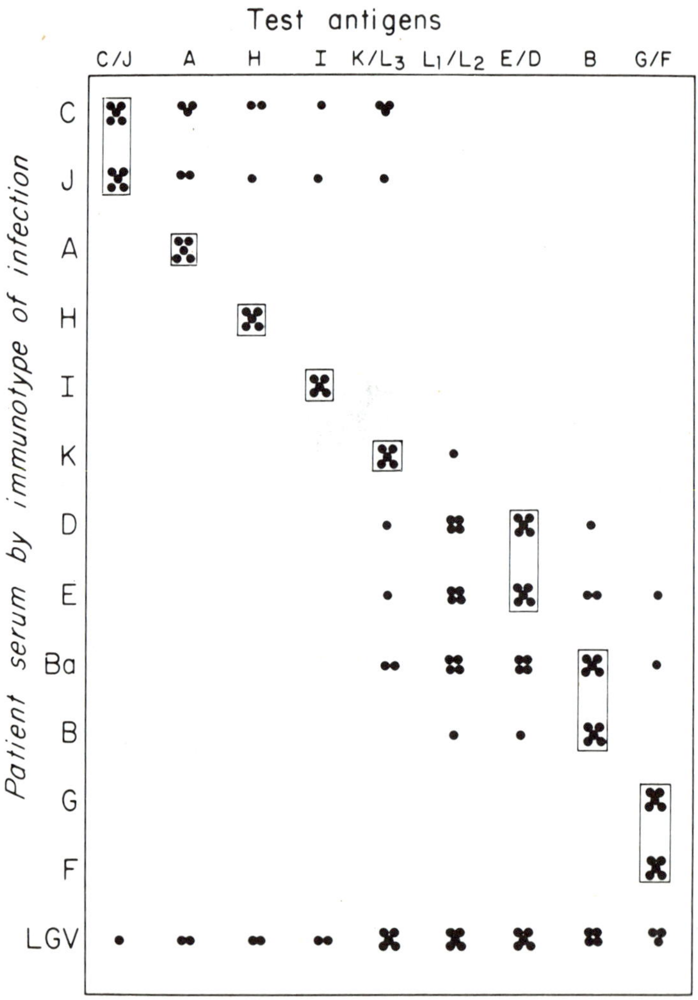

FIG. 1. *Schematic expression of reaction patterns in the micro-IF test of patient sera derived from infection with different immunotypes of C. trachomatis. Homologous reaction is expressed by five dots. Difference in number of dots shows extent of cross-reaction in terms of titer difference from homologous titer in test with twofold dilutions of serum.*

body response in human serum and tears, the types have tended to fall into two large groups, one in which type C is the senior strain and the other with type B the senior strain. The C group includes "CJ," J, A, H, I, K, and L_3. The B complex which has previously been recognized in mouse antiserum includes B, Ba, D, E, L_1, and L_2. G and F are related to the B complex but show little cross-reaction in human serum. The K/L_3 closely related strains form a bridge between the groups since all the LGV strains cross-react; although K antigen reacts readily with C antisera, it may also react with B complex antisera.

Table 2 gives examples of type-specific micro-IF antibody patterns in patient sera, with eight test antigens.

Table 3 shows the highest serum micro-IF antibody titer responses for 165 patients with either ocular or genital infection proven by isolation. The results are divided by infection with trachoma or LGV types, by whether trachoma types are ocular or genital infections, and further by sex. Table 3 shows that with genital trachoma infection females showed on average a fivefold-higher antibody response than males. Nineteen percent (11/57) of the males failed to show antibody in paired sera compared with none of the females (0/49). There was no difference between sexes for ocular infections.

All of the 12 proven cases of LGV showed high micro-IF antibody titers. Laboratory diagnosis for LGV by serology alone has not been possible on the basis of monotypic antibody reaction. However, extremely high antibody titer response to the B complex antigens or to K/L_3 with considerable heterotypic reactions is suggestive of LGV (14).

Simplification of Test for Serodiagnosis

Several developments have allowed further simplification of the micro-IF test for routine serodiagnosis: (i) The number of test antigens has been reduced from 15 to 8 as described above. (ii) HeLa 229 cell culture-grown organisms are used as antigen (Kuo, Wang, and Grayston, this volume, p. 328);

TABLE 2. *Examples of type-specific micro-IF antibody patterns in sera of patients with infection proven by isolation of C. trachomatis organisms*

Patient sex, age (years)	Isolate type/strain/origin	Antibody vs. test antigens							
		C/J	A	H	I	K	E/D	B	G/F
f, 5	C/TW-337/OT	*512*	32	128	32	32	0[a]	0	0
m, 25	J/UW-121/Ur	*64*	32	8	8	0	0	0	0
m, 28	H/DK-41/Ur	0	0	*64*	0	0	0	0	0
f, 20	I/UW-29/Cx	0	0	0	*512*	0	0	0	0
f, 21	K/UW-71/Cx	0	8	0	0	*256*	0	0	0
f, 19	E/UW-25/Cx	0	0	0	0	16	*128*	16	8
m, 29	D/UW-98/Ur	0	0	0	0	8	*64*	16	0
f, 9	Ba/DK-54/OT	0	0	0	0	128	256	*512*	0
m, 10	B/TW-334/OT	0	0	0	0	0	0	*64*	0
f, 23	B/UW-20/Cx	0	0	0	0	8	0	*128*	0
f, 17	G/DK-7/Vag	0	0	0	0	0	0	0	*1,024*
m, 35	F/UW-160/Ur	0	0	0	0	0	0	0	*64*

[a] Titer less than 8.

TABLE 3. *Antibody response by titer in the micro-IF test for 165 patients with C. trachomatis infection proven by isolation of the organisms*

	No. of patients with isolate(s)					
	Trachoma types				LGV types	
	Ocular infection		Genital infection			
Micro-IF titer[a]	Male	Female	Male	Female	Male	Female
8,192					1	
2,048				2	2	1
1,024	3	1		1	4	1
512		5	1	3	1	1
256	2	7	3	6	1	
128	5	11	5	14		
64	2	5	12	10		
32	1		11	9		
16	1	2	10	4		
8	1	1	4			
<8			11			
Total no.	15	32	57	49	9	3
Geometric mean titer	128	142	21	101		
	61				1,149	

[a] All sera tested with twofold dilutions and titrated to end point.

the previously used yolk sac-grown organisms are more difficult and time-consuming to grow and prepare for use as antigen. (iii) Use of three fourfold dilutions of serum and tears (instead of six or more twofold dilutions) has been shown to provide adequate type specificity and titer change measurement.

Demonstration of Recent Infection with Anti-IgM Conjugate

In the micro-IF test it has been shown that IgM antibody is associated with early response to infection with trachoma organisms (13, 14, 25; Bowie et al., this volume, p. 19). Two typical examples are shown in Table 4. IgM antibody appeared first, lasted for 1 to 2 months, and was replaced by IgG antibody. Concurrent tests for IgM and IgG antibodies may be done with anti-human IgM or IgG conjugate without the necessity of serum fractionation of immunoglobulins.

Table 5 shows examples of the simplified micro-IF test in which immunoglobulin specific conjugates are used with sera of nongonococcal urethritis (NGU) patients or their sexual consorts. The venereal disease clinic patients have often had more than one *C. trachomatis* infection. The usefulness of tests for IgM antibody in demonstrating (i) type specificity and (ii) recent infection is shown in these series of results. Recent infection is indicated by the presence of IgM antibody in all seven persons. In patients 2, 4, and 5, confirmatory isolation was not obtained. In patient 2, despite the early disappearance of G/F antibody and the persistence of E/D antibody (original antigenic sin [25]), recent infection with the G/F type is indicated by the IgM response.

TABLE 4. *Micro-IF antibodies in specific immunoglobulin fractions in acute and convalescent sera from patients with proven infections*

Patient age, sex	Isolation type	Date	IgMAG antibody vs.:								IgM antibody vs.:								IgG antibody vs.:							
			C	A	H	I	K	E/D	B	G/F	C	A	H	I	K	E/D	B	G/F	C	A	H	I	K	E/D	B	G/F
19, f	H	5-19-71	·[a]	·	·	·	·	·	·	·	·	·	·	·	·	·	·	·	·	·	·	·	·	·	·	·
		6-2-71	·	·	128	·	·	·	·	·	·	·	128	·	·	·	·	·	·	·	·	·	·	·	·	·
		2-22-72	·	·	128	·	·	·	8	8	·	·	·	·	·	·	·	·	·	·	64	·	·	·	·	·
18, f	D	3-17-71	·	·	·	·	·	·	·	·	·	·	·	·	·	·	·	·	·	·	·	·	·	·	·	·
		3-31-71	·	·	·	·	16	256	8	·	·	·	·	·	8	32	·	·	·	·	·	·	·	·	·	·
		5-12-71	·	·	·	·	·	32	·	·	·	·	·	·	·	·	·	·	·	·	·	·	·	16	·	·

[a] Titer less than 8.

TABLE 5. *Examples of the results of the simplified micro-IF test with sera of male NGU patients and sexual contacts*

Patient[a] (sex)	Isolation (type)	Date	IgMAG antibody vs.:								IgM antibody vs.:								IgG antibody vs.:							
			C/J	A	H	I	K	E/D	B	G/F	C/J	A	H	I	K	E/D	B	G/F	C/J	A	H	I	K	E/D	B	G/F
1. (m)	+(I)	6-12-75	·[b]	·	·	·	·	·	·	·																
	–	6-24-75	·	·	·	32	·	·	·	·	·	·	·	32	·	·	·	·	·	·	·	32	·	·	·	·
2. (m)	NT	5-22-73	·	·	·	·	8	128	32	·	·	·	·	·	·	·	·	·	·	·	·	·	32	128	32	·
		6-25-73	·	·	·	·	8	32	32	32	·	·	·	·	·	·	·	32	·	·	·	·	8	128	32	32
		7-17-73	·	·	·	·	8	32	8	8	·	·	·	·	·	·	·	·	·	·	·	·	8	32	32	·
		12-29-73	·	·	·	·	·	32	·	·	·	·	·	·	·	·	·	·	·	·	·	·	8	32	8	·
3. (m)	+(GF)	12-16-74	8	·	·	·	8	32	32	128	·	·	·	·	·	·	·	32	8	·	·	·	8	32	32	32
4. (m)	–	4-20-72	128	·	·	·	·	·	·	·	8	·	·	·	·	·	·	·	128	·	·	·	·	·	·	·
	–	5-18-72	128	·	·	·	·	·	·	·																
5. (f)	–	4-26-72	128	8	·	8	32	·	·	·	8	·	·	·	·	·	·	·	128	8	·	8	32	·	·	·
	–	5-10-72	128	8	·	8	32																			
6. (m)	+(F)	12-21-72	·	·	·	·	·	·	·	32	·	·	·	·	·	·	·	8	·	·	·	·	·	·	·	32
	+(D)	6-28-73	·	·	·	·	32	32	32	32	·	·	·	·	·	8	·	·	·	·	·	·	32	32	32	32
7. (f)	+(F)	12-27-72	·	·	·	·	·	·	·	128	·	·	·	·	·	·	·	128	·	·	·	·	·	·	·	128

[a] Bracketed numbers indicate patients who were sexual consorts.
[b] Titer less than 8.

A similar IgM response occurred in patient 3, from whom type G/F organism was isolated. In patient 6 the IgM antibody response and the sequential isolation of two types mutually confirm two separate infections.

Fortuitously timed paired sera showing fourfold trachoma antibody titer rises are infrequently obtained in venereal disease. The use of tests for IgM antibody increases the opportunity for serological evidence of recent infection.

Table 6 compares micro-IF serological results and isolation of the organisms for diagnosis of trachoma infection in first-episode NGU patients reported by Bowie et al. (this volume, p. 19). There were 59 male patients from whom repeat specimens of urethral swabs and serum were collected at an interval of 10 to 14 days. The isolation and serology were carried out independently in different laboratories. The simplified micro-IF test was applied to screen all sera for antibody. Sera with antibody were tested for IgM and IgG antibodies. Of the 59 patients, 22 had antibody which was shown to be present in the IgM fraction and/or to demonstrate fourfold or greater titer rise. Isolation was obtained from 19 of these 22 patients. No antibody was demonstrated in 32 of the patients, and isolation was also negative. Antibody not present in the IgM fraction and without titer rise was found in five patients, one of whom was isolation positive. Thus recent infection, demonstrated serologically, was confirmed by isolation in 19 of 22 patients, and infection demonstrated by isolation was confirmed to be recent serologically in 19 of 20.

Use of Tears to Demonstrate Trachoma Antibody

Ocular infection with *C. trachomatis* remains a major cause of human disease. We have found that demonstration of tear micro-IF antibody is a practical method of diagnosis because of its simplicity and reproducibility and because of the stability of antibody. Table 7 shows examples of antibodies in tears from selected subjects. Tears are collected on filter-paper strips (25; Grayston et al., this volume, p. 113). The simplified micro-IF test was used with three fourfold dilutions starting with the filter-paper eluate, which is approximately a 1:10 dilution of tears. The first three examples are from the family trachoma studies on Taiwan reported by Grayston et al. (this volume, p. 113). The next two examples are from cases of eye infection occurring in Florida. These specimens

TABLE 6. *Laboratory diagnosis of C. trachomatis infection in first-episode male NGU patients: comparison of the results of micro-IF serology and isolation of the organism*

Micro-IF test			Isolation[a]		Total
Antibody	Titer rise	IgM	+	−	
+	+	+	9	2	11 } 22
	+	−	3	0	3 }
	−	+	7	1	8 }
	−	−	1	4	5
−			0	32	32
Total			20	39	59

[a] Isolation utilizing IUdR-treated McCoy cells.

TABLE 7. *Examples of C. trachomatis micro-IF antibodies in the eluate of tear-saturated filter-paper strips*

Patient sex, age, area	Eye disease diagnosis	Isolation type	Date	IgMAG antibody titer vs. immunotype: C/J	A	H	I	K	E/D	B	G/F
m, 7, Taiwan	Trachoma	+(D)	1-10-75	·[a]	·	·	·	10	160	40	·
f, 17, Taiwan	Trachoma	+(C)	1-10-75	160	10	·	·	40	·	·	·
f, 8, Taiwan	Trachoma	+(B)	1-9-73	·	·	·	·	·	·	40	·
m, 19, Florida	Trachoma	NT[b]	9-23-73	·	·	·	·	·	10	40	·
f, 20, Florida	Trachoma	NT	4-8-74	·	·	·	·	·	·	·	40
m, NB,[c] Seattle	Normal	0	8-8-75	·	·	·	·	·	·	·	·
(born 6-24-75)		0	11-5-75	·	·	·	·	·	40	·	·
		0	12-10-75	·	·	·	·	·	40	·	·
f, NB, Seattle,	"Sticky eye"	0	8-12-75	·	·	·	·	·	·	·	·
(born 7-29-75)		0	9-10-75	·	·	·	·	·	·	·	·
		0	10-20-75	·	·	·	·	·	·	·	40
		0	12-11-75	·	·	·	·	·	·	·	10

[a] Negative with undiluted eluate, which is approximately 1:10 dilution of original tear.
[b] Not tested.
[c] Newborn from mother with positive isolation from the cervix.

were exposed to room temperature for 2 to 3 weeks before arriving at our laboratory. Serum specimens were tested in both cases, and the results were similar to the patterns of tear antibody. The last two cases were from the prospective studies of neonatal infection reported by Alexander et al. (this volume, p. 148). Both babies were born from mothers with isolation-proven infection in the cervix. Isolation attempts were negative. Asymptomatic infection of the eye was suggested by the appearance of type-specific antibody in tears 3 to 5 months after birth.

Serological Epidemiology

Table 8 demonstrates the usefulness of the simplified micro-IF screening test in serological surveys. All the children had some relationship to respiratory disease. They were from the pneumonia studies of Foy et al. (26) and the virus watch studies of Fox et al. (27). The adult group included normal (no disease) persons and patients attending venereal disease clinics and their sexual consorts. About one-quarter of the adult population with antibodies were found to have antibodies against multiple immunotypes. E/D was the most frequent type, followed by G/F. This distribution corresponds closely to the type distribution of genital isolates described above (Table 1). Ten percent of the children had antibody. The immunotype distribution was different from either group of adults, with the C group (C/J, A, H, I, and K) predominating.

DISCUSSION

From 300 *C. trachomatis* strains typed in the past 5 years, we have found none that did not fit immunologically with earlier strains. This suggests that most

TABLE 8. *Results of trachoma micro-IF screening test by antibody type with sera from several Seattle population groups*

Group	Total no. tested	No. with antibody type: C/J	A	H	I	K	B	E/D	G/F	Multiple	No. with antibody
Children (<15 yrs)	210	1	5	3	4	3	1	2	1	2	22 (10%)
Adults											
Normal	346	6		4	5	2	8	32	11	21	89 (26%)
Venereal disease and contacts ..	788	24	1	15	17	22	11	202	81	125	498 (63%)

immunotypes have been described. With the further simplification of the micro-IF test described here, it complements cell culture isolation in practicality for diagnosis of *C. trachomatis* infection. The test for IgM antibody is particularly useful in sexually active adults where antibodies to multiple immunotypes may be found in routine tests. Tear specimens are easier to obtain in areas with endemic trachoma, and tear antibodies are more closely related than serum antibody to active disease (Grayston et al., this volume, p. 113). The finding of trachoma antibodies in young children with pneumonia and the respiratory diseases requires further study to see whether the antibodies are associated with respiratory disease or with eye or genital tract infection.

ACKNOWLEDGMENTS

This study was supported by U.S. Public Health Service research grants 5-RO1-EY-00219 from the National Eye Institute and AI-12192 from the National Institute of Allergy and Infectious Diseases.

REFERENCES

1. Grayston JT, Wang SP: New knowledge of chlamydiae and the diseases they cause. J Infect Dis 132:87–105, 1975
2. Holmes KK, Handsfield HH, Wang SP, et al: Etiology of nongonococcal urethritis. N Engl J Med 292:1199–1205, 1975
3. Gordon FB, Harper AL, Quan AL, et al: Detection of Chlamydia (Bedsonia) in certain infections of man. I. Laboratory procedures: comparison of yolk sac and cell culture for detection and isolation. J Infect Dis 120:451–462, 1969
4. Kuo CC, Wang SP, Wentworth BB, et al: Primary isolation of TRIC organisms in HeLa 229 cells treated with DEAE-dextran. J Infect Dis 125:665–668, 1972
5. Wentworth BB, Alexander ER: Isolation of *Chlamydia trachomatis* by use of 5-iodo-2-deoxyuridine-treated cells. Appl Microbiol 27:912–916, 1974
6. Blyth WA, Taverne J: Cultivation of TRIC agents: a comparison between the use of BHK-21 and irradiated McCoy cells. J Hyg 72:121–128, 1974
7. Hobson D, Johnson FWA, Rees E, et al: Simplified method for diagnosis of genital and ocular infections with Chlamydia. Lancet 2:555–556, 1974
8. Croy, TR, Kuo CC, Wang SP: Comparative susceptibility of eleven mammalian cell lines to infection with trachoma organisms. J Clin Microbiol 1:434–439, 1975
9. Nichols RL, Bobb AA, Haddad NA, et al: Immunofluorescent studies of microbiologic epidemiology of trachoma in Saudi Arabia. Am J Ophthalmol 63:1372–1408, 1967

10. McComb DE, Nichols RL: Antibody type specificity to trachoma in eye secretions of Saudi Arab children. Infect Immun 2:65–68, 1970
11. Hanna L, Jawetz E, Nabli B, et al: Titration and typing of serum antibodies in TRIC infections by immunofluorescence. J Immunol 108:102–107, 1972
12. Dwyer RSC, Treharne JD, Jones BR, et al: Results of microimmunofluorescence tests for detection of type-specific antibody in certain chlamydial infections. Br J Vener Dis 48:452–459, 1972
13. Philip RN, Casper EA, Gordon FB, Quan AL: Fluorescent antibody responses to chlamydial infection in patients with lymphogranuloma venereum and urethritis. J Immunol 112:2126–2134, 1974
14. Wang SP, Grayston JT: Human serology in *Chlamydia trachomatis* infection with microimmunofluorescence. J Infect Dis 130:388–397, 1974
15. Wang SP, Grayston JT, Alexander ER, et al: Simplified microimmunofluorescence test with trachoma-lymphogranuloma venereum (*Chlamydia trachomatis*) antigens for use as screening test for antibody. J Clin Microbiol 1:250–255, 1975
16. Wang SP, Grayston JT: Immunologic relationship between genital TRIC, lymphogranuloma venereum, and related organisms in a new microtiter indirect immunofluorescence test. Am J Ophthalmol 70:367–374, 1970
17. Wang SP: A microimmunofluorescence method. Study of antibody response to TRIC organisms in mice, p 273–388. *In* RL Nichols (ed), Trachoma and related disorders caused by chlamydial agents. Excerpta Medica, Amsterdam, 1971
18. Alexander ER, Wang SP, Grayston JT: Further classification of TRIC strains from ocular trachoma and other sources by the mouse toxicity prevention test. Am J Ophthalmol 63:1469–1478, 1967
19. Wang SP, Grayston JT: Classification of TRIC and related strains with microimmunofluorescence, p 305–321. *In* RL Nichols (ed), Trachoma and related disorders caused by chlamydial agents. Excerpta Medica, Amsterdam, 1971
20. Wang SP, Grayston JT, Gale JL: Three new immunologic types of trachoma-inclusion conjunctivitis organisms. J Immunol 110:873–879, 1973
21. Kuo CC, Wang SP, Grayston JT, et al: TRIC type K, a new immunologic type of *Chlamydia trachomatis*. J Immunol 113:591–596, 1974
22. Wang SP, Grayston JT: *Chlamydia trachomatis* immunotype J. J Immunol 115: 1711–1716, 1975
23. Wang SP, Kuo CC, Grayston JT: A simplified method for immunological typing of trachoma-inclusion conjunctivitis-lymphogranuloma venereum organisms. Infect Immun 7:356–360, 1973
24. Philip RN, Casper EA, Ormsbee RA, et al: Microimmunofluorescence test for the serological study of Rocky Mountain spotted fever and typhus. J Clin Microbiol 3:51–61, 1976
25. Wang SP, Grayston JT: Local and systemic antibody response to trachoma eye infection in monkeys, p 217–232. *In* RL Nichols (ed), Trachoma and related disorders caused by chlamydial agents. Excerpta Medica, Amsterdam, 1971
26. Foy HM, Grayston JT, Kenny GE, et al: Epidemiology of *Mycoplasma pneumoniae* infection in families. J Am Med Assoc 197:859–866, 1966
27. Hall CE, Cooney MK, Fox JP: The virus watch program. I. Infection and illness experience of virus watch families during a community wide epidemic of echovirus type 30 aseptic meningitis. Am J Public Health 60:1456–1465, 1970

Serological Responses to Chlamydial Ocular and Genital Infections in the United Kingdom and Middle East

JOHN D. TREHARNE, R. J. DINES,
AND S. DAROUGAR

The micro-immunofluorescence (micro-IF) test has divided subgroup A *Chlamydiae* (1), *Chlamydia trachomatis* agents (2), into a number of distinct serotypes (3, 4). By use of this classification, type-specific antibodies have been demonstrated in sera from patients with chlamydial infection (5–11). This test, however, is both time consuming and expensive, and therefore a simple and sensitive modification is described.

The modified serodiagnostic test in which four antigen pools are used is only slightly less sensitive than the full test in which up to 14 antigens are used (Treharne et al., J. Clin. Pathol., in press). Furthermore, this division into four antigen pools corresponds closely with the clinical and epidemiological types of chlamydial infection (12) and requires only a simple additional titration for the determination of type-specific antibody to individual serotypes.

This paper reports tests with the modified micro-IF test on 531 sera and tears from patients with a wide variety of chlamydial infections. In addition, it attempts to correlate the presence, the titer, and in some cases the immunoglobulin class of antichlamydial antibody with the presence of chlamydial agent or active clinical disease in the patient.

MATERIALS AND METHODS

Chlamydial antigens. The following antigens were used.

1. Hyperendemic trachoma isolates of *C. trachomatis*: A/HAR-1/OT, B/TW-5/OT, C/UW-1/OT.
2. Paratrachoma isolates of *C. trachomatis*: D/IC-Cal-8/ON, E/DK-20/ON, F/MRC-301/GU, G/IOL-238/R, H/UW-4/GCx, I/UW-12/GU, J/UW-36/GCx, K/UW-31/GCx.
3. Lymphogranuloma venereum (LGV) isolates of *C. trachomatis*: L1/440/LN, L2/434/Bu, L3/404/LN.
4. Selected *Chlamydia* subgroup B isolates of *C. psittaci*: 33-L, isolated from a patient with LGV (13); IOL-395/0, isolated from a patient with follicular conjunctivitis; A-22/18, isolated from a sheep with enzootic abortion; A-10, isolated from the eye of a guinea pig with inclusion conjunctivitis (14).

In addition to using these representative serotypes as individual antigens, sera and tears were initially screened against the four antigen pools listed in Table 1.

Preparation of antigens. Antigens were prepared from the yolk sacs of live chick embryos as previously described (15). Pooled antigens were prepared as described by Treharne et al. (in press).

TABLE 1. *System of pooled chlamydial antigens used for screening sera or tear fluids*

Pool	Antigens
1	Hyperendemic trachoma TRIC agent serotypes A, B, and C
2	Paratrachoma TRIC agent serotypes D, E, F, G, H, I, and K
3	Lymphogranuloma venereum serotypes L1, L2, and L3
4	*C. psittaci* isolates 33-L, IOL-395/0, A-10, and A-22/18

Sera. All sera were absorbed before use with an equal volume of 40% (wt/vol) uninfected yolk sac.

Tears. Fluid was collected from the conjunctival sac of patients by use of a dry cellulose sponge measuring approximately 5 by 2 by 1 mm. When saturated, this sponge absorbed approximately 0.05 ml of fluid. After collection, the sponge was transferred to a tightly stoppered capsule and stored at −180°C until processed. These tears were not absorbed with normal yolk sac before titration.

Micro-IF test. Initially, all sera and tears were titrated in screening dilutions of 1:8, 1:32, and 1:128 against the four antigen pools. Subsequently, complete titrations were carried out against single selected antigens of those sera or tears shown to be positive by the screening titration. Certain sera were tested with a goat anti-human IgM (μ-chain specific) fluorescein-labeled conjugate (Hyland Division, Travenol Laboratories Inc.). All other sera and tears were tested with a goat anti-human immunoglobulin fluorescein-labeled conjugate (Hyland Division, Travenol Laboratories Inc.). The other technical procedures for the micro-IF test have been described previously (6, 15).

RESULTS

Results obtained by using the micro-IF test with the complete range of antigens to detect type-specific antibody in sera of 2,262 patients are given in Table 2. The results shown are those in which sera gave clear-cut reactions to a single antigen only, that is to say, at least a two- to fourfold-higher titer to a single antigen than to

TABLE 2. *Geographical distribution and clinical origin of serological responses of 2,262 patients with proven or suspected C. trachomatis infections*

Origin	Micro-IF type-specific antibody detected												
	A	B	C	D/E	F	G	H	I	J	K	L1	L2	L3
Geographical area													
N. Africa: Tunisia, Egypt	71	12	7	1	—	—	—	—	—	—	—	—	—
Asia: Iran, Afganistan	210	574	127	3	1	3	2	—	—	1	—	—	—
Europe: United Kingdom	2	3	3	840	132	154	6	3	9	86	2	8	2
Clinical site and type of disease													
1. Ocular infections													
a. Hyperendemic trachoma	281	586	134	2	—	—	—	—	—	—	—	—	—
b. Paratrachoma	1	1	—	273	31	26	1	1	—	2	—	—	—
2. Genital or lymph node infections													
a. Paratrachoma	1	2	3	569	102	131	7	2	9	85	—	—	—
b. LGV	—	—	—	—	—	—	—	—	—	—	2	8	2
Totals	283	589	137	844	133	157	8	3	9	87	2	8	2

any other cross-reacting antigen. The majority of sera from hyperendemic trachoma in North Africa or Asia were shown to possess type-specific antibody to the A, B, or C trachoma-inclusion conjunctivitis (TRIC) serotypes. Conversely, sera from patients with ocular or genital infections of paratrachoma in Europe and the United Kingdom generally possess type-specific antibody to TRIC types D, E, F, G, H, I, J, and K. All sera from infections clinically diagnosed as LGV contained high levels of type-specific antibody directed to one of the three LGV serotypes.

Table 3 shows the mean percent reactions of human sera from serotyped isolation-positive chlamydial infections against the four antigen pools listed in Table 1. It can be seen that in order to determine the type-specific antibody present in any individual serum, further titrations would be required against only the limited selections of single-serotype antigens listed in the right-hand column. These pooled antigens were used to screen sera and tears from patients suffering from a wide variety of diseases, some of which were proven chlamydial infections. Type-specific antibody was eventually determined by use of the whole range of individual micro-IF antigens.

Table 4 shows the comparison of titers of antibody in sera and tears of school children with hyperendemic trachoma in Southern Tunisia. It is evident that higher titers occur in those patients from whom *C. trachomatis* is isolated. However, 70% of isolation-negative patients have serum antibody, whereas only 20% of isolation-negative patients have antibody levels of 1:8 or greater in their tears.

TABLE 3. *Percent cross-reactions, with antigen pools, of 98 sera from patients with typed chlamydial isolates*

Type of isolate	No. of sera	Subgroup A *Chlamydia* (*C. trachomatis*)			Subgroup B *Chlamydia*, pool 4[d]	Single antigens[e]
		Pool 1[a]	Pool 2[b]	Pool 3[c]		
A	11	100				(A, C)
B	8	100	6–13	6–13		—
C	10	100				(A, C)
D	10	0–6	100	6–25		(D, E, F, G)
E	12	0–13	100	13–25		(D, E, F, G)
F	6		100	6–25		(D, E, F, G)
G	8		100	13–50		(D, E, F, G)
H	5		100			(H, I)
I	3	0–6	100			(H, I)
K	4		100	50–100		—
L1	2		0–13	100		(L1, L3)
L2	7	0–13	25–50	100		—
L3	2		0–25	100		(L1, L2)
Subgroup B	10				100	Not known

[a] Hyperendemic trachoma serotypes.
[b] Paratrachoma serotypes.
[c] Lymphogranuloma venereum serotypes.
[d] *C. psittaci*-type isolates.
[e] Single antigens against which further titrations would be necessary to define the infecting serotypes.

TABLE 4. *Micro-IF antibody titers in tears and sera of Tunisian schoolchildren*

Specimen	No. of patients	Cell culture isolation	Geometric mean titer	Percent positive[a]
Sera	21	+	1:40	90
Sera	73	−	1:9	70
Tears	20	+	1:13	75
Tears	51	−	1:2	20

[a] Patients with titers of 1:8 or greater.

In Table 5, geometric mean titers of antibody in sera and tears are compared with the clinical intensity of conjunctival inflammatory disease in these Tunisian schoolchildren. A good correlation of high antibody levels in tears (1:43) in a high percentage of patients (70%) with severe imflammatory disease is demonstrated. These antibody levels diminish to levels of 1:4 in only 17% of patients when the disease becomes less active. Antibody in sera, however, can be detected in a high percentage of patients (81%) and at fairly high levels (1:25), even in the mild stage of the disease.

High levels of antibody (1:93) were detected in all of 15 isolation-positive patients presenting with ocular paratrachoma in the United Kingdom. A further 10 patients with isolation-positive paratrachoma due to *C. trachomatis* were followed for 6 months (Table 6). The levels of antichlamydial IgM antibody in the sera, and the percentage of patients in which this immunoglobulin was detected, correlated well with the presence of agent in the conjunctiva. Levels of antichlamydial IgG remained high in all patients, even in the absence of detectable *C. trachomatis* infection in the conjunctiva.

Sera from parents of 16 babies with ophthalmia neonatorum due to TRIC agent were tested for the presence of type-specific antibodies to *Chlamydia* (Table 7). All 16 mothers and one female consort of the father of family number 14 had antibody to *Chlamydia*. Of the 17 women, 16 showed highest levels of antibody to the paratrachoma types; the mother of family 16 had the highest level of type-specific antibody to serotype B of the hyperendemic trachoma group. The mother in family 3 developed equal levels of type-specific antibody to both type C (hyperendemic) and to type J (paratrachoma). Only 9 of the 16 fathers (56%) developed antibody to *Chlamydia*. In each of these 9, the type specificity of the antibody was the same as that found in the mother.

TABLE 5. *Comparison of geometric mean antibody titers (G.M.T.) in sera and tears with the activity of clinical disease in Tunisian school children with trachoma*

Clinical activity	Positive sera[a]		Positive tears[a]	
	G.M.T.	Percent	G.M.T.	Percent
Severe	1:51	82	1:43	70
Moderate	1:37	88	1:12	31
Mild	1:25	81	1:4	17

[a] Patients with titers of 1:8 or greater.

TABLE 6. *Geometric mean titers of antichlamydial IgG and IgM antibody in 10 ocular paratrachoma patients in the United Kingdom from whom C. trachomatis was isolated*

Days (approx) after presenting	Antibody	Geometric mean titer[a]	Patients isolation positive[a] (%)	Patients with positive sera (%)
0	IgM	1:19	100	60
	IgG	1:32	100	80
30	IgM	1:6	20	40
	IgG	1:223	20	100
90	IgM	1:2	0	20
	IgG	1:181	0	100
180	IgM	0	0	0
	IgG	1:147	0	100

[a] Patients with titers of 1:8 or greater.

Table 8 shows the geometric mean titers of patients in the United Kingdom presenting with genital infections. Those patients with proven *C. trachomatis* isolation-positive infections showed good levels of antichlamydial antibody, the highest levels (1:66) being found in women with cervicitis. The percentage of isolation-positive patients with antibody levels of 1:8 or greater was also highest in women with cervicitis (77%). Only very low levels of antibody, in a small percentage of women, could be detected in the absence of *C. trachomatis*. Of 11 men with postgonococcal urethritis, 8 had antichlamydial IgG antibody and also had IgM antichlamydial antibody in their first serum specimen. Geometric mean titers of antichlamydial antibody in men with *C. trachomatis* isolation-positive postgonococcal urethritis and men with nonspecific urethritis were very similar.

Figure 1 summarizes the results obtained with the pooled antigen system on 531 sera from persons with proven or suspected chlamydial infection. In hyperendemic trachoma, the titers of sera and tears were significantly higher in isolation-positive cases than in those from whom *Chlamydia* was not isolated. In all cases, sera reacted most strongly with the hyperendemic trachoma antigen pool; this was subsequently confirmed by titration against single antigens.

Relatively high antibody levels (1:114), predominantly to paratrachoma antigens, were detected in all mothers of babies with ophthalmia neonatorum due to TRIC agent. Fewer fathers had detectable levels of antibody (56%), and, although the type specificity of antibody paralleled that of the mothers, the geometric mean titer was much lower.

Sera from patients presenting in the United Kingdom with ocular infection due to *Chlamydia* showed high geometric mean titers (1:93) to the paratrachoma antigen pool.

In men with *Chlamydia* isolation-positive, nonspecific urethritis, antibody to *Chlamydia* could be detected in only 54% (19 of 35); the geometric mean titer was 1:37 in those 19 men who developed antibody, and these sera reacted most strongly with the pooled paratrachoma antigens.

TABLE 7. *Micro-IF reciprocal titers to the pooled antigens and to homologous single chlamydial antigens, with sera from 16 parents of babies with TRIC isolation-positive ophthalmia neonatorum*

Family no.	Parental status	TRIC agent isolation	Antigen pool 1 (A, B, C)	Antigen pool 2 (D, E, F, G, H, I, K)	Antigen pool 3 (L1, L2, L3)	Antigen pool 4 (*C. psittaci* isolates)	Single-antigen titration (highest titer)	Micro-IF type reaction pattern
1	Mother	+	16	128	32	—	256	G
	Father	+	—	16	16	—	16	G
2	Mother	+	8	256	64	—	256	D–E
	Father	+	—	64	16	—	64	D–E
3	Mother	+	512	256	256	8	512	C–J
	Father	–	—	—	—	—	—	—
4	Mother	+	8	32	32	—	32	K
	Father	–	—	—	—	—	—	—
5	Mother	+	—	32	32	—	32	F
	Father	–	—	16	16	—	8	F
6	Mother	+	—	64	64	—	64	K–L3
	Father	–	—	—	—	—	—	—
7	Mother	+	—	128	128	8	128	F
	Father	–	—	32	32	8	64	F
8	Mother	+	16	64	64	—	128	K
	Father	+	8	128	128	—	256	K
9	Mother	–	—	16	16	—	16	K–L3
	Father	–	—	64	64	—	32	K–L3
10	Mother	+	—	128	128	8	256	D–E
	Father	–	—	16	16	—	8	D–E
11	Mother	+	64	256	128	8	256	D–E
	Father	–	—	8	8	—	8	D–E
12	Mother	+	—	32	16	—	32	D–E
	Father	–	—	8	—	—	8	—
13	Mother	+	8	128	128	—	256	D–E
	Father	–	—	—	—	—	—	—
14	Mother	+	—	128	32	—	128	F
	Female consort	–	—	64	32	—	128	F
	Father	+	—	—	—	—	—	—
15	Mother	–	—	128	32	—	64	K–L3
	Father	–	—	—	—	—	—	—
16	Mother	–	256	32	16	—	256	B
	Father	–	128	32	16	—	64	B

Eight sera from isolation-positive cases of lymphogranuloma venereum from the United Kingdom and the United States developed high geometric mean antibody levels (1:512) to chlamydial antigens in the LGV pool. These sera, however, showed strong cross-reactions with the paratrachoma antigen pool.

TABLE 8. *Micro-IF serological responses in genital infections in the United Kingdom*

Category	Sex	*C. trachomatis* isolated	No. of patients	Geometric mean titer[a]	Positive patients[a] (%)	Type-specific antibody found
Cervicitis	F	+	117	1:66	77	D/E, F, G, H, J, and K
	F	−	79	1:2	14	D/E, F, G, and K
Nonspecific urethritis	M	+	35	1:37	51	D/E, F, and K
	M	−	30	1:2	10	D/E and K
Postgonococcal urethritis	M	+	25	1:32	44	D/E, F, G, H, and K
	M	−	56	1:2	8	D/E, F, G, I, and K/L3

[a] Patients with a 1:8 titer or greater.

Finally, all 30 sera from patients with clinically diagnosed psittacosis developed high levels of antibody (geometric mean titer, 1:256) to the selected antigens in the subgroup B pool, with no cross-reactions at a 1:8 level with the other antigen pools.

DISCUSSION

The detection of type-specific antibodies to chlamydial infections by use of the micro-IF test has been shown by many workers to be a practicable technique (5–11). To date, however, some 12 TRIC agent serotypes and 3 LGV serotypes have been described; hence, the work required to titrate, or even screen, sera against such a range of antigens is time consuming and thus limits the usefulness of the existing test in large-scale serodiagnosis.

Recently, a micro-IF test was described which is capable not only of detecting type-specific antibody to subgroup A chlamydial infections but also, by use of appropriate antigens, of clearly distinguishing subgroup A chlamydial infections from subgroup B (*C. psittaci*) infections (8). A further modification of the micro-IF test in which three antigen pools of *C. trachomatis* and one of *C. psittaci* are used has been described by Treharne et al. (J. Clin. Pathol., in press). The present study indicates that this modified test provides a practicable, routine, and simplified serodiagnostic test which offers considerable savings in time, material, and costs.

Levels of antibody from sera and tears of school children with hyperendemic trachoma in Southern Tunisia correlated well with the isolation of *C. trachomatis* from the conjunctiva, the highest levels being found in those patients who were isolation positive. It would seem, however, that measurable levels of serum antibody persist in the absence of clinical or microbiological signs of infection. The presence of tear antibody, on the other hand, appears closely related to the presence of agent in the conjunctiva and to the inflammatory activity of the clinical disease and thus may provide a meaningful criterion of present or recent ocular infection due to *C. trachomatis*.

High levels of antichlamydial antibody were found in the sera of patients with ocular paratrachoma in the United Kingdom (1:93). These levels are higher than the levels we have found in hyperendemic ocular trachoma (1:40), probably because 80% of the females and approximately 50% of the males with ocular infection in the United Kingdom have concurrent genital infections due to *C. trachomatis* (16), thus providing an additional antigenic stimulation.

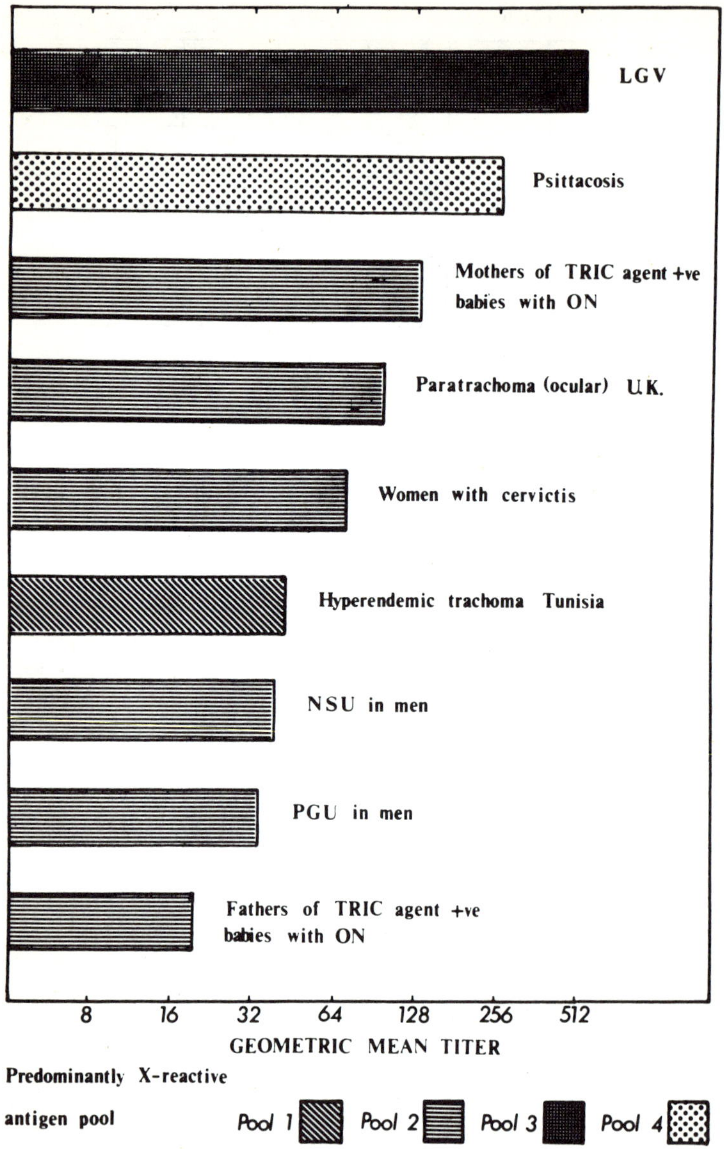

FIG. 1 *Predominant antibody type and geometric mean titers of Chlamydia isolation-positive patients determined by the modified micro-IF test. NSU = nonspecific urethritis; PGU = postgonococcal urethritis; ON = ophthalmia neonatorum.*

It has been suggested, on the evidence of experimental animal infections, that IgM-specific antichlamydial antibodies appear early in chlamydial infections, are transient, and are not recalled unless the reinfection is with a heterologous serotype (10). Thus, their presence may indicate a recent primary infection or a reinfection with a heterologous serotype of *C. trachomatis*. We were able to demonstrate IgM-specific antichlamydial antibody in the first serum specimen of 6 of 10 patients with ocular paratrachoma in the United Kingdom. The presence of this antibody was only short-lived, whereas a greater than fourfold rise in the titer demonstrated for IgG-specific antichlamydial antibody persisted at a high level for 6 months in the absence of *C. trachomatis*. This suggests that the demonstration of IgM-specific antichlamydial antibody may be a useful adjunct in the serodiagnosis of active chlamydial infections (9).

IgM-specific antichlamydial antibody was also demonstrated in the first serum specimen of 8 of 11 men presenting with postgonococcal urethritis. This IgM antibody was less cross-reactive than IgG-specific antichlamydial antibody, although titers were generally lower and the antichlamydial type specificity of antibody produced in individual patients was always the same, irrespective of the specific immunoglobulin class.

The type specificity of antibody detected in both mothers and fathers of babies with ophthalmia neonatorum also correlated well, although the mothers generally had much higher levels of antibody than the fathers. It was predictable that higher antibody levels would be found in the mothers as they all must have had genital infections with *C. trachomatis* at the time the babies were born. In fact, 13 of 16 mothers were isolation positive, whereas *C. trachomatis* could be isolated from only 4 of 16 fathers. Furthermore, in addition to having a larger area of infection, women often remain asymptomatic for long periods of time, thus providing a larger and longer antigenic stimulation than usually occurs in males, and this leads to higher antibody levels.

In chlamydial isolation-positive patients the lowest levels of type-specific serum antibody were found in men presenting with genital infections. Isolation-positive patients with hyperendemic trachoma in Southern Tunisia had slightly higher levels (1:40) in their serum. Women presenting with cervicitis, mothers of babies with ophthalmia neonatorum due to TRIC agent, and adult patients in the United Kingdom presenting with ocular paratrachoma infection due to TRIC agent gave the highest levels of antichlamydial antibody to *C. trachomatis* serotypes. The titer of chlamydial antibody does appear, however, to correlate with the nature of the infection: highest levels of antibody are produced by the more invasive LGV and *C. psittaci* organisms.

Antibody could be detected in chlamydial isolation-positive patients in most instances. The main exceptions were in isolation-positive males with nonspecific or postgonococcal urethritis and in fathers of babies with TRIC agent ophthalmia neonatorum. This may suggest that in these men infection was mild, or of short duration, thus providing little antigenic stimulus. In common with others (6, 11), we have noted more broadly cross-reactive sera in LGV patients, although some broadly cross-reactive sera have come from patients with infections due to paratrachoma serotypes. A fourfold or higher titer of antibody to any individual serotype has been accepted as a positive indication of infection by that serotype, and this has been borne out by numerous replicate titrations on single serum specimens from *Chlamydia*-positive patients where isolates have been serotyped.

ACKNOWLEDGMENTS

We acknowledge the encouragement and advice of Barrie R. Jones throughout the course of this study. We also thank in particular the many clinicians from Moorfields Eye Hospital, the

Whitechapel Clinic, the London Hospital, and other hospitals for their help in the collection of sera and tears examined in this study.

This work was supported by a grant from the Department of Health and Social Security and by grants from the Wellcome Trust, the World Health Organization, and an anonymous donor.

REFERENCES

1. Gordon FB, Quan AL: Occurrence of glycogen in inclusions of the psittacosis-lymphogranuloma venereum-trachoma agents. J Infect Dis 115:186–196, 1965
2. Page LA: Proposal for the recognition of two species in the genus *Chlamydia*. Int J Syst Bacteriol 18:51–66, 1968
3. Wang SP, Grayston JT: Immunologic relationship between genital TRIC, lymphogranuloma venereum and related organisms in a new microtiter indirect immunofluorescence test. Am J Ophthalmol 70:367–375, 1970
4. Wang SP, Grayston JT, Gale JL: Three new types of trachoma-inclusion conjunctivitis organisms. J Immunol 110:873–879, 1973
5. Briones OC, Hanna L, Jawetz E, et al: Type-specific antibodies in human Chlamydia trachomatis infections of the eye. J Immunol 113:1262–1270, 1974
6. Dwyer RStC, Treharne JD, Jones BR, et al: Results of micro-immunofluorescence tests for the detection of type-specific antibody in certain chlamydial infections. Br J Vener Dis 48:452–459, 1972
7. Hanna L, Jawetz E, Nabli B, et al: Titration and typing of serum antibodies in TRIC infections by immunofluorescence. J Immunol 108:102–107, 1972
8. Jones BR, Treharne JD: Micro-immunofluorescence type-specific serological tests for chlamydial infection. Proc R Soc Med 67:735–736, 1974
9. Philip RN, Casper EA, Gordon FB, et al: Fluorescent antibody responses to chlamydial infection in patients with lymphogranuloma venereum and urethritis. J Immunol 112:2126–2134, 1974
10. Wang SP, Grayston JT: Local and systemic antibody response to trachoma eye infection in monkeys, p 217–232. In RL Nichols (ed), Trachoma and related disorders caused by chlamydial agents. Excerpta Medica, Amsterdam, 1971
11. Wang SP, Grayston JT: Human serology in Chlamydia trachomatis infection with microimmunofluorescence. J Infect Dis 130:388–397, 1974
12. Jones BR, The prevention of blindness from trachoma. Trans Ophthalmol Soc UK 95:16–33, 1975
13. Schachter J, Meyer KF: Lymphogranuloma venereum. II. Characterization of some recently isolated strains. J Bacteriol 99:636–638, 1969
14. Murray ES: Guinea pig inclusion conjunctivitis virus. 1. Isolation and identification as a member of the psittacosis-lymphogranuloma-trachoma group. J Infect Dis 114:1–12, 1964
15. Treharne JD, Katzenelson E, Davey SJ, et al: Comparison of serotyping subgroup A chlamydial isolates by a one-way cross-reaction and by a two-way cross-reaction immunofluorescence test, p 289–304. *In* RL Nichols (ed), Trachoma and related disorders caused by chlamydial agents. Excerpta Medica, Amsterdam, 1971
16. Darougar S, Jones BR, Kinnison JR, et al: Advances in the diagnostic isolation of Chlamydia including TRIC agents from eye, genital tract and rectum. Br J Vener Dis 48:416–420, 1972

Single-Antigen Indirect Immunofluorescence Test for Screening Venereal Disease Clinic Populations for Chlamydial Antibodies

SHIRLEY J. RICHMOND AND E. OWEN CAUL

Indirect immunofluorescence (IF) methods have recently been developed which will detect chlamydial antibodies in sera from patients infected with ocular or genital trachoma-inclusion conjunctivitis (TRIC) agent strains of *Chlamydia trachomatis*. The micro-immunofluorescence (micro-IF) test of Wang and Grayston (1), which uses as antigen the elementary bodies of the known serotypes of *C. trachomatis* propagated in the yolk sac of embryonated eggs, has been most widely used (2–7). More recently, a single-antigen IF test has been described (8) in which irradiated McCoy cells infected with genital strains of *C. trachomatis* are grown in wells on slides coated with polytetrafluoroethylene (PTFE). In this system the whole chlamydial inclusion acted as the antigen in an indirect IF test, and propagation of the different *C. trachomatis* serotypes in eggs was avoided. This test was capable of detecting antibodies to the group antigen, as well as type-specific antibodies to *C. trachomatis* serotypes. With the use of a single *C. trachomatis* strain (TRIC serotype E) as antigen, chlamydial antibodies were detected in 82 of 116 women (71%) attending venereal disease (VD) clinics, compared with 21 of 58 women (36%) attending a Family Planning Association Clinic, who had been less exposed to the risk of acquiring sexually transmitted infections than the VD clinic group.

In the present work, studies with this single-antigen IF test have been extended to include a larger number of patients attending VD clinics and to relate antibody findings to the clinical information available on these patients, in an attempt to assess the practical value of such a screening test, both in individual patients and for epidemiological studies on VD clinic populations.

MATERIALS AND METHODS

Antigen for IF Test

T181, a TRIC agent strain of *C. trachomatis*, serotype E, was used as antigen throughout this work.

Preparation of Antigen

Antigen was prepared on PTFE-coated slides as previously described (8), except that McCoy cells treated with cytochalasin B (9) were used for growing the organism instead of irradiated McCoy cells.

Single-Antigen IF Test and Complement Fixation Test

The single-antigen IF test and the complement fixation (CF) test were carried out as previously described (8).

Sera Tested

A total of 245 sera from a previous study of 279 women attending VD clinics (10) were used. These sera were collected at the time isolation of chlamydiae was attempted and were stored at −20°C. All sera were screened for the presence of chlamydia-specific immunoglobulins with the use of a fluorescein-conjugated rabbit anti-human globulin (Wellcome Research Laboratories, Beckenham, England). Any serum with a total immunoglobulin titer ≥8 was also tested for chlamydia-specific IgM and for complement-fixing antibodies to the chlamydial group antigen. The specificity of the fluorescein-conjugated goat anti-human IgM (Behringwerke, Marburg-Lahn, Germany) was tested as described previously (8). Preliminary results on some of these sera have already been reported (8); these results are included in the overall results presented here.

Sixty-nine sera were from men attending VD clinics with nongonococcal urethritis (NGU), on whom isolation studies had previously been carried out (11). Acute-phase sera only were tested from these patients, Since a number of these sera had been frozen and thawed several times before the present test, a procedure which may destroy IgM, they were tested only for total chlamydial immunoglobulins.

Thirty-nine sera from 18 patients with clinical and serological evidence of psittacosis were tested by CF, and by IF for total chlamydial immunoglobulins.

RESULTS

Chlamydial Antibodies Detected in IF Test

Chlamydia-specific antibodies were detected in 67 of 76 women (88%) from whom chlamydiae were isolated (culture-positive women) and in 83 of 169 women (49%) from whom chlamydiae were not isolated (culture-negative women; $P < 0.01$; Table 1). Similar results were obtained in sera from men with NGU: here 21 of 24 culture-positive men (88%) and only 19 of 45 culture-negative men (42%) had chlamydial antibodies ($P < 0.01$; Table 1). In women with chlamydial antibodies, the geometric mean titer was higher than in men (Table 2), an observation similar to that made in previous studies when the micro-IF test was used (6, 7), and in each sex the mean titer was higher in the culture-positive patients than in the culture-negative patients. Among the culture-positive women, 22 of 76 (29%) had specific IgM in their sera compared with 11 of 169 (7%) culture-negative women ($P < 0.001$). Chlamydia-specific IgM was found only in patients for whom the total immunoglobulin titer was ≥64.

TABLE 1. *Chlamydial antibodies in patients attending venereal disease clinics*

Group studied	No. of reactive sera/no. of sera tested		
	Culture-positive patients	Culture-negative patients	Total patients
Women	67/76 (88%)	83/169 (49%)	150/245 (61%)
Men with NGU	21/24 (88%)	19/45 (42%)	40/69 (58%)

TABLE 2. *Geometric mean titers in patients with chlamydial antibodies*

Group studied	Culture-positive patients	Culture-negative patients	Total patients
Women	74.7	43.6	55.4
Men with NGU	29.0	19.2	23.0

In the culture-negative female patients there was a tendency for the proportion of patients with chlamydial antibodies to increase with increasing age up to about 25 years, though this trend was not statistically significant ($0.05 < P < 0.1$; Table 3).

Comparison of Antibodies Detected by IF and CF Tests

There was no significant difference in the chlamydial antibody titer measured by IF and by CF in sera from patients with psittacosis (Fig. 1), which suggests that the IF test will detect antibodies to the chlamydial group antigen. However, when the CF and IF titers in sera from female patients attending VD clinics were compared, many sera were reactive by IF that were negative by CF (Fig. 2). Thus, in 26 sera (18%) the CF and IF titers were not significantly different, in 33 sera (23%) the CF titer was significantly (>fourfold) lower than the IF titer, and in 86 sera (59%) there were no detectable complement-fixing antibodies. Assuming that all sera with IF titers <8 were also <8 by CF test, group-specific antibodies were present in 38 of 73 sera (52%) from culture-positive patients compared with only 21 of 164 sera (13%) from culture-negative patients ($P < 0.001$).

Chlamydial Antibodies Related to Clinical Data

Results were analyzed to see whether chlamydial antibodies detected by IF in sera from women attending VD clinics were affected by the presence or absence of other genital tract pathogens in these patients, by pregnancy, or by NGU in the male consort (Table 4). The distribution of antibodies in these different categories of patients did not differ significantly from the antibody distribution in the group as a whole. In men with NGU, 12 of 15 culture-positive patients (80%) whose symptoms had lasted 7 days or less before serum was obtained already had antibodies in their serum, and antibodies were present in all 7 culture-positive men whose symptoms had been present for >7 days (Table 5).

TABLE 3. *Chlamydial antibodies in culture-negative female patients in relation to age*

Age group (years)	No. of patients with antibody	Total no. of patients	Percent with antibody
14–17	9	28	32
18–21	27	61	44
22–25	22	37	60
>25	25	45	55

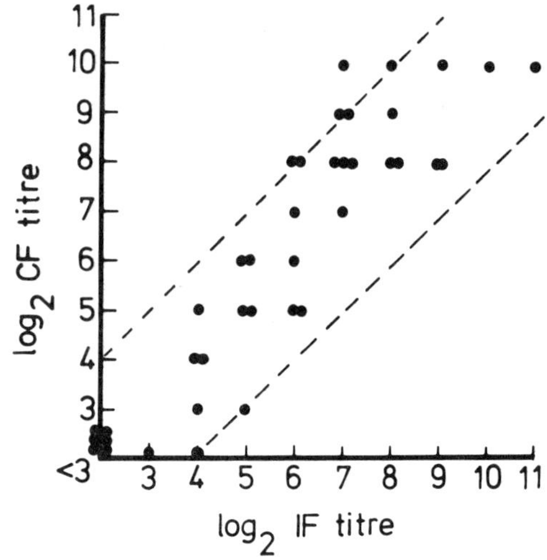

FIG. 1. *Immunofluorescence (IF) and complement fixation (CF) antibody titers in 39 sera from 18 patients with psittacosis. Sera on or between dotted lines: not >fourfold difference in IF and CF titers.*

DISCUSSION

The single-antigen IF test provides a simple, sensitive method of screening large numbers of sera for both group- and type-specific chlamydial antibody. It is interesting to compare our results by this test with those obtained by the micro-IF test in previous studies with comparable populations. In the present studies 88% of culture-positive men with NGU had chlamydial antibodies compared with 86% (5), 83% (4), and 70% (7) detected by the micro-IF test. In culture-negative men with NGU, we found 52% with antibody, and frequencies of 24% (5), 44% (7), and 76% (4) have previously been found by micro-IF. It appears, therefore, that in such epidemiological surveys the single-antigen IF test is as sensitive as the micro-IF test in detecting chlamydial antibodies, despite the fact that only the E serotype of *C. trachomatis* was used in the former test. This conclusion has recently been confirmed by Thomas and her colleagues (12); they identified 119 sera (obtained from patients attending a VD clinic) which reacted in the micro-IF test with at least one *C. trachomatis* serotype, and 111 (93%) of these sera reacted with serotype E.

The data illustrated in Fig. 1 showed that the single-antigen IF test could detect antibodies to the chlamydial group antigen, and that for this it was of comparable sensitivity to the CF test. In the VD clinic populations studied in the present work, antibodies were detected by IF more frequently and often at a higher titer than the complement-fixing antibodies. This was probably due to the frequent presence of type-specific antibodies to genital TRIC agent strains of *C. trachomatis* in these populations, which were detected in the IF but not in the CF test. A comparison of IF and CF titers in one serum therefore allows one to assess the proportion of type- and group-specific chlamydial antibodies present in that particular serum (Fig. 2).

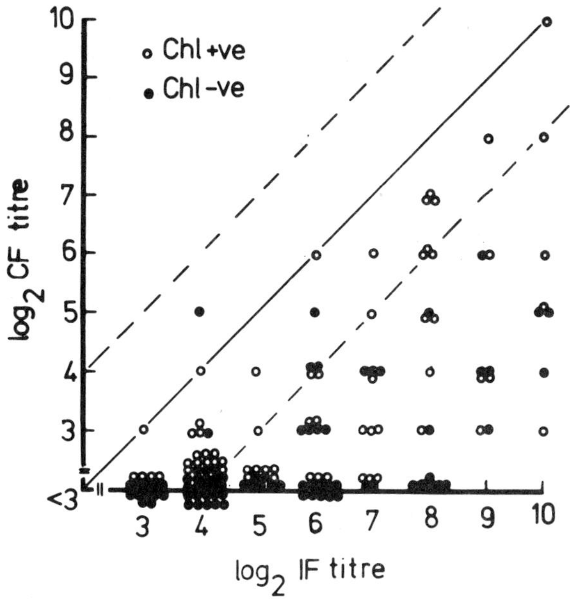

FIG. 2. *Immunofluorescence (IF) and complement fixation (CF) antibody titers in 145 sera reactive by IF. Single serum specimens were collected from female patients attending venereal disease clinics at the time isolation of chlamydiae was attempted. (○) Culture-positive patients. (●) Culture-negative patients. Sera on solid line: IF titer = CF titer. Sera on or between dotted lines: not >fourfold difference in IF and CF titers.*

A number of differences could be detected between the culture-positive and culture-negative patients within the VD clinic groups studied in the present work. About 90% of culture-positive patients, both male and female, had antibodies in their sera at the time of isolation, significantly more than the 50% of culture-negative patients with antibody. This antibody is therefore probably due to the current infection in a substantial number of these culture-positive patients. It is not known how long antibodies detected by IF take to appear in

TABLE 4. *Chlamydial antibodies in female patients in relation to other genital tract pathogens, pregnancy, and NGU in the male consort*

Clinical category	No. of reactive sera/no. of sera tested	
	Culture-positive patients	Culture-negative patients
Consorts of men with NGU	9/13 (69%)	10/42 (42%)
Women with gonorrhea	29/31 (94%)	6/15 (40%)
Women with other genital tract pathogens[a]	14/15 (93%)	26/51 (51%)
Women with no other genital tract pathogens	24/30 (80%)	51/103 (50%)
Pregnant patients	6/7 (86%)	14/27 (52%)

[a] *Candida albicans* and/or *Trichomonas vaginalis* and/or *Herpesvirus hominis.*

TABLE 5. *Chlamydial antibodies in men with nongonococcal urethritis in relation to duration of symptoms*

Men studied	No. of reactive sera/no. of sera tested	
	Symptoms for 0–7 days	Symptoms for >7 days
Culture positive	12/15 (88%)	7/7 (100%)
Culture negative	13/30 (43%)	4/12 (33%)

primary genital chlamydial infections in humans; in experimental chlamydial infections established in the urethra of baboons (13), antibodies (detected by the micro-IF test) did not appear until about 3 weeks after infection; chlamydiae were reisolated from the urethra both before the appearance of serum antibodies and after seroconversion. It is therefore unlikely that many of the culture-positive patients with antibody were in the acute primary stage of their infection at the time the chlamydial isolation was made. The absence of antibodies in a small proportion of culture-positive patients may be due to the presence of type-specific antibodies to a strain which does not cross-react with the TRIC E serotype used in this work. However, a similar proportion of patients without antibody has been found in surveys in which the micro-IF test has been used, which is capable of detecting antibodies to all the different known TRIC serotypes. It is likely, therefore, either that these culture-positive patients without antibody are in the early stages of a primary genital chlamydial infection or that the infection has not caused sufficient mucosal damage to provoke a humoral response; subsequent serum specimens from such patients should yield valuable information on the serum antibody response to these infections.

About 50% of culture-negative patients attending VD clinics also had chlamydial antibodies in their sera (Table 1). These antibodies are presumably due either to a previous chlamydial infection or to a current infection in which isolation of chlamydiae was not achieved. Group-specific antibodies were found significantly less frequently in culture-negative patients than in culture-positive patients, perhaps because type-specific antibodies persist longer than group-specific antibodies after a genital chlamydial infection.

The frequency of chlamydial antibodies in these culture-negative patients is probably a reflection of the extent to which this population has been exposed to genital chlamydial infections in the past. Similarly, the increase in the proportion of culture-negative female patients with antibodies with increasing age (Table 3) may be an indication of the increased chance of previous exposure to genital chlamydial infections in the older patients.

Although screening single serum specimens from patients for chlamydial antibodies by the single-antigen IF test does yield useful epidemiological information on the extent to which a particular population has been exposed to genital chlamydial infections (8), such a test is of little value in the assessment of individual patients, at any rate in groups as frequently exposed to these infections as VD clinic populations. For, although chlamydial antibodies detected by IF are suggestive of a *C. trachomatis* infection, this may be either a past or a current infection. If we assume that our culture technique detected the majority of active chlamydial infections, then serum antibodies, even in high titer, are not diagnostic of an active infection, since such antibodies may also be found in culture-negative patients. Neither is the presence of chlamydia-specific IgM a reliable indication of an active infection; although it was found significantly more often in culture-positive than in culture-negative patients, it was detected in fewer than one-third of all culture-positive patients. Moreover, the antibody distribution in the different diagnostic categories of patients was not significantly different from the findings in the group as a whole (Table 4). Testing single serum specimens for chlamydial antibodies by IF is therefore of very limited diagnostic value in the individual patients.

ACKNOWLEDGMENT

We are grateful to Hélène Mair for providing 15 sera from patients with psittacosis.

REFERENCES

1. Wang SP, Grayston JT: Immunologic relationship between genital TRIC, lymphogranuloma venereum, and related organisms in a new microtiter indirect immunofluorescence test. Am J Ophthalmol 70:367–375, 1970
2. Dwyer RStC, Treharne JD, Jones BR, et al: Results of micro-immunofluorescence tests for the detection of type-specific antibody in certain chlamydial infections. Br J Vener Dis 48:452–458, 1972
3. Hanna L, Jawetz E, Nabli B, et al: Titration and typing of serum antibodies in TRIC infections by immunofluorescence. J Immunol 108:102–107, 1972
4. Philip RN, Casper EA, Gordon FB, et al: Fluorescent antibody responses to chlamydial infection in patients with lymphogranuloma venereum and urethritis. J Immunol 112: 2126–2134, 1974
5. Reeve P, Gerloff RK, Casper E, et al: Serological studies on the role of *Chlamydia* in the etiology of non-specific urethritis. Br J Vener Dis 50:136–139, 1974
6. Wang SP, Grayston JT: Human serology in *Chlamydia trachomatis* infection with microimmunofluorescence. J Infect Dis 130:388–397, 1974
7. Holmes KK, Handsfield HH, Wang SP, et al: Etiology of nongonococcal urethritis. N Engl J Med 292:1199–1205, 1975
8. Richmond SJ, Caul EO: Fluorescent antibody studies in chlamydial infections. J Clin Microbiol 1:345–352, 1975
9. Sompolinsky D, Richmond S: Growth of *Chlamydia trachomatis* in McCoy cells treated with cytochalasin B. Appl Microbiol 28:912–914, 1974
10. Hilton AL, Richmond SJ, Milne JD, et al: *Chlamydia* A in the female genital tract. Br J Vener Dis 50:1–10, 1974
11. Richmond SJ, Hilton AL, Clarke SKR: Role of *Chlamydia* subgroup A in non-gonococcal and post-gonococcal urethritis. Br J Vener Dis 48:437–444, 1972
12. Thomas BJ, Reeve P, Oriel JD: Simplified serological test for antibodies to *Chlamydia trachomatis*. J Clin Microbiol 4:6–10, 1976
13. Digiacomo RF, Gale JL, Wang SP, et al: Chlamydial infection of the male baboon urethra. Br J Vener Dis 51:310–313, 1975

Microscopic Demonstration of Chlamydial Inclusions by Giemsa, Iodine, or Immunofluorescence Stains

LAVELLE HANNA

The first chlamydial infection to be recognized as a clinical entity was trachoma, an ocular disease described by Hippocrates and referred to in early Egyptian records. The demonstration in 1907 by Halberstaedter and Prowazek of inclusion bodies in Giemsa-stained epithelial scrapings (first in experimentally infected apes and then in humans) offered the first laboratory procedure for support of clinical diagnosis. During the next few years, investigators found similar inclusions in scrapings from babies with nongonococcal ophthalmia neonatorum, the clinical entity "inclusion blennorrhea" was described, and inclusions were found in female genital epithelium and in cases of nongonococcal urethritis in men. Subsequently, experimental infection of monkeys with material from cervix, urethra, or infant's eye resulted in follicular conjunctivitis and confirmed the relationship among the infections. In 1936, Rice showed that in inclusions the chlamydial particles were in a glycogen-like matrix which stained reddish-brown with iodine. These early studies have been reviewed comprehensively by Jawetz (1).

After T'ang et al. (2) reported the isolation in chick embryo yolk sac of chlamydiae from patients with trachoma, other laboratory tests became available. Yolk sac-adapted strains could be grown in large quantities in the laboratory, and this permitted development of improved methods of serological diagnosis. Immunofluorescence was first applied by Nichols et al. (3) and Hanna et al. (4) for the detection of inclusions in scrapings.

These three staining methods—Giemsa, iodine, and immunofluorescence—remain the most commonly applied techniques for direct demonstration of chlamydial agents. Accepted techniques for each are given below (5), and some of the advantages and limitations are listed briefly.

GIEMSA

Giemsa stain is prepared by dissolving 0.5 g of the powdered stain in 33 ml of glycerol at 55 to 60°C for 1.5 to 2 h. To this is added 33 ml of absolute methanol, acetone-free. The solution is mixed thoroughly and allowed to sediment; it is then stored at room temperature as stock. Giemsa stain is also available commercially as a solution. Dilutions of the stock stain are made with neutral distilled water or buffered water (pH 6.8 to 7.2) in a ratio of 1 part of stock Giemsa solution to 40 or 50 of diluent.

The smear is air-dried, fixed with absolute methanol for at least 5 min, and again dried. It is then covered for 1 h with the diluted Giemsa stain, freshly prepared each day. The slide is then rinsed rapidly in 95% ethanol to remove excess dye, dried, and examined for the presence of inclusions.

Staining for 30 min with 10% Giemsa may be preferred for cell culture preparations.

Giemsa is the standard reference stain. The intracytoplasmic inclusion consists of a vesicle containing the microcolony of chlamydiae in a matrix. The vesicle is bounded by a membrane derived from host cell membrane. It is usually seen as a crescent or helmet-shaped structure cupping the nucleus, in an epithelial cell, as shown in Fig. 1a. Seen by high magnification, it consists of particles ranging from the small (about 300 nm) reddish-purple elementary body to the larger (1 μm) bluish-purple initial body. The two types of particles are frequently present in the same inclusion. Because it is the reference standard, any laboratory worker studying chlamydiae should familiarize himself with the appearance of these Giemsa-staining inclusions. In addition to the presence or absence of the pathognomonic inclusions, the cytological pattern of the inflammatory reaction can be observed, and also the presence of any bacteria.

Among other advantages of Giemsa staining, (i) scrapings spread on slides and air-dried can be held for hours unfixed, and almost indefinitely after fixation with methanol; (ii) the stain itself is commercially available, and if the water for diluent is kept at a constant pH, staining will be uniform and reproducible; and (iii) no equipment other than a standard microscope is needed.

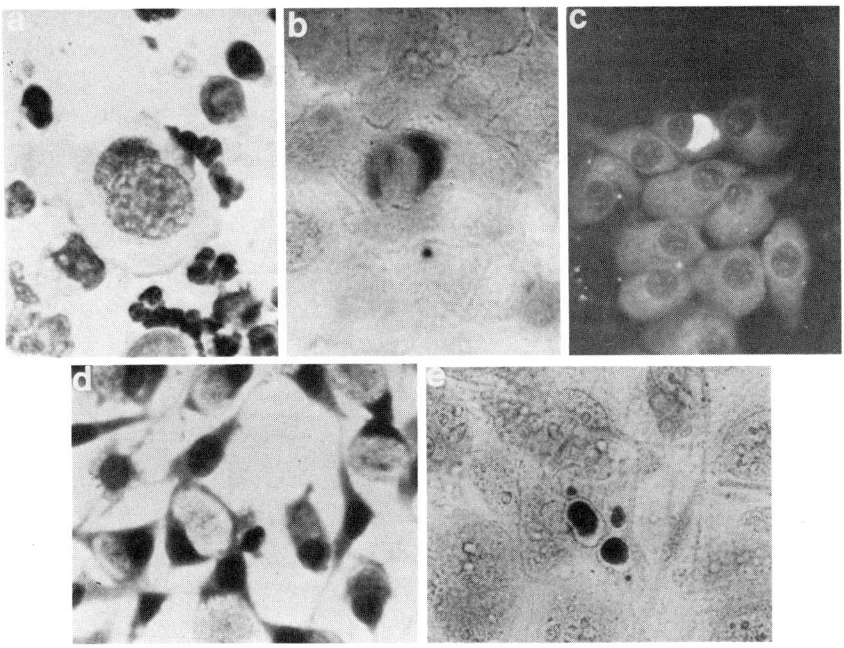

FIG. 1. *Typical intracytoplasmic inclusions of Chlamydia trachomatis. (a) Giemsa-stained crescent-shaped inclusion with many leukocytes in conjunctival scraping. (b) Iodine-stained inclusion in conjunctival scraping. (c) Immunofluorescent inclusion in conjunctival scraping. (d) Giemsa-stained oval inclusions in cell culture. (e) Iodine-stained oval inclusions in cell culture.*

Excellent examples of chlamydial inclusions and of sources of confusion are shown in the "WHO Guide to the Laboratory Diagnosis of Trachoma" (6). The main problems are in misidentifying other similar structures:

1. Extruded nuclei are irregular masses seen in the cytoplasm but attached to the nucleus, and resembling the nucleus in color texture.
2. Pigment granules are greenish-black in color, irregular in size, and often seen cupping the nucleus in a pattern similar to the chlamydial inclusion.
3. Clusters of granules in goblet cells are less distinct and more diffuse than chlamydial inclusions.

IODINE

Scrapings are air-dried, fixed in absolute methanol, and flooded with Lugol's iodine or 5% iodine in 10% potassium iodide for 3 to 5 min. They are examined as wet mounts. The glycogen-like matrix within the inclusion takes the stain, as shown in Fig. 1b. These reddish-brown masses are easily distinguished from nonstaining cells, even in thick smears, and specimens can be screened very quickly. Unfortunately, the iodine-staining matrix may be demonstrated during only a part of the reproductive cycle. As a result, if the iodine-stained preparation is decolorized with methanol and restained with Giemsa, the iodine-staining structures can be reidentified and confirmed, but additional Giemsa-staining inclusions may also be noted.

IMMUNOFLUORESCENCE

For immunofluorescence, sera from rabbits or guinea pigs hyperimmunized with yolk sac-grown chlamydial antigen (LGV antigens such as LB-1 [LGV-II] show broad cross-reactivity and are recommended) can be fluorescein labeled for use in a direct test or can be used with the appropriate fluorescein-labeled antiglobulin.

The scraping is air-dried, fixed in cold acetone (4°C) for at least 10 min, and rinsed in phosphate-buffered saline, pH 7.2 (PBS); it can be stored at 4°C for 2 to 3 days or at −20°C for months or years. At the time of examination, the scraping is thawed and air-dried. For the direct test, fluorescein-labeled serum diluted in PBS is added, and the slide is incubated in a moist chamber at 37°C. Plastic or glass petri dishes, 14 to 15 cm in diameter, containing filter paper saturated with water provide convenient small chambers. For larger numbers of slides, the bottoms of flat trays may be covered with wet paper towels, and other trays are inverted over the tops. After 30 min, the labeled serum is decanted and the slide is rinsed twice in PBS, 10 min each, with occasional shaking. The saline is shaken off, 90% glycerol plus 10% saline is added, and cover glasses are placed over the scrapings. For the indirect test, the unlabeled specific antiserum is added as described above; after the first series of rinses the slides are air-dried, and an appropriate dilution of labeled antiglobulin is added for a second period of incubation at 37°C for 30 min. The slides are again rinsed in PBS, with shaking, for two 10-min periods and then rinsed quickly in distilled water; glycerol-saline is added, and cover glasses are placed in position. The entire area of the scraping is examined under oil at a magnification of about ×500 with an appropriate light source (e.g., ultraviolet light). Inclusions are frequent in very active or severe infections, but rare in chronic infections.

Immunofluorescence stain appears to be the most sensitive method in current use for the demonstration of inclusions, and it is highly specific (3, 7–9). The increased sensitivity is due, at least in part, to the ease with which the microscopist can examine the entire specimen. The typical *Chlamydia trachomatis* inclusion is a sharply defined mass, fluorescing with a yellow-green light, clearly within the cytoplasm of an epithelial cell, not attached to the nucleus but in the same plane of focus, and resembling a Giemsa-stained inclusion in shape (Fig. 1c).

In our laboratory we have followed a volunteer experimentally infected 11 times during the period 1961 to 1966 (10). During the past 10 years, there have been no further inoculations, and clinical activity has been minimal or absent. Conjunctival scrapings and serum for antibody titrations have been collected regularly thoughout this period. He continues to show small numbers of inclusions at intervals, sometimes for weeks at a time, and he continues to have specific antibodies demonstrable by microimmunofluorescence.

With immunofluorescence, as with Giemsa-stained preparations, an experienced observer can note the accompanying cytology. Polymorphonuclear leukocytes can be identified, eosinophils can often be differentiated, and the presence of bacteria can be noted. However, the technique requires reagents not yet commercially available, fluorescence microscopes (recent models of standard microscopes may be more adaptable), and an experienced observer. Preparations must be fixed within a few hours of collection. They must be stored at −20°C or below for storage of more than 2 to 3 days, to avoid damage to the antigens. Once the slides have been frozen, they must be kept at a constant temperature until examined. If cells are not spread in a thin layer, one may not be able to see through the mass. Strands of mucus fluoresce brightly and may be confused with inclusions.

Indirect immunofluorescence is more practical than direct immunofluorescence for most laboratories. Preparation of fluorescein-conjugated antichlamydia immune serum is laborious and is suited mainly for research laboratories. Specific antisera can be produced in rabbits, guinea pigs, or other laboratory animals. Some investigators have chosen to use human sera with high levels of anti-lymphogranuloma venereum (LGV) antibody. Fluorescein-conjugated antisera directed against gamma globulins of laboratory animals or humans are commercially available.

DIRECT DEMONSTRATION OF CHLAMYDIAE ACCORDING TO ANATOMIC SITE

In diseases due to *C. psittaci* or in lymphogranuloma venereum, the agent is rarely demonstrated in infected sites by direct microscopy. The ease with which epithelial cells can be scraped from the conjunctiva contributed largely to the clarification of chlamydial infections of the eye. This is a clean area, and microscopic structures can be identified with relative ease. The choice between Giemsa staining and immunofluorescence depends upon availability of experienced personnel, proper microscopes, and reagents.

The transitional cells of the urethra are similar to conjunctival cells, and either Giemsa staining or immunofluorescence can be satisfactory. The patient, however, may feel more reluctant to submit to a urethral scraping than to a conjunctival scraping.

In examining cervical material microscopically, grave problems are encountered. Iodine stain cannot be used in genital tract scrapings because of the high glycogen content of many cells. In a cervical scraping, many of the cells are squamous, and inclusions will be found only in transitional cells. It may be impossible to differentiate Giemsa-stained inclusions from the bacteria and debris present, or to see through mucus plugs. The cytological picture is of little help, since the presence of neutrophils may be related to the stage of the menstrual cycle rather than to an inflammatory process. Immunofluorescence may be complicated by the normal bacterial flora which adhere to cells. Recognition of inclusions in Papanicolaou-stained preparations has been reported (11) but has not been generally accepted.

Use of the immunoferritin technique in examination of specimens from the genital tract has been reported recently (12; Richmond and Ashley, this volume, p. 272).

FACTORS CONTRIBUTING TO DIRECT DEMONSTRATION OF CHLAMYDIAL INCLUSIONS

In attempting direct demonstration by microscopy of chlamydial inclusions in conjunctival scrapings, the severity of the disease has been a factor. In some areas of Saudi Arabia, for example, where trachoma is endemic and severe, Nichols et al. have reported counts of thousands of inclusions per scraping, with an average of about 50 inclusions in those inclusion positive (13). In this situation any technique will be satisfactory, and there seems little reason for more than Giemsa staining of scrapings. In scrapings from American Indians with trachoma, on the other hand, no Giemsa-stained inclusions have been noted for over a decade, and only by immunofluorescence has it been possible to demonstrate the occasional inclusions in children who, nevertheless, do have active disease (14). Between these extremes, Yoneda et al. (15) have reported that about 10% of conjunctival smears from children with chronic trachoma in southern Tunisia were inclusion positive by Giemsa, with a geometric mean of three inclusions among those positive. In acute neonatal inclusion conjunctivitis, the scraping from the baby is by definition inclusion positive, with up to 10% of the cells infected. Here, again, a Giemsa-stained scraping is sufficient for laboratory diagnosis. If the same spectrum exists in genital infections, particularly among asymptomatic individuals, demonstration of agent by *any* laboratory procedure may sometimes be very difficult.

Irrespective of the method chosen, proper specimen collection is essential. Inclusions will be found only in epithelial cells, and a proper sample should contain a minimum of 1,000 such cells, spread as nearly as possible in a monolayer. If the greater part of the material collected is pus, the specimen must be considered unsatisfactory.

Thus far, I have discussed Giemsa, iodine, and immunofluorescence staining only in terms of clinical specimens. I would like to point out that these techniques also provide the means of locating and identifying the agent isolated in cell culture. The first two methods, Giemsa and iodine, have been applied more often, and the choice is based on individual laboratory preference. Inclusion counts based on iodine-stained preparations may be lower than those based on Giemsa-stained preparations, but the iodine-stained preparations can be examined much more rapidly. In cell culture the inclusions are likely to be found some distance from the nucleus in the cytoplasm, and they tend to be round rather than crescent shaped (Fig. 1d, e). In Giemsa-stained preparations, large, ballooning vacuoles are noted, often with relatively few chlamydiae within the vacuole. It should be noted, parenthetically, that in cell culture the presence of a high concentration of serum, with a high lipid content, may result in iodine-stained light brown masses within noninfected cells.

ACKNOWLEDGMENT

This study was supported by U.S. Public Health Service grant EY 00186 from the National Eye Institute.

REFERENCES

1 Jawetz E: Agents of trachoma and inclusion conjunctivitis. Annu Rev Microbiol 18:301–334, 1964
2. T'ang FF, Chang HL, Huang YT, et al: Studies on the etiology of trachoma with special reference to isolation of the virus in chick embryo. Chin Med J 75:429–447, 1957
3. Nichols RL, McComb DE, Haddad N, et al: Studies on trachoma. II. Comparison of fluorescent antibody, Giemsa, and egg isolation methods for detection of trachoma virus in human conjunctival scrapings. Am J Trop Med Hyg 12:223–229, 1963
4. Hanna L, Okumoto M, Thygeson P, et al: TRIC agents isolated in the United States. X. Immunofluorescence in the diagnosis of TRIC agent infection in man. Proc Soc Exp Biol Med 119:722–728, 1965
5. Hanna L, Schachter J, Jawetz E: Chlamydiae (psittacosis-lymphogranuloma venereum-trachoma group), p 795–804. *In* EH Lennette, EH Spaulding, JP Truant (ed), Manual of clinical microbiology, 2nd ed. American Society for Microbiology, Washington, D.C., 1974
6. Hanna L et al: (Participant) Guide to the laboratory diagnosis of trachoma. World Health Organization Symposium, Geneva, 1975, p 16–28
7. Hanna L: An evaluation of the fluorescent antibody technic in the diagnosis of trachoma and inclusion conjunctivitis. Rev Int Trach 45:345–356, 1968
8. Schachter J, Hanna L, Tarizzo ML, et al: Relative efficacy of different methods for laboratory diagnosis in chronic trachoma in the United States, p 469–475. *In* RL Nichols (ed), Trachoma and related disorders caused by chlamydial agents. Excerpta Medica, Amsterdam, 1971
9. Schachter J, Dawson CR, Balas S, et al: Evaluation of laboratory methods for detecting acute TRIC agent infection. Am J Ophthalmol 70:375–380, 1970
10. Hanna L, Dawson CR, Briones O, et al: Latency in human infections with TRIC agents. J Immunol 101:43–50, 1968
11. Naib AM: Exfoliative cytopathology. Little, Brown and Co., Boston, 1970, p 78–79
12. Ashley CR, Richmond SJ, Caul EO: Identification of the elementary bodies of *Chlamydia trachomatis* in the electron microscope by an indirect immunoferritin technique. J Clin Microbiol 2:327–331, 1975
13. Nichols RL, Bobb AA, Haddad NA, et al: Immunofluorescent studies of the microbiologic epidemiology of trachoma in Saudi Arabia. Am J Ophthalmol, Ser. 3, 63:1372–1408, 1967
14. Dawson CR, Ostler HB, Hanna L, et al: Tetracyclines in the treatment of chronic trachoma in American Indians. J Infect Dis 124:255–263, 1971
15. Yoneda C, Dawson CR, Daghfous T, et al: Cytology as a guide to the presence of chlamydial inclusions in Giemsa stained conjunctival smears in severe endemic trachoma. Br J Ophthalmol 59:116–124, 1975

Diagnosis of Genital Chlamydial Infections by Immune Electron Microscopy

SHIRLEY J. RICHMOND AND CHARLES R. ASHLEY

At the present time, definitive diagnosis of a genital chlamydial infection depends on isolating the organism from the genital tract. Diagnoses by direct demonstration of the organism in clinical material may provide useful alternative techniques, which would allow an assessment of the efficiency of the isolation procedure.

This paper describes and discusses an immune electron microscopy method for detecting extracellular elementary bodies (EB) of *Chlamydia trachomatis*. Ferritin staining of EB was achieved by an indirect immunoferritin technique. The principle of this technique is similar to that of indirect immunofluorescence. Since ferritin molecules are electron dense, ferritin-conjugated immunoglobulins allow antigen-antibody reactions to be identified in an electron microscope (1, 2), in a way similar to that in which fluorescein-conjugated immunoglobulins allow detection of these reactions by fluorescence microscopy.

Details of the indirect immunoferritin technique have already been published elsewhere (3), but briefly the test was carried out as follows. Well-washed, methanol-fixed preparations of EB on formvar-coated grids were incubated with a rabbit antiserum, R367, raised against a trachoma-inclusion conjunctivitis (TRIC) E serotype of *C. trachomatis* (4). This antiserum contained antibodies directed against the chlamydial group antigen (3, 5). After reaction of this antiserum with the EB, the grids were washed thoroughly and incubated with a ferritin-conjugated goat anti-rabbit globulin (GIBCO Bio-cult Laboratories Ltd., Paisley, Scotland). Grids were again washed thoroughly and then examined under an electron microscope.

Ferritin-stained particles were defined as particles between 200 and 500 nm in diameter, with ferritin attached over the whole surface of the particle, clearly in excess of the amount of ferritin attached to the background. Such ferritin-stained particles were easily identified in the supernatant fluids from McCoy cell cultures heavily infected with genital chlamydiae (Fig. 1A). Ferritin staining did not occur when a control rabbit serum was used in place of R367 (Fig. 1B); neither did it occur when the EB were unfixed or treated with mild fixatives such as glutaraldehyde, which did not disrupt the limiting envelope of the particle. The failure of EB with intact outer envelopes to stain with ferritin suggests that the chlamydial group antigen is not exposed on the outer surface of the particle.

A theoretical advantage of this indirect immunoferritin technique over culture methods of diagnosing chlamydial infections is that it may be possible to demonstrate ferritin-stained EB when the organisms are no longer viable. To investigate this possibility, pooled cells and supernatant fluids from several McCoy cell cultures, heavily infected with a genital TRIC agent strain of *C. trachomatis* (serotype E), were left at room temperature. Samples were withdrawn at various times and stored at −70°C. Subsequently, samples were

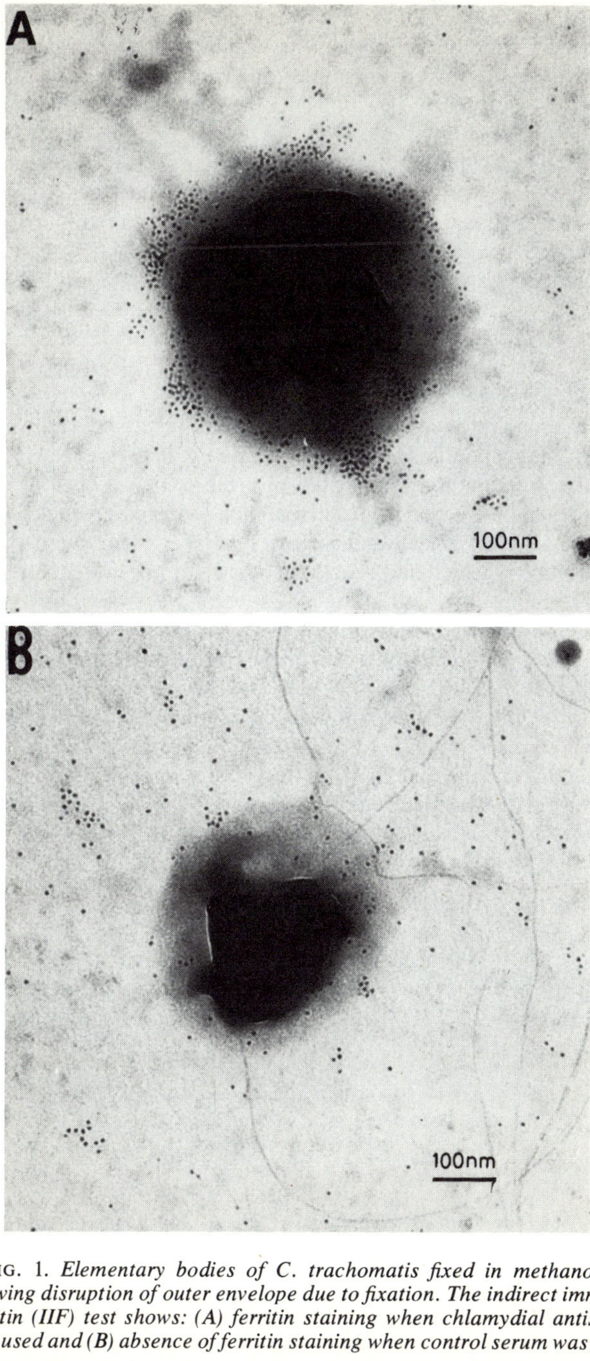

FIG. 1. *Elementary bodies of C. trachomatis fixed in methanol and showing disruption of outer envelope due to fixation. The indirect immunoferritin (IIF) test shows: (A) ferritin staining when chlamydial antiserum was used and (B) absence of ferritin staining when control serum was used.*

inoculated into McCoy cell cultures as well as tested by the indirect immunoferritin test. As would be expected, the viability of the organisms declined rapidly within a few days. Ferritin-stained particles, however, were detected by immune electron microscopy in samples kept for as long as 36 days at room temperature (Fig. 2).

Having established, first, that cell culture-grown EB could be ferritin stained and, second, that this staining occurred when EB were no longer viable, clinical specimens were examined by the indirect immunoferritin technique to see whether EB could be recognized in these as well as in cell culture material. A specimen was considered positive in the indirect immunoferritin test if a ferritin-stained particle was found, but no quantitative assessment was made of the numbers of such particles in each specimen. If no ferritin-stained particles were seen after scanning the grid under an electron microscope for 25 min, the specimen was recorded as negative. Results obtained by the indirect immunoferritin technique were compared with the results of culturing the same specimens for chlamydiae in McCoy cells treated with cytochalasin B (6).

The clinical specimens used consisted of urethral or cervical swabs obtained from patients attending venereal disease clinics, which were placed in medium containing 10% sorbitol (7). Two categories of specimens were examined: (i) 43 "fresh" specimens which had been kept at 4°C for not more than 24 h and had then been frozen at −70°C and stored for several weeks before they were cultured and tested by immune electron microscopy; (ii) 75 specimens which were sent to the laboratory by mail. These had not been kept at 4°C and many of them had been more than 24 h in transit before they reached the laboratory, where they were stored at −70°C. Table 1 shows that culture techniques were slightly, though not significantly, more sensitive than the indirect immunoferritin technique for detecting chlamydiae in the "fresh" specimens, and that there was a good percentage match between the two methods. (The percentage match = the percentage of co-positive and co-negative specimens.) This association between

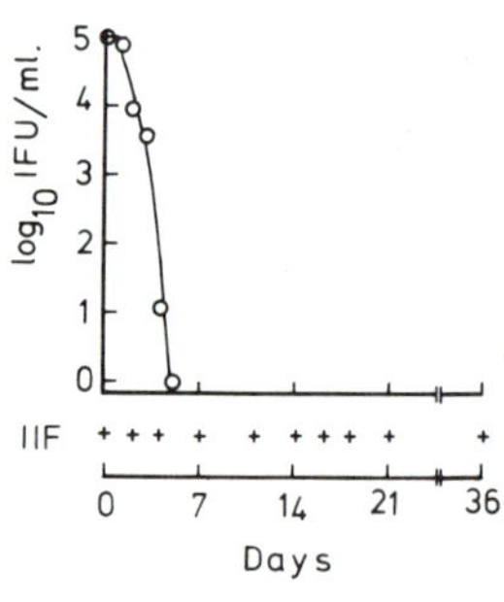

FIG. 2. *Detection of chlamydiae by indirect immunoferritin (IIF) technique and by isolation in McCoy cell cultures. Cell culture-grown elementary bodies (EB) of C. trachomatis, held at room temperature, were tested at intervals for viability in McCoy cell cultures and for ferritin staining by the IIF technique.* (○) *Log$_{10}$ inclusion-forming units per milliliter grown in McCoy cells treated with cytochalasin B.* (+) *Ferritin-stained EB detected by IIF technique.*

TABLE 1. *Comparison of results obtained in indirect immunoferritin (IIF) test with results obtained by isolation of chlamydiae in McCoy cell cultures*

No. of clinical specimens	No. of clinical specimens				Percentage match[a]
	Positive IIF test		Negative in IIF test		
	Chlamydiae isolated	Chlamydiae not isolated	Chlamydiae not isolated	Chlamydiae isolated	
43 "fresh"	11	2	25	5	84
75 mailed	7	13	53	2	80

[a] Percentage match = percentage of co-positive and co-negative specimens.

the indirect immunoferritin test and culture results was highly statistically significant ($P < 0.001$). By contrast, whereas the number of positive specimens detected by the indirect immunoferritin test in the specimens that had been mailed was similar to that found in the "fresh" specimens, the isolation rate was significantly reduced ($P < 0.01$). The percentage match between the two techniques was still, however, good in these specimens (Table 1), and the association between the two techniques was again highly significant ($P < 0.001$).

These results show that direct visualization of EB in an electron microscope by the indirect immmunoferritin technique is a feasible method of detecting chlamydiae in clinical specimens obtained from the genital tract. Provided the chlamydiae in the specimens are viable, the indirect immunoferritin technique is of comparable sensitivity to culture methods for diagnosing chlamydial infections. However, when transport of clinical specimens to the laboratory is delayed, so that the viability of the chlamydiae in the specimens is reduced, the indirect immunoferritin test is a significantly more sensitive method of diagnosing chlamydial infections than culture techniques ($P < 0.05$).

Both techniques detect extracellular chlamydial particles. However, it is theoretically possible that a chlamydial infection may exist where the organisms are present predominantly in their intracellular form, producing no or only very low levels of extracellular particles. Such infections would not be detected either by the indirect immunoferritin or culture techniques, but could be demonstrated by finding intracellular chlamydial antigens in urethral or cervical epithelial cells. Work is therefore in progress at present in our laboratory to see how culture results correlate with the presence of intracellular chlamydial antigens demonstrated by fluorescent-antibody staining of exfoliated endocervical cells.

REFERENCES

1. Singer SJ: Preparation of an electron-dense antibody conjugate. Nature (London) 183:1523–1524, 1959
2. Baxandall J, Perlmann P, Afzelius BA: A two-layer technique for detecting surface antigens in the sea urchin egg with ferritin-conjugated antibody. J Cell Biol 14:144–151, 1962
3. Ashley CR, Richmond SJ, Caul EO: Identification of the elementary bodies of *Chlamydia trachomatis* in the electron microscope by an indirect immunoferritin technique. J Clin Microbiol 2:327–331, 1975

4. Sowa J, Collier LH, Sowa S: A comparison of the iodine and fluorescent antibody methods for staining trachoma inclusions in the conjuctiva. J Hyg 69:693–708, 1971
5. Richmond SJ, Caul EO: Fluorescent antibody studies in chlamydial infections. J Clin Microbiol 1:345–352, 1975
6. Sompolinsky D, Richmond SJ: Growth of *Chlamydia trachomatis* in McCoy cells treated with cytochalasin B. Appl Microbiol 28:912–914, 1974
7. Richmond SJ: The isolation of *Chlamydia* subgroup A (*Chlamydia trachomatis*) in irradiated McCoy cells. J Med Lab Technol 31:7–9, 1974

Detection of *Chlamydia trachomatis* Infections by a Radioisotopic Enzyme Test

HOWARD M. JENKIN AND VICTOR S. C. FAN

This report reviews our current knowledge of glycogen, glycogen synthetase, and specific aspects of lipid metabolism of chlamydial agents infecting various cells in vitro.

Fan and Jenkin (1) showed that glycogen was accumulated into HeLa cells infected with *Chlamydia trachomatis* strain TW-3, whereas cells infected with *C. psittaci* strain meningopneumonitis (MN) or normal cells did not show this reaction when [U-^{14}C]glucose was used as a substrate. Inclusions of *C. trachomatis* in infected cells stained with iodine (2) or the periodate-Schiff reagent (3) showed a positive reaction. The stained material has been presumed to be glycogen accumulated in the inclusions, thereby showing a direct relationship between the histochemical and chemical reactions. Jenkin and Makino (unpublished data) found plasmalogens in chlamydia-infected cells, which would also give a positive reaction in inclusions when stained with iodine or the periodate-Schiff reagent. Fatty acids and aldehydogenic moieties are reactive sites for these stains. Jenkin and Fan (4) showed that *C. trachomatis* strain lymphogranuloma venereum (LGV), isolate 440LN, accumulated glycogen for 24 to 48 h after infection of HeLa cells; the glycogen then disappeared, which also correlates with the disappearance of the iodine-positive staining reaction of these cells. It is proposed that the glycogen synthetase activity of *C. trachomatis*-infected cells will not disappear and may prove to be a sensitive measure of chlamydial infections associated with nonspecific urethritis.

A more difficult test to distinguish chlamydia-infected cells from normal cells and cells infected with other organisms could be based on our knowledge that incorporation of [U-^{14}C]serine and its accumulation into the phosphatidylethanolamine (PE) fraction is greater in *C. trachomatis*-infected cells (5) than in normal cells. This particular phospholipid has an unusually high amount of branched fatty acids (40%) in the second position of the glycerol backbone of the lipid (6), not reported to our knowledge in other biological systems.

MATERIALS AND METHODS

Cells. HeLa 229 cells (7) were cultivated in 32-ounce (about 960-ml) prescription bottles with Eagle minimal essential medium (8) supplemented with 5% fetal calf serum at pH 7.2 (MEM FC_5) as growth medium. The cells were incubated at 37°C and were transferred every 3 to 4 days, with 0.05% trypsin (1:250; Difco) used to remove the cells. Almost complete monolayers of HeLa cells were formed after 18 h of incubation when the inoculum used was 4×10^5 cells per 1-ounce French square bottle for the experimental samples.

Monkey kidney cells (LLC-MK-2; 9) were cultivated as monolayers in 32-ounce prescription bottles. Minimal essential medium supplemented with 5% calf serum (MEM CS_5) was used to cultivate the cells, which were incubated at

37°C and subcultured every 5 to 7 days. The procedures used for subculturing the cells, cultivating infected cells, and determining the viability counts were identical to those described by Jenkin and Anderson (10).

Organisms. *C. trachomatis* strain LGV (440LN) was isolated by Schachter and Meyer (11). The organism was kindly supplied by J. Schachter from the fifth passage in chick embryo yolk sacs and was further passed two times in white Leghorn chick embryo yolk sacs (12). The yolk sac suspensions of LGV were used to initiate infection in monkey kidney cells and to prepare a stock inoculum of LGV grown in cell culture. The LGV cells used in the experiments were obtained from homogenates of infected monkey kidney cells after 5 to 11 passages in these cells.

C. trachomatis strain TW-3 (12) in the 21st yolk sac passage was propagated as described by Grayston et al. (13); 33% homogenates of strain TW-3 were prepared by diluting whole infected yolk sacs 1:2 (wt/vol) in sucrose-potassium glutamate (14). Fresh or frozen suspensions of the organisms were used to infect HeLa cells.

C. psittaci strain MN (14) was derived from the 84th yolk sac passage and 3 further passages in the allantoic cavity of chick embryos. The allantoic fluid inoculum of strain MN for HeLa cell infection was prepared as described by Jenkin (15). All strains were tested for sterility by use of blood agar slants, thioglycolate broth, Trypticase soy broth, and *Mycoplasma* plates (Hyland, Los Angeles, Calif.).

Enriched propagating medium for LGV growth in MK-2 cells. MEM CS_5 was enriched to obtain a higher yield of organisms for preparation of inoculum than was readily achieved by standard procedures. This was accomplished by changing the glucose concentration from 1,000 mg/liter to 2,700 mg/liter, doubling the concentration of the amino acids and vitamins, and adding 10% fetal calf serum (International Scientific Industries, Inc., Cary, Ill.). Hanks balanced salt solution (8) was maintained at 1× concentration.

Plaque assay medium. Plaque assay medium was prepared as described by Fan and Jenkin (16).

Cell culture inoculum. The infection of monkey kidney cells with *C. trachomatis* strain LGV was performed as described by Fan and Jenkin (16).

Infectivity titrations. Infectivity titrations were performed by the techniques described by Jenkin and Fan (4).

Lipid extraction, separation, and analyses. Lipid was analyzed by the methods described by Fan and Jenkin (5).

Phosphorus determination. Phosphorus was analyzed according to Fan and Jenkin's modification (16) of the technique of Rouser et al. (17).

Labeling of PE. Normal and LGV-infected MK-2 cells were labeled with DL-[3-^{14}C]serine according to the technique of Fan and Jenkin (16).

Glycogen isolation. Glycogen was isolated from HeLa cells by the method of Fan and Jenkin (4).

Preparation and assay of crude glycogen synthetase. Crude glycogen synthetase was prepared and assayed by the method described by Jenkin and Fan (4).

Protein assay. Protein analysis of the cell lysate supernatant fluid was performed by the method of Lowry et al. (18).

Reagents and isotopes. [U-^{14}C]glucose (specific activity, 309 mCi/mmol) was purchased from Amersham/Searle Co. (Des Plaines, Ill.). Uridine diphosphate (UDP) [U-^{14}C]glucose (specific activity, 237 mCi/mmol) and adenosine diphosphate (ADP) [U-^{14}C]glucose (specific activity, 228 mCi/mmol) were purchased from New England Nuclear Corp. (Boston, Mass.).

Shell fish glycogen and ADP glucose, dipotassium salt, were purchased from Calbiochem (Los Angeles, Calif.). Glucose-6-phosphate and UDP glucose were obtained from Sigma Chemical Co. (St. Louis, Mo.).

RESULTS AND DISCUSSION

[U-^{14}C]glucose labeling of normal and infected HeLa cells. As the first step to show that staining of *C. trachomatis* inclusions with iodine and the periodate-Schiff reagent were indeed associated with glycogen, infected and normal HeLa cells were pulse-labeled every 24 h with [U-^{14}C]glucose, and the glycogen was extracted at different time intervals of infection with *C. trachomatis* strain TW-3 and *C. psittaci* strain MN. As illustrated in Fig. 1, the labeled glycogen increased between

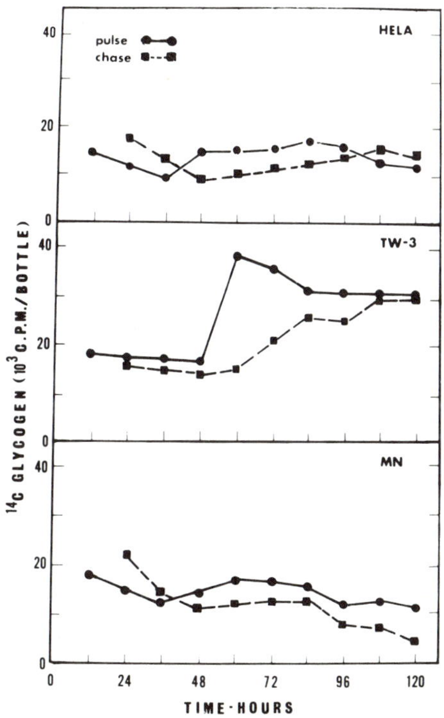

FIG. 1. *Incorporation of [U-^{14}C]glucose into glycogen of normal HeLa cells and HeLa cells infected with Chlamydia strains TW-3 and MN. Cells were exposed to [U-^{14}C]glucose (2×10^6 cpm/bottle) for the first 12 h of incubation; they were then washed, unlabeled glucose was added, and the cells were incubated for another 12 h. From 12 h to the end of the experiment, cells were pulsed with [U-^{14}C]glucose for 24 h, washed, and chased for 12 h. Each point represents the radioactivity at the end of the pulse or chase period from the average of triplicate samples. These data represent typical results of two independent experiments.*

48 and 72 h after inoculation and remained elevated for the rest of the infection in TW-3-infected cells. MN-infected cells and normal HeLa cells at 48 h after infection showed no increase in glycogen, but the normal cells showed some increase in glycogen on continued incubation from 48 to 96 h. In labeled LGV-infected cells, labeled glycogen rose sharply at about 36 h after infection but dropped precipitously by 48 h (Fig. 2). A comparison of the two experiments showed differing patterns of glycogen metabolism among the three chlamydial agents and normal HeLa cells.

The reduction or loss of iodine-staining characteristics of *C. trachomatis* inclusions paralleled the gradual decrease of glycogen in TW-3 infections and the abrupt loss of this characteristic in LGV infections. The time of appearance and disappearance of iodine-positive material in FL cells infected with *C. trachomatis* TE-55, as reported by Bernkopf et al. (19), was similar to the glycogen pattern of LGV isolate 440L.

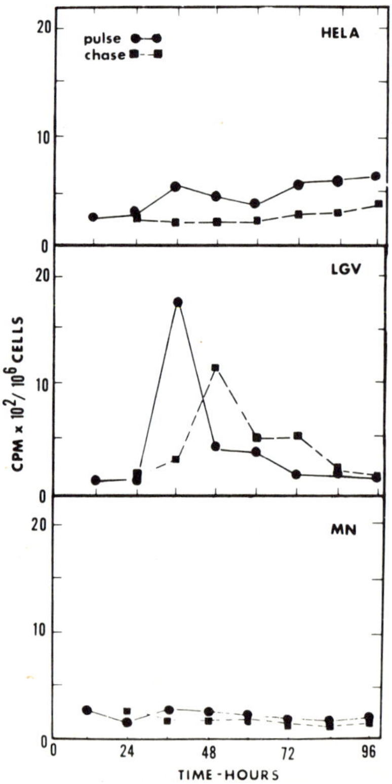

FIG. 2. *Incorporation of [U-¹⁴C]glucose into glycogen of normal HeLa cells and HeLa cells infected with Chlamydia strains MN and LGV. Cells were exposed to [U-¹⁴C]glucose (2 × 10⁵ cpm/bottle) and labeled for 12-h pulses; they were then washed and chased for 12 h. Each point represents the radioactivity at the end of the pulse or chase period from the average of triplicate samples. These data represent typical results of two independent experiments.*

Glycogen synthetase activity with labeled ADP glucose or UDP glucose as substrates. A number of enzymes such as glycogen synthetase, phosphorylases, and amylases (glucosidases) are associated directly with glycogen metabolism. Leloir and Cardini (20) and Villar-Palasi and Larner (21) reported that UDP glucose is a substrate for glycogen synthesis in mammalian cells, whereas Greenberg and Preiss (22) and Recondo and Leloir (23) reported ADP glucose as the substrate for bacteria and plants.

Glycogen synthetase activity was investigated to study one major enzyme associated with glycogen metabolism in chlamydial infections. The time course of glycogen synthesis from ADP glucose substrate, with the use of a crude enzyme preparation from HeLa cells infected for 36 h with LGV (Fig. 3), shows that enzyme activity was constant with time in the presence of excess substrate.

Normal and infected cell lysates were examined for glycogen synthetase activity at various times (Table 1). HeLa cells infected with TW-3 and LGV showed a rise in enzyme activity when ADP glucose was used as a substrate, parallel to the incorporation of [^{14}C]glucose into glycogen. MN-infected cells showed little activity in the presence of ADP glucose or UDP glucose. Normal HeLa cells utilized UDP glucose as a substrate for glycogen metabolism, as was expected.

The fact that different substrates are utilized by lysates of normal and infected cells suggests that *C. trachomatis* agents have a glycogen synthetase enzyme, independent of the host. Since the enzyme found is associated with a bacterial-type glycogen synthetase, it is not likely that the enzyme is host derived.

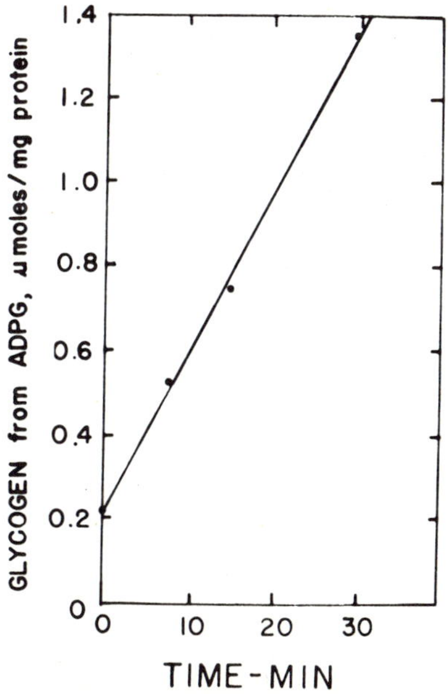

FIG. 3. *Linearity of initial rate of glycogen synthetase. A 100-µl amount of crude enzyme was added to the mixture (see text), which was incubated at 37°C to start the reaction. These data are representative of results from at least 40 independent experiments.*

TABLE 1. *Glycogen synthetase activity in normal and chlamydia-infected HeLa cells*[a]

Time (h)	LGV-infected		TW-3-infected		Normal cells		MN-infected	
	ADPG	UDPG	ADPG	UDPG	ADPG	UDPG	ADPG	UDPG
16	4.9	0.1	0.5	0.8	<0.1	0.8	<0.1	0.4
24	16.7	0.5	2.5	1.2	<0.1	1.0	<0.1	1.6
36	37.5	6.1	2.1	1.0	0.3	1.0	0.4	0.7
48	2.8	0.7	4.2	3.1	<0.1	1.1	<0.1	<0.1
72	0.7	0.5	2.5	1.7	<0.1	2.1	<0.1	<0.1

[a] Enzyme activity was measured with adenosine diphosphate glucose (ADPG) and uridine diphosphate glucose (UDPG) as substrate and is expressed as nanomoles of substrate incorporation per minute per milligram of protein.

This observation perhaps can be used to advantage in examining cervical and vaginal wall scrapings for chlamydial agents. Once the enzyme is produced, it should remain active in the tissue, although the glycogen-staining characteristics of infected cells may disappear, particularly when certain LGV strains are implicated in nonspecific urethritis infections. Specimens from a scraping containing 10^5 to 10×10^5 cells should be sufficient to use with a highly active ADP [^{14}C]glucose in 2- to 4-h pulses. The glycogen can be isolated, and the ^{14}C activity can be directly

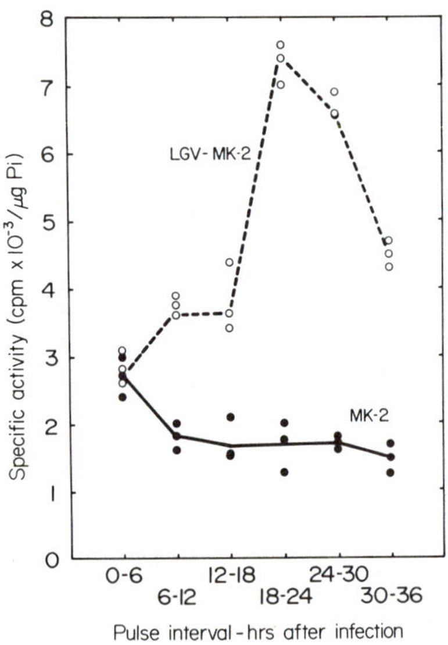

FIG. 4. *Incorporation of* DL-[3-^{14}C]*serine into phosphatidylethanolamine of normal and LGV-infected monkey kidney cells. Normal and LGV-infected MK-2 cells (2 × 10^5 cells/2-ounce prescription bottle) were incubated with* DL-[3-^{14}C]*serine in MEM CS*$_5$ *(0.1 µCi/ml) every 6 h after infection. Phosphatidylethanolamine was isolated after incubation. The deacylated portion of the lipid was measured for its radioactivity and phosphorus content. The specific activity is expressed as counts per minute per microgram of inorganic phosphorus.*

measured. The practicality of this test remains to be proven by workers who have access to human specimens.

Incorporation of DL-[3-^{14}C]serine into PE of normal and LGV-infected MK-2 cells. Fan and Jenkin (5) showed that major changes in the lipid metabolism of LGV-infected MK-2 cells occurred during the course of infection. An increase of PE and decreases in phosphatidylcholine, phosphatidylserine, and phosphatidylinositol were found 24 to 36 h after infection. To learn more about the mode of action of lipid metabolic events in the LGV-infected and normal MK-2 cells, we performed a tracer study using DL-[3-^{14}C]serine. The label was followed into the PE in pulse and chase experiments to determine turnover of this key phospholipid. Figure 4 shows that the labeled serine was incorporated into PE with a high peak activity 18 to 24 h after infection. The normal cells showed a low magnitude of incorporation of the label into PE during the 36-h observation period.

Figure 5 illustrates the lack of turnover of PE in LGV-infected cells, once peak activity was observed, during 36 h of infection.

On the basis of the difference in biosynthetic pathways of lipid metabolism of mammalian cells and bacteria (Fig. 6), a phosphatidylserine decarboxylase should be present in the bacterial cells, and therefore infected mammalian cells should be able to convert phosphatidylserine to PE. With prelabeled phosphatidylserine in which the serine is labeled in the 1 position used as substrate, it could be possible to measure the increase in [^{14}C]CO_2 evolved from phosphatidylserine when mixed

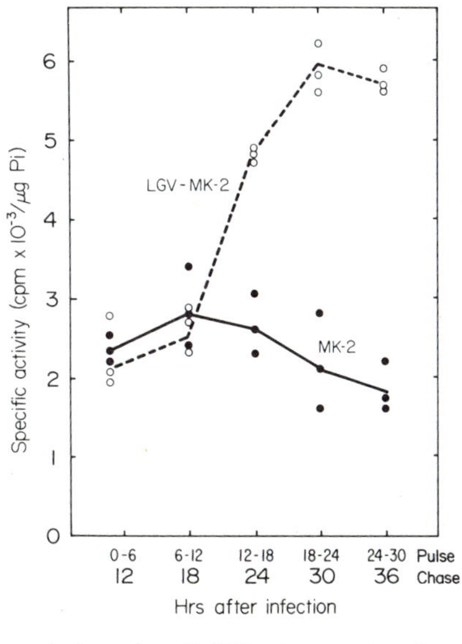

FIG. 5. *Accumulation of DL-[3-^{14}C]serine in phosphatidylethanolamine (PE) of normal and LGV-infected monkey kidney cells. Normal and LGV-infected cells were prepared and incubated with DL-[3-^{14}C]serine as described in Fig. 1. Labeled cells were further incubated in the absence of labeled serine for another 6 h. PE samples were isolated and deacylated. The deacylated portion of the lipid was measured for its radioactivity and phosphorus content. The specific activity is expressed as counts per minute per microgram of inorganic phosphorus.*

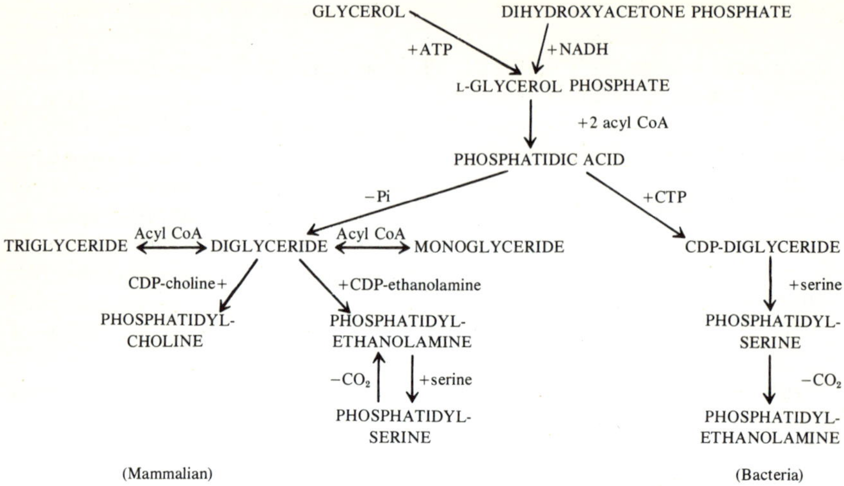

FIG. 6. *Biosynthetic pathways of lipids in mammalian tissue and bacteria.*

with chlamydia-infected cells from human specimens. DL-[3-^{14}C]serine could also be used to label PE, which can be isolated from infected cells by means of a simple lipid extraction. Since the activity of infected cells is two to four times higher than that of normal cells, one could use this type of test clinically in conjunction with other diagnostic tools. The problem with this assay is that further investigations must be performed to observe whether other microbial flora found in the genital tract may mask the reactions observed experimentally with *C. trachomatis* strains.

In summary, two enzyme systems, glycogen synthetase and possibly phosphatidylserine decarboxylase, which are peculiar to bacterial agents and perhaps specifically to *C. trachomatis*, have been speculated to be useful tools to aid in the diagnosis of chlamydial infections of the genital tract. This conclusion is based on experimental data obtained by use of normal HeLa and MK-2 cell culture systems and cells infected with *C. trachomatis* and *C. psittaci*.

ACKNOWLEDGMENTS

This work was supported by Public Health Service research grants HE 08214 from the Program Projects Branch, Extramural Programs, National Heart Institute, and AI 07747NAIAD from the National Institute for Allergy and Infectious Diseases. It was also aided by support to a Student Fellow from Fight for Sight, Inc., financed by a grant from Burroughs Wellcome Co. (USA), Inc., to Fight for Sight, Inc., New York, N.Y., the World Health Organization, and The Hormel Foundation.

REFERENCES

1. Fan VSC, Jenkin HM: Glycogen metabolism in chlamydia-infected HeLa cells. J Bacteriol 104:608–609, 1970
2. Rice CE: Carbohydrate matrix of the epithelial cell inclusion in trachoma. Proc Soc Exp Biol Med 33:317–318, 1935
3. Gordon FB, Quan AL: Occurrence of glycogen in inclusions of the psittacosis-lymphogranuloma venereum-trachoma agents. J Infect Dis 115:186–196, 1965

4. Jenkin HM, Fan VSC: Contrast of glycogenesis of *Chlamydia trachomatis* and *Chlamydia psittaci* strains in HeLa cells, p 52–59. *In* RL Nichols (ed), Trachoma and related disorders caused by chlamydial agents. Excerpta Medica, Amsterdam, 1971
5. Fan VSC, Jenkin HM: Biosynthesis of phospholipids and neutral lipids of monkey kidney cells (LLC-MK-2) infected with *Chlamydia trachomatis* strain lymphogranuloma venereum. Proc Soc Exp Biol Med 148:351–357, 1975
6. Makino S, Jenkin HM, Yu HM, et al: Lipid composition of *Chlamydia psittaci* grown in monkey kidney cells in defined medium. J Bacteriol 103:62–70, 1970
7. Scherer WF, Syverton JT, Gey GO: Studies on propagation in vitro of poliomyelitis viruses. IV. Viral multiplication in a stable strain of human malignant epithelial cells (strain HeLa) derived from epidermoid carcinoma of cervix. J Exp Med 97:695–709, 1953
8. Eagle H: Amino acid metabolism in mammalian cell cultures. Science 130:432–437, 1959
9. Hull RN, Cherry WR, Tritch OJ: Growth characteristics of monkey kidney cells strains LLC-MK1, LLC-MK2 (NCTC-3196) and their utility in virus research. J Exp Med 115:903–918, 1962
10. Jenkin HM, Anderson LE: The effect of oleic acid on the growth of monkey kidney cells (LLC-MK2). Exp Cell Res 59:6–10, 1970
11. Schachter J, Meyer KF: Lymphogranuloma venereum. II. Characterization of some recently isolated strains. J Bacteriol 99:636–638, 1969
12. Wang SP, Grayston JT: Egg infectivity assay of trachoma virus. Proc Soc Exp Biol Med 115:587–591, 1964
13. Grayston JT, Wang SP, Johnston PB: Etiologic studies of trachoma on Taiwan. Proc Soc Exp Biol Med 103:596–599, 1960
14. Francis T, Magill TP: An unidentified virus producing acute meningitis and pneumonitis in experimental animals. J Exp Med 100:575–580, 1938
15. Jenkin HM: Continuous passage of agents of trachoma in cell culture. I. Characteristics of TW-3 and Bour strains of trachoma cultivated in serial passage in HeLa 229 cells. J Infect Dis 116:390–399, 1966
16. Fan VSC, Jenkin HM: Lipid metabolism of monkey kidney cells (LLC-MK-2) infected with *Chlamydia trachomatis* strain *lymphogranuloma venereum*. Infect Immun 10:464–470, 1974
17. Rouser G, Fleischer S, Yamamoto A: Two dimensional thin layer chromatographic separation of polar lipids and determination of phospholipids by phosphorus analysis of spots. Lipids 5:494–496, 1970
18. Lowry, OH, Rosebrough NJ, Farr AL, et al: Protein measurement with the Folin phenol reagent. J Biol Chem 193:265–275, 1951
19. Bernkopf H, Mashiah P, Becker Y: Correlation between morphological and biochemical changes and the appearance of infectivity in FL cultures infected with trachoma agent. Ann NY Acad Sci 98:62–81, 1962
20. Leloir LF, Cardini CE: Biosynthesis of glycogen from uridine diphosphate glucose. J Am Chem Soc 79:6340–6341, 1957
21. Villar-Palasi C, Larner J: A uridine coenzyme-linked pathway of glycogen synthesis in muscle. Biochim Biophys Acta 30:449, 1958
22. Greenberg E, Preiss J: The occurrence of adenosine diphosphate glucose-glycogen transglucosylase in bacteria. J Biol Chem 239:4314–4315, 1964
23. Recondo E, Leloir LF: Adenosine diphosphate glucose and starch synthesis. Biochem Biophys Res Commun 6:85–88, 1961

Tissue Culture Procedures for the Isolation of *Chlamydia trachomatis* from Patients with Nongonococcal Genital Infections

DEREK HOBSON

At the time of the last major symposium on chlamydial infections (1), tissue culture (TC) procedures had only recently begun to replace inoculation of the chick embryo yolk sac as a feasible method of isolating *Chlamydia trachomatis* directly from the infected patient. The purpose of this review is to describe the advances in laboratory practice over the past 6 years and to discuss the practical problems that still need to be solved before TC methods can be applied universally and routinely to investigate all new patients with nongonococcal genital infections (nongonococcal urethritis [NGU], etc.).

HISTORICAL DEVELOPMENT OF TC PROCEDURES

At first sight, it is difficult to understand why it took so long to design TC techniques to isolate *C. trachomatis*, or why the methods still in use are more complex and time consuming than those used in routine virological diagnosis. It had been known since 1935 that *C. psittaci* grew well in TC (2), and by 1960, soon after the first isolation of *C. trachomatis* in eggs (3, 4), it was found that certain strains could readily be propagated to high titer in HeLa (5), FL (6), and BSC-1 cells (7). Cultural conditions did not seem to be exacting, and adsorption of the agent to TC cells presented no greater problem than that of enteroviruses. Multiple growth cycles occurred, even under agar, with formation of plaques (6).

By hindsight, these early successes may have done more to confuse than to help solve today's problems in isolating *C. trachomatis* directly from the patient. Many of the egg-adapted strains of *C. trachomatis* used in early TC studies differ in laboratory behavior and antigenic constitution from any of the subsequently isolated serotypes. The circumstances which led to the emergence of these "fast" strains remain unknown, but they developed enhanced virulence for chick embryos and serological relationship with the JH strain of lymphogranuloma venereum (8). Results obtained with such laboratory variants in TC cannot be extrapolated with any assurance that they will help in designing better procedures for isolating genital strains of *C. trachomatis*.

MODERN TC PROCEDURES

Methods devised by Gordon and his colleagues in the United States (9, 10), and simplified by Darougar and co-workers in England (11), were first brought into large-scale diagnostic use 7 years ago as a result of collaboration between the two groups of workers (12). Various modifications of these methods have subsequently been introduced, but they continue to share many of the basic features of Gordon's original technique. Their common aspects can be summarized as follows:

1. A defined number of TC cells freshly obtained from stock culture bottles is suspended in liquid growth medium containing gentamicin, or vancomycin plus streptomycin, to prevent bacterial contamination. Penicillin, chloramphenicol, or tetracyclines must not be used for this purpose because chlamydiae are sensitive to them.

2. The cells are dispensed into flat-bottomed glass or plastic capped vials containing a flying cover slip and are incubated until an even and just-confluent monolayer of well-spread cells has formed. These cell cultures may need special pretreatment or maintenance to control their growth rate within narrow limits.

3. The clinical specimen or standard strain of *C. trachomatis* is then inoculated. For reasons to be discussed later, effective contact of the infective chlamydial particle with the TC monolayer needs to be augmented, usually by centrifugation.

4. After incubation at 35 to 37°C for 48 to 72 h, each successful infective unit will have produced one intracytoplasmic inclusion filled with progeny particles.

5. At this time, the cover slip is removed, fixed, stained with iodine, Giemsa, or carmine (13), and examined for inclusions by microscopy (×400).

COMMON PROBLEMS IN TC PROCEDURES

Since current methods draw largely on the same basic procedure, they all suffer from similar limitations and give rise to similar practical problems.

1. Many TC lines used in routine virological diagnosis are insensitive to *C. trachomatis* infection, or require special treatment to make them susceptible.

2. The transformed or malignant TC lines in widest use for isolating *C. trachomatis* divide rapidly and repeatedly, with a lack of contact inhibition between the cells. Unrestrained overgrowth may prevent recognition of inclusions because the TC cells become packed, overlaid, and deficient in cytoplasm, and they may actually inhibit chlamydial growth by exhausting the medium of nutrients essential for the agent's development.

3. The uptake of *C. trachomatis* by TC cells is inefficient, in sharp contrast to *C. psittaci*, lymphogranuloma venereum, or the "fast" strains. Unless special adsorption conditions are adopted, less than 1% of all the infective particles in a clinical specimen may produce detectable inclusions.

4. In all current techniques, each input infective particle of *C. trachomatis* yields only one final inclusion (14). Although this inclusion may burst and liberate progeny particles during incubation, there is rarely any evidence of daughter inclusions produced by secondary growth cycles. Hence, there is no cumulative easily scored cytopathic effect. Every inoculated culture must be stained and examined microscopically at the optimal time in the single growth cycle. It is difficult to obtain high-titer yields to *C. trachomatis* from patients to determine their serotype or antibiotic susceptibility.

Some of these shortcomings have already been partly overcome, but often by empirical maneuvers which have not helped to identify the cause of the original problem and have sometimes added further problems. Much more basic research work is needed to make the routine laboratory diagnosis of genital infections due to chlamydiae as simple and precise as the diagnosis of gonorrhea. The outstanding difficulties in reaching this target are given more detailed consideration below.

CHOICE OF TC CELL LINE

In recent years, the supposition has arisen that the McCoy cell line first used by Gordon (9, 14) is virtually unique in its susceptibility to *C. trachomatis*. In fact, many other cell lines of human or animal origin can support the growth of this agent (15–17), and the Seattle stock of HeLa 229 seems to be at least as sensitive as the McCoy line (16). It is interesting that McCoy, originally used because it was thought to be a line of human synovial cells, is now known (14, 15) to be closely related in karyotype and antigenic constitution to the mouse fibroblastic tumor cell line L929, which has long been used for the laboratory growth of *C. psittaci* (18). This finding could simplify future research into the exact nutritional requirements of *C. trachomatis* in TC, since the precise amino acid and vitamin needs of L cells themselves have long been identified (19, 20).

Since the bulk of world experience in growing *C. trachomatis* over the past 13 years has been with McCoy cells, there could be advantages in continuing its use as a reference TC line, to facilitate exchange of information and standardization of techniques. However, the McCoy cultures now used have had varying passage histories in a variety of media in many different laboratories for many years. Any polyploid cell line maintained under such imprecisely specified conditions for so long is likely to have become increasingly heterogeneous, but these sublines have never been compared directly in a single reference laboratory in terms of their identity or their susceptibility to *C. trachomatis*, or their freedom from contamination with mycoplasmas or viruses. Furthermore, no attempt has yet been reported to reclone any McCoy line to select for greater sensitivity to *C. trachomatis* infection, although such procedures have proved useful in virus studies, e.g., enhancing the sensitivity of Chang conjunctival cells to myxoviruses (21). In this context, it is pertinent that some families of HeLa cells do not share the high sensitivity of HeLa 229 to *C. trachomatis* (16) and that the sensitivity of the Seattle variety of HeLa 229 is not always matched by HeLa 229 maintained elsewhere (Kuo, personal communication). Similarly, stocks of BHK-21 held in different laboratories have given discrepant results on infection with *C. trachomatis* (15, 16).

It is far from certain, however, that malignant mouse fibroblasts such as the McCoy line can offer the ideal "milieu intèrieure" for such a restricted pathogen as *C. trachomatis*, which shows marked specificity in vivo for columnar or transitional epithelium of humans and the higher primates only. Many of the problems of current isolation techniques may not be simply procedural, but may be due to fundamental defects in the nature of the cells employed.

The susceptibility of other cell lines or primary cell cultures deserves further investigation, but it will be important to test each under its own optimal conditions of medium, temperature, pH, etc., and not merely to infect with *C. trachomatis* under the cultural conditions which happen to have been favorable for McCoy or HeLa 229 cells (16).

TC CONDITIONS

The growth of chlamydiae, as prokaryotic parasites of eukaryotic cells, is dependent on the nutritional status of the host cell. It has been known for 20 years that *C. psittaci* can infect but remain latent in L cells maintained in inadequate medium (18). The growth of *C. trachomatis* (T'ang strain) in FL cells was prevented by omitting arginine, histidine, and leucine from the TC medium, and three other strains would not grow in FL cells in Eagle medium unless extra serine was added (22). Removal of calf serum from the medium also prevented chlamydial growth (23). A specific requirement for isoleucine has been demonstrated for *C. psittaci* growing in L cells; competition between host cell and parasite for this amino acid could be prevented by treating the culture with cycloheximide (24), which inhibits protein synthesis of eukaryotic TC cells but does not affect the metabolism of prokaryotic chlamydiae (25, 26).

It is not surprising, therefore, that inconsistent isolation rates of *C. trachomatis* from genital infections may be reported from different laboratories using different sublines of McCoy cells grown in different TC media and different sources of serum to produce monolayers of different cell density and age, often treated with various metabolic inhibitors to control their growth rate. In addition to the constitution of the medium, differences in the temperature and pH of incubation and in CO_2 concentration can all affect the growth of *C. trachomatis* in McCoy cells (27, 28).

REPLICATING VERSUS NONREPLICATING TC

For many years, *C. psittaci* has been grown in conventional replicating cell cultures like those used in virological diagnosis. However, the first successful isolations of *C. trachomatis* from NGU patients were in McCoy cells which had been heavily irradiated to prevent their replication before the specimens were inoculated (10–12), and by 1974, as Blyth and Taverne (15) observed, "a virtually unsubstantiated belief that irradiated McCoy cells have some special advantage for the isolation of TRIC agents has become widespread amongst workers investigating the incidence of these agents in the genital tract." During the past 2 years, the practical problems in exposing each working batch of McCoy cells to 4,000 to 6,000 rads of X-irradiation has led other workers to investigate alternative methods of preventing cell replication, using antimetabolic chemicals such as iododeoxyuridine (IUdR) (29), cytochalasin B (30), or cycloheximide (Ripa and Mårdh, p. 323, this volume).

The common final effect of all these materials is to produce monolayers of large nondividing cells with abundant cytoplasm, which persist virtually unchanged throughout 48 to 72 h of incubation after infection with *C. trachomatis*. Inclusions tend to be larger and more easily recognizable than in untreated cells. There may be a significant increase in the number of inclusions produced by a given inoculum, but some of this may be due to improved precision of counting, rather than to real improvement in the efficiency of plating.

The antireplicative effect of these chemicals is not simply radiomimetic; it may be the result of a complex of activities against several different biochemical pathways of the TC cell. The mode of action varies from one material to another and may not be reproducible from one commercial batch to another of the same material. The effect may vary from one cell line to another and in different stocks of the same cell line. Thus, the infection-enhancing effect of irradiation and IUdR for McCoy cells cannot be demonstrated with HeLa 229 cells (16), and different batches of IUdR have given discrepant results (Ripa and Mårdh, p. 323, this symposium). Furthermore, the growth of certain serotypes of *C. trachomatis* in McCoy cells is not enhanced by IUdR (16). To date, the only quantitative biochemical evaluation of the precise mode of action of any antireplicative agent has been with cycloheximide in L cells infected with *C. psittaci* (24–26).

The satisfactory use of replicating cultures of BHK (15), McCoy (28, 31), and HeLa 229 (32) cells in the diagnosis of chlamydial infections indicates that the apparent necessity for antireplicative agents can be obviated by close attention to TC conditions. Important factors include the number and condition of the cells explanted and the age of the monolayer at the time of infection (16, 33), the control of pH, CO_2 concentration, and temperature (27, 28), and the provision of nutrients essential for chlamydial growth in concentrations above the basic minimal requirements of the TC cell itself (22, 24, 28).

Irradiation, IUdR, and cycloheximide can act as inducers of latent virus infections (e.g., with C-type particles) carried within many cell lines. Their exact dosage and time of addition must be carefully controlled; e.g., IUdR must be removed before infection, to prevent inhibition of chlamydial deoxyribonucleic acid replication. Thus, these materials add to the complexity of already complicated procedures, and all constitute potential biohazards to laboratory staff. Detailed investigation of their basic mode of action, to confirm that these agents do more than provide an elaborate empirical corrective to empirical but unbalanced TC systems, might usefully point out simpler and more rational future improvements in TC isolation procedures.

ADSORPTION OF INFECTIVE PARTICLES

Failure to isolate *C. trachomatis* from clinical specimens is more often due to inefficient adsorption and penetration into the TC cell than to incapacity of the cell to support chlamydial growth within its cytoplasm. This may be an artifact of the current procedures, since natural infections of humans are readily transmissible by brief contact.

It is not yet proven whether uptake of *C. trachomatis* by TC cells involves electrostatic interactions (34) or chlamydia-specific sialic acid receptors on the cell surface, similar to those for myxoviruses (35), or phagocytosis without an adsorption phase (36). With any of these possible mechanisms, the efficiency of plating of *C. trachomatis* may be affected by the pH, ionic strength, and viscosity of the medium, by the time and temperature of interaction, and by changing the medium shortly after inoculation (28). Similarly, the presence in clinical specimens of nonviable chlamydial particles, mucus, bacterial products, enzymes, or toxic chemicals applied in local therapy may block or inactivate the process of uptake. With the present lack of detailed knowledge, the best that can be done is to increase the effective collision between chlamydial particles and TC cells by centrifugation (37) or to modify electrostatic effects and increase pinocytosis by diethylaminoethyl (DEAE)-dextran (27, 38, 39).

The number of inclusions in TC freshly inoculated with a standard inoculum of *C. trachomatis* increases linearly with the force and the duration of centrifugation (28, 40, 41). Inclusion counts are higher after centrifugation at 33°C than at lower temperatures (27). In diagnostic practice, 2,500 to 3,000 × *g* for 1 h at 33 to 35°C is sufficient to ensure maximal isolation rates from clinical specimens (28, 42).

Pretreatment of TC monolayers with DEAE-dextran before inoculating *C. trachomatis* gives increased inclusion counts in HeLa 229 cells (32) and irradiated

McCoy cells (27; Rota, p. 314, this volume) when used in conjunction with centrifugation, but is much less effective when used alone (27). Different sources of DEAE-dextran may give discrepant results, and material of the molecular weight used in the reported studies may no longer be commercially available (Ripa and Mårdh, p. 323, this volume).

RECOGNITION OF POSITIVE SPECIMENS

Despite all recent advances in TC procedures, *C. trachomatis* usually fails to produce repeated cycles of reinfection in the monolayer. This may be an artifact of imperfect TC conditions, related to the inefficiency of adsorption-penetration, but it is also possible that the TC cells become refractory after the first cycle, either by age effects or by acquiring defense mechanisms such as interferon, or by blockage from abnormal or inactivated first-cycle progeny.

Since the TC monolayer, at best, will show only one inclusion per input viable particle, weakly positive clinical specimens will be falsely scored as negative unless the time and method of examination of the culture are well chosen. The inclusion count is usually maximal after 48 to 60 h of incubation but falls rapidly after 72 to 84 h as infected cells burst and drop off the cover slip (Johnson et al., p. 000, this volume). Different methods of staining the monolayer may give discrepant results. Iodine may be far less efficient in demonstrating inclusions in TC than in direct scrapings from trachoma lesions (Hanna, p. 266, this volume), since the glycogen matrix, on which iodine staining depends, is inconstant in amount and in time in the growth cycle in TC (16, 43; Johnson et al., p. 309, this volume). Giemsa-stained monolayers may be difficult to examine because of minimal color differentiation of cell nuclei, cytoplasm, and chlamydiae. This is especially so in crowded monolayers of cells with scanty cytoplasm or in nonreplicating treated McCoy cells, where nuclear reduplication and fragmentation tend to occur. This problem can be overcome by modification of the Giemsa stain (13) combined with dark-ground microscopy, where stained inclusions show intense yellow fluorescence against a dull green-brown background of TC cells.

DISCUSSION

Despite all the ongoing technical problems and the variety of TC isolation procedures adopted to overcome them in different laboratories, remarkably similar isolation rates of *C. trachomatis* have been reported from epidemiological surveys in clinics in many different regions of Britain and the United States. It already seems clear that approximately 40% of males with NGU and 30% of their female contacts have chlamydial infections. An important target for future research is to determine whether this represents the true role of chlamydia in NGU, etc., or whether an even higher incidence might be revealed by further improvements in the sensitivity of TC isolation procedures. Such an assessment requires independent evidence of infection, obtained by direct identification of chlamydiae by light or electron microscopy (Richmond and Ashley, p. 272, this volume), by detection of chlamydial antigens or enzymes (Jenkin and Fan, p. 277, this volume) in the patients' secretions, or by investigating the patients' antibody responses (Wentworth, p. 299, this volume).

The epidemiological information already available is certain to stimulate justifiable clinical demands for nationwide routine screening of all new cases of NGU, etc., for possible *C. trachomatis* infection, and for the follow-up of infected patients after treatment. In Britain alone this would probably entail examining 250,000 specimens or more each year. It is doubtful whether the present complex and labor-intensive TC methods could cope with such a regular service commitment, on an even greater scale than current laboratory involvement in the control of gonorrhea. Thus, the main objective over the next few years must be to learn how to simplify TC isolation techniques further without losing their present sensitivity, and to investigate alternative methods of making at least a provisional diagnosis of *C. trachomatis* infection.

In this context, it cannot be emphasized too often that most of the expense and time spent in perfecting laboratory procedures can be vitiated by failure to ensure, in advance, that the right sort of specimen is taken from the right site at the right stage of the disease, before treatment, by an experienced clinician, and transmitted in the right transport medium with the minimal delay to a laboratory that can process it expeditiously.

REFERENCES

1. Nichols RL (ed): Trachoma and related disorders caused by chlamydial agents. Excerpta Medica, Amsterdam, 1971
2. Bland JOW, Canti RG: Growth and development of psittacosis virus in tissue culture. J Pathol Bacteriol 40:231–241, 1935
3. T'ang FF, Chang CL, Huang YT, et al: Studies on the etiology of trachoma with special reference to isolation of the virus in chick embryo. Chin Med J 75:429–447, 1957
4. Jones BR, Collier LH, Smith CH: Isolation of virus from inclusion blennorrhoea. Lancet 1:902–905, 1959
5. Furness G, Graham DM, Reeve P: The titration of trachoma and inclusion blennorrhoea viruses in cell culture. J Gen Microbiol 23:613–619, 1960
6. Bernkopf H, Mashiah P: The growth cycle of a trachoma agent in FL cell cultures. J Immunol 88:570–571, 1962
7. Bernkopf H: A plaque test for TRIC agents. Am J Ophthalmol 63:1206–1207, 1967
8. Wang SP, Grayston JT: Studies on the identity of the 'fast' egg-killing chlamydia strains, p 322–336. *In* RL Nichols (ed), Trachoma and related disorders caused by chlamydial agents. Excerpta Medica, Amsterdam, 1971
9. Gordon FB, Magruder GB, Quan AL, et al: Cell cultures for detection of trachoma virus from experimental simian infections. Proc Soc Exp Biol Med 112:236–242, 1963
10. Gordon FB, Quan AL: Isolation of the trachoma agent in cell culture. Proc Soc Exp Biol Med 118:354–359, 1965
11. Darougar S, Kinnison JR, Jones BR: Simplified irradiated McCoy cell culture for isolation of chlamydia, p 63–70. *In* RL Nichols (ed), Trachoma and related disorders caused by chlamydial agents. Excerpta Medica, Amsterdam, 1971
12. Gordon FB, Harper A, Quan AL, et al: Detection of chlamydia (Bedsonia) in certain infections of man. I. Laboratory procedures: comparison of yolk sac and cell culture for detection and isolation. J Infect Dis 120:451–462, 1969
13. Johnson FWA: A comparison of staining techniques for demonstrating group A chlamydia in tissue culture. Med Lab Technol 32:233–238, 1975
14. Gordon FB, Dressler HR, Quan AL, et al: Effect of ionizing irradiation on susceptibility of McCoy cell cultures to *Chlamydia trachomatis*. Appl Microbiol 23:123–129, 1972

15. Blyth WA, Taverne J: Cultivation of TRIC agents: a comparison between the use of BHK-21 and irradiated McCoy cells. J Hyg 72:121–128, 1974
16. Croy JR, Kuo CC, Wang SP: Comparative susceptibility of eleven mammalian cell lines to infection with trachoma organisms. J Clin Microbiol 1:434–439, 1975
17. Hobson D, Johnson FWA, Byng RE: Isolation of Chlamydia trachomatis in tissue cultures of human thyroid. Lancet 2:643, 1976
18. Morgan HR, Bader JP: Latent viral infection of cells in tissue culture. IV. Latent infection of L cells with psittacosis virus. J Exp Med 106:39–44, 1957
19. Eagle H: The specific amino acid requirements of a mammalian cell strain L in tissue culture. J Exp Med 102:839–852, 1955
20. Eagle H: The minimum vitamin requirements of the L and HeLa cells in tissue culture, the production of specific vitamin deficiencies and their cure. J Biol Chem 214:595–600, 1955
21. Sugiura A, Kilbourne ED: Genetic studies of influenza viruses. 2. Plaque formation by influenza viruses in a clone of a variant human heteroploid cell line. Virology 26:478–488, 1965
22. Ossowski L, Becker Y, Bernkopf H: Amino-acid requirements of trachoma strains and other agents of the PLT group in cell cultures. Isr J Med Sci 1:186–193, 1965
23. Becker Y: The molecular biology of trachoma agent. Isr J Med Sci 8:1134–1137, 1972
24. Hatch TP: Competition between *Chlamydia psittaci* and L cells for host isoleucine pools: a limiting factor in chlamydial multiplication. Infect Immun 12:211–220, 1975
25. Alexander JJ: Separation of protein synthesis in meningopneumonitis agent from that of L cells by differential susceptibility to cycloheximide. J Bacteriol 95:327–332, 1968
26. Alexander JJ: Effect of infection with the meningopneumonitis agent on deoxyribonucleic acid and protein synthesis by its L cell host. J Bacteriol 97:653–657, 1969
27. Rota TR, Nichols RL: *Chlamydia trachomatis* in cell culture. Appl Microbiol 26:560–565, 1973
28. Johnson FWA, Hobson D: Factors affecting the sensitivity of replicating McCoy cells in the isolation and growth of chlamydia A (TRIC agents). J Hyg 76:441–451, 1976
29. Wentworth BB, Alexander ER: Isolation of *Chlamydia trachomatis* by use of 5 iodo-2-deoxyuridine-treated cells. Appl Microbiol 27:912–916, 1974
30. Sompolinsky D, Richmond S: Growth of *Chlamydia trachomatis* in McCoy cells treated with cytochalasin B. Appl Microbiol 28:912–914, 1974
31. Hobson D, Johnson FWA, Rees E, et al: Simplified method for diagnosis of genital and ocular infections with chlamydia. Lancet 2:555–556, 1974
32. Kuo CC, Wang SP, Wentworth BB, et al: Primary isolation of TRIC organisms in HeLa 229 cells treated with DEAE-dextran. J Infect Dis 125:665–668, 1972
33. Richmond SJ: Growth of Chlamydia trachomatis in cell culture. Lancet 1:865, 1976
34. Becker Y, Hochberg E, Zakay-Rones Z: Interaction of trachoma elementary bodies with host cells. Isr J Med Sci 5:121–124, 1969
35. Kuo CC, Wang SP, Grayston JT: Effect of polycations, polyanions and neuraminidase on the infectivity of trachoma-inclusion conjunctivitis and lymphogranuloma venereum organisms in HeLa cells: sialic acid residues as possible receptors for trachoma-inclusion conjunctivitis. Infect Immun 8:74–79, 1973
36. Friis RR: Interaction of L cells and *Chlamydia psittaci*: entry of the parasite and host responses to its development. J Bacteriol 110:706–721, 1972
37. Weiss E, Dressler HR: Centrifugation of rickettsiae and viruses onto cells and its effect on infection. Proc Soc Exp Biol Med 103:691–695, 1960
38. Rota TR, Nichols RL: Infection of cell cultures by trachoma agent. Enhancement by DEAE-D. J Infect Dis 124:419–421, 1971
39. Harrison MJ: Enhancing effect of DEAE-dextran on inclusion counts of an ovine Chlamydia (Bedsonia) in cell culture. Aust J Exp Biol Med Sci 48:207–213, 1970

40. Reeve P, Taverne J: Strain differences in the behavior of TRIC agents in cell cultures. Am J Ophthalmol 63:1167–1173, 1967
41. Darougar S, Cubitt S, Jones BR: Effect of high speed centrifugation on the sensitivity of irradiated McCoy cell culture for the isolation of chlamydia. Br J Vener Dis 50:308–312, 1974
42. Reeve P, Owen J, Oriel JD: Laboratory procedures for the isolation of Chlamydia trachomatis from the human genital tract. J Clin Pathol 28:910–914, 1975
43. Gordon FB, Quan AL: Occurrence of glycogen in inclusions of the psittacosis-lymphogranuloma venereum-trachoma agents. J Infect Dis 115:186–196, 1965

Isolation of *Chlamydia* from Ocular Infections

S. DAROUGAR, R. M. WOODLAND, T. FORSEY, S. CUBITT, J. ALLAMI, AND B. R. JONES

Chlamydia trachomatis is responsible for a variety of ocular infections, namely, trachoma-inclusion conjunctivitis (TRIC) ophthalmia neonatorum, inclusion conjunctivitis, TRIC punctate keratoconjunctivitis (1), and trachoma of sexually transmitted origin in developed countries as well as hyperendemic trachoma of eye-to-eye transmission in developing countries.

The technique of culture in irradiated McCoy cells (2) provides a sensitive and practical method for the primary isolation of *C. trachomatis*. This is three to four times more sensitive than isolation in the yolk sac of fertile hen eggs (2–4) or the direct detection of inclusions in conjunctival scrapings.

In our laboratory, modification of this technique resulted in the development of a simplified one-passage culture technique (5). This rapid test requires only two tubes of cells for each clinical specimen and 2 to 3 days to obtain results. Its sensitivity is similar to that of the original three-passage technique.

Over the past 10 years, we have investigated the effect of various factors on the collection of specimens and the isolation of *C. trachomatis* in cell culture.

COLLECTION OF SPECIMENS

Isolation from conjunctival swabbing produced four times the number of inclusions as compared with isolation from the standard technique of scraping (6). Swabbing provides a simple and rapid method of specimen collection, free from the undesirable sequelae of scraping such as irritation, bleeding, and conjunctival reactions.

STORAGE AND TRANSPORT

A portable liquid nitrogen refrigerator (−180°C) provides a practical means of storage and transport of specimens if inoculation of fresh specimens is not possible. The use of plastic ampoules, with screw caps and silicone rubber washers (Sterilin Ltd.), containing 2SP (2) for storage and transportation of specimens in liquid nitrogen has obviated the hazards of breakage and explosion associated with glass containers.

Studying the effect of different freezing temperatures for storage of clinical specimens, we found that after 1 week of storage those specimens snap-frozen and stored in liquid nitrogen (−180°C) yielded only half as many inclusions as specimens stored at −70°C in a mechanical refrigerator. No further marked differences in the numbers of inclusions were observed when specimens were stored at −70 or −180°C for a period of 1 to 6 months. Storage at −20°C proved to be unsatisfactory (Darougar et al., in preparation). This result indicates that snap-freezing in liquid nitrogen may reduce the infectivity of chlamydial particles and hence the isolation rate.

EFFECT OF PRESERVATIVES

The preservative effects of serum, lactalbumin hydrolysate, dimethyl sulfoxide, and glycerol on the infectivity of *Chlamydia* were assessed. The addition of serum to transport medium (3%, vol/vol) gave marked protection against damage caused by snap-freezing and freezing and thawing of specimens (Darougar et al., in preparation). The other compounds showed no preservative effect.

CENTRIFUGATION

Centrifugation of clinical specimens at 1,800 × *g* for 1 h at 18°C increased the sensitivity of irradiated McCoy cells for the isolation of *Chlamydia* (2). The number of inclusions was four times higher when the centrifugation temperature was increased to 35°C. Centrifugation at 15,000 × *g* for 1 h at 35°C produced the highest rate of isolation (7). Centrifugation of clinical specimens at 15,000 × *g* produced an equal number of inclusions whether a horizontal swing-out rotor or a fixed-angle rotor was used. Generally, fixed-angle rotors have a greater capacity and thus increase the number of specimens that can be processed with a high-speed centrifuge.

STAINING TECHNIQUES

In our laboratory, inoculated monolayers are stained with Giemsa after approximately 60 h of incubation and are examined under dark-field illumination with a Zeiss RA microscope at ×200 magnification. Bright, yellow-golden inclusions contrast well with the dark background; screening of cover slips is rapid, each taking 2 to 3 min, and is more accurate and less fatiguing than examination under bright-field illumination.

The iodine staining technique is simple, rapid, and useful for detecting mature inclusions of *C. trachomatis* in cell culture (8). We use this technique routinely to classify chlamydial isolates into subgroup A (*C. trachomatis*) and subgroup B (*C. psittaci*).

Staining with fluorescent antibody (FA) is more sensitive than using Giemsa for detecting chlamydial inclusions in conjunctival scrapings (9–12). We have assessed the sensitivity of the FA technique for detecting chlamydial inclusions in irradiated McCoy cells. Using an indirect FA staining technique (4), we detected inclusions after 24 h of incubation. However, in most cases, these were too small to fulfill criteria for positive identification. After 30 h of incubation, inclusions stained with this technique were larger, with a defined morphology, and were thus easier to identify.

A study was carried out to compare the sensitivity of the FA stain with that of Giemsa for detection of inclusions in cell cultures. Inoculating 41 clinical specimens into irradiated McCoy cells, we found no marked differences in the rate of chlamydial isolation or the number of inclusions produced in the monolayers, whether they were stained by FA after approximately 40 h of incubation or by Giemsa after 60 h of incubation.

The immunoperoxidase staining technique has been used to study the developmental cycle of *C. psittaci* in the peritoneal macrophages of infected mice (13). We have compared the sensitivity of immunoperoxidase staining (peroxidase-antiperoxidase method) (14) with that of FA and Giemsa for detecting

TABLE 1. *Isolation of C. trachomatis from active untreated cases of chlamydial ocular infections*

Ocular infection	No. of patients	No. positive
Ophthalmia neonatorum	53	48 (90.5%)
Inclusion conjunctivitis or TRIC punctate keratoconjunctivitis	732	660 (90.1%)
Endemic trachoma (sexually transmitted origin)	25	18 (72%)
Hyperendemic trachoma (eye-to-eye transmission)		
Tunisia (school children)[a]	70	23 (32.8%)
Iran (preschool children)	147	58 (39.4%)

[a] This work was carried out in collaboration with C. R. Dawson and T. Daghfous.

chlamydial inclusions in irradiated McCoy cells. In the first experiment, eight laboratory isolates of feline keratoconjunctivitis agent (*C. psittaci*) were inoculated. The number of inclusions detected was 56% and 61% higher with immunoperoxidase and FA staining, respectively, than with Giemsa. In the second experiment, 50 conjunctival specimens collected from cats suffering from chlamydial keratoconjunctivitis (Darougar et al., p. 186, this volume) were inoculated into irradiated McCoy cells. The number of positive chlamydial cultures detected by the immunoperoxidase staining technique was 38 (76%) compared with 19 (38%) obtained with Giemsa.

ISOLATION OF *C. TRACHOMATIS*

The simplified one-passage technique of culture in irradiated McCoy cells is being used in our laboratory for routine laboratory diagnosis of chlamydial ocular and genital infections. This technique also provides microbiological back-up for clinical, epidemiological, and therapeutic studies of chlamydial infections. Other modifications include the collection of specimens by swabbing, addition of fetal calf serum to transport medium (3%, vol/vol), and transportation and storage of clinical specimens in a portable liquid nitrogen refrigerator. All clinical specimens are centrifuged at 2,500 × *g* for 1 h at 35°C except those collected from cases of hyperendemic trachoma in Iran, which are centrifuged at 15,000 × *g*.

The results of isolation of *C. trachomatis* from ocular infections are shown in Tables 1 and 2.

TABLE 2. *Isolation of C. trachomatis in relation to the severity of hyperendemic trachoma*

Grade of severity	Percentage of isolation	
	Tunisia	Iran
Severe	72.7	68
Moderate	29.2	20
Mild	23.3	16
Trivial	0	12

REFERENCES

1. Jones BR: Ocular syndromes of TRIC virus infection and their possible genital significance. Br J Vener Dis 40:3–18, 1964
2. Gordon FB, Harper IA, Quan AL, et al: Detection of Chlamydia (Bedsonia) in certain infections in man. I. Laboratory procedures: comparison of yolk sac and cell culture for detection and isolation. J Infect Dis 120:451–462, 1969
3. Darougar S, Dwyer RStC, Treharne JD, et al: Comparison of laboratory methods of diagnosis of chlamydial infections, p 445–460. *In* RL Nichols (ed), Trachoma and related disorders caused by chlamydial agents. Excerpta Medica, Amsterdam, 1971
4. Darougar S, Treharne JD, Dwyer RStC, et al: Isolation of TRIC agent (Chlamydia) in irradiated McCoy cell culture from endemic trachoma in field studies in Iran. Comparison with other laboratory tests for detection of Chlamydia. Br J Ophthalmol 55:591–599, 1971
5. Darougar S, Kinnison JR, Jones BR: Simplified irradiated McCoy cell culture for isolation of Chlamydia, p 63–70. *In* RL Nichols (ed), Trachoma and related disorders caused by chlamydial agents. Excerpta Medica, Amsterdam, 1971
6. Darougar S, Jones BR: Conjunctival swabbing for the isolation of TRIC agent (Chlamydia). Br J Ophthalmol 55:585–590, 1971
7. Darougar S, Cubitt S, Jones BR: Effect of high-speed centrifugation on the sensitivity of irradiated McCoy cell culture for the isolation of Chlamydia. Br J Vener Dis 50:308–312, 1974
8. Gordon FB, Quan AL: Occurrence of glycogen in inclusions of psittacosis-lymphogranuloma venereum-trachoma agents. J Infect Dis 115:186–196, 1965
9. Hanna L, Okumoto M, Thygeson P, et al: TRIC agent isolated in the United States. X. Immunofluorescence in the diagnosis of TRIC agent infections in man. Proc Soc Exp Biol Med 119:722–728, 1965
10. Nichols RL, McComb DE, Haddad N, et al: Comparison of fluorescent antibody, Giemsa and egg isolation methods for detection of trachoma virus in human conjunctival scrapings. Am J Trop Med Hyg 12:223–229, 1963
11. Nichols RL, Bobb AA, Haddad NA, et al: Immunofluorescent studies of the microbiologic epidemiology of trachoma in Saudi Arabia. Am J Ophthalmol 63:1372–1408, 1967
12. Surman PG, Hardy D, Howarth WH: The immunofluorescent technique applied to trachomatous eye smears in aboriginal school children in South Australia. Am J Ophthalmol 63(Suppl):1361–1372, 1967
13. Mields W, Hentschke J, Becker W, et al: The immunoperoxidase method for detection of viral and chlamydial antigens. II. Demonstration of development cycle of Chlamydia psittaci in peritoneal macrophages of infected mice. Zentralbl Veterinaermed 21:48–57, 1974
14. Dougherty RM, Marucci AA, Distefano HS: Application of immunohistochemistry to study of avian leukosis virus. J Gen Virol 15:149–162, 1972

Sensitivity of Cell Culture for Isolation of *Chlamydia trachomatis* from Genital Sources

BERTTINA B. WENTWORTH

Although it seems obvious that cell culture serves better than embryonated eggs for the isolation of *Chlamydia trachomatis* from genital sources (1–3), it is difficult to determine the sensitivity of our present methods for diagnosis of chlamydial infections. It also seems impossible, as yet, to determine the optimal isolation system. In the few cases where more than one cell system has been used to test the same specimens, there has been little difference between systems. Although all the other cell systems tested appear to give slightly better results than the originally described irradiated McCoy cells (4), that is, HeLa 229 (5), BHK-21 (6), and 5-iodo-2′-deoxyuridine (IUdR)-treated McCoy cells (7), none of these gave significantly more isolations than the irradiated cells, and all increases were small.

Given that all present methods have yielded approximately the same results, how can one determine whether cell culture is a sensitive tool for isolation of genital *Chlamydia*? If one looks at isolations from males, studies of nongonococcal urethritis (NGU) have shown isolation rates from as low as 5% (8) to almost 80% (9), according to the population studied. The larger studies show isolation from about 40% of NGU patients and give rise to a question concerning other etiologies for this syndrome (9–13). Are the other half of these cases due to different etiological agents, or are these really cases in which we have failed to isolate a chlamydial agent by present methods?

The classical methods for evaluation of sensitivity of bacterial isolation have been three: (i) multiple cultures of the same site at the same time, (ii) sequential cultures of the same site over time, and (iii) comparison of isolation with serological evidence of infection.

There appear to be no data which bear on the first criterion. Criterion (ii) is complicated by the use of antibiotic therapy, but in a Seattle treatment study chlamydiae were recovered 8 to 10 days after the initial culture from all 10 untreated NGU patients who had initially positive cultures, though from none of 10 treated cases cultured within 3 days after completion of 7 days of therapy (14).

Criterion (iii) is amenable to study by the micro-immunofluorescence test (15, 16). In two Seattle NGU studies, this test was used to evaluate antibody response, and either irradiated or IUdR-treated McCoy cells were used for isolation (13; W. R. Bowie et al., in preparation). In the first study (Tables 1 and 2), 36 isolations were obtained from 86 patients whose paired sera were examined for significant changes in antibody, that is, fourfold or greater increases in titer or seroconversion from less than 1:8. The overall rate of antibody prevalence in this group of patients was high; 23 of 36 (64%) patients yielding isolates had antibody in their initial sera, and 25 of 50 (50%) patients without isolations were also seropositive. There were 11 significant changes in antibody titer among the 36 patients with chlamydial isolates; 4 were rises in titer, and the remaining 7 were seroconversions among the 13 patients without antibody in their initial sera. Among the 50 patients without isolations, there were 2 patients whose sera showed significant

TABLE 1. *Nongonococcal urethritis: changes in antibody status versus isolation*

	Chlamydial isolation[a]					
	Positive[b]			Negative[c]		
Study	Significant change	Stable	Seronegative	Significant change	Stable	Seronegative
A	11 (30)	19 (53)	6 (17)	2 (4)	24 (48)	24 (48)
B	12 (63)	5 (26)	2 (19)	3 (8)	7 (17)	30 (75)
Combined	23 (42)	24 (44)	8 (14)	5 (6)	31 (34)	54 (60)

[a] Data are expressed as number (percentages in parentheses) of paired sera showing significant fourfold or greater change in antibody titer or seroconversion.
[b] Study A, $n = 36$; study B, $n = 19$; combined, $n = 55$.
[c] Study A, $n = 50$; study B, $n = 40$; combined, $n = 90$.

changes in titer. One had a rise in titer and one had a seroconversion from a titer of less than 1:8. If these two, added to the 36 isolations, represent a true rate of 38 chlamydial etiologies among these patients, isolation has revealed 95% of the cases.

In a more recent study (Tables 1 and 2), there were 22 isolations from 62 NGU patients. The patients in this study were specially selected to be less than 36 years of age, without history of previous NGU or gonorrhea, and with symptoms of less than 1 month's duration. This selection resulted in a population with much lower antibody prevalence (28%) than that found among the patients of the first study. Of the 22 patients with chlamydial isolates, 10 (45%) had antibody, versus 20% of those 40 without isolates. Twelve of 19 patients with isolates and paired sera had significant changes in antibody titer; 4 showed fourfold or greater changes in titer, and 8 were seroconversions in patients without initial antibody. There were an additional 3 patients with significant changes in antibody titer among the 40 patients without isolations, so that isolation demonstrated 88% of these cases.

Since significant changes in antibody titer were not observed among all the patients from whom chlamydiae were isolated, we must consider that some of those patients without isolations and with stable antibody titers may still have NGU of chlamydial etiology, which both serology and isolation have failed to reveal. This argument would add a possible 24 cases to the first study and leave 24 who remained seronegative. However, in the second study, only 7 of the 37 patients without

TABLE 2. *Nongonococcal urethritis: serology and chlamydial isolations*

Data	Isolation +, serology +	Isolation −, serology +
Study A	11/36 (31%)	2/50 (4.0%)
Study B	12/22 (55%)	3/40 (7.5%)
Combined data	23/58 (42%)	5/90 (5.5%)

isolations showed stable titers in serial sera, and no antibody could be demonstrated in sera from the other 30, even though from 2 to 5 serial sera (with a mean of 3.3 sera per patient) were examined. Either these 30 patients were not infected with *Chlamydia* or they did not respond to infection with antibody production.

The combined data from these two studies (Table 2) show a sensitivity of 92% for isolations versus serology, and for neither of the studies, nor for the combined data, are the differences statistically significant. If the antibody studies represent the truth about chlamydial infection, it is obvious that chlamydiae are not responsible for those one-half of NGU cases not yielding isolations in cell culture.

Isolations from females are more difficult to evaluate, and cell culture may be less effective for demonstrating cervical than for demonstrating urethral infection. This is due to the extent to which other organisms in cervical specimens interfere with isolation of chlamydiae. Contaminated cultures have resulted in unsatisfactory tests in about 10% of some NGU studies (10, 11), but the use of gentamicin appears to have reduced contamination among cervical cultures from more than 20% (17) to 6 to 8% (18, 19). In a Seattle study, 385 cervical specimens were examined for *Chlamydia*, *Mycoplasma*, *Trichomonas*, *Candida*, gonococci, cytomegalovirus, and *Herpesvirus hominis*, and 44% of the specimens yielded two or more of these agents (20). *C. trachomatis* was recovered from 22% of the specimens, and there was no evidence that chlamydiae were less prevalent than expected in specimens yielding other organisms than in specimens not yielding other isolates. Although *C. trachomatis* was isolated from 27% of the patients with cervical abnormality versus 11% of controls, *Mycoplasma hominis*, *Ureaplasma urealyticum*, cytomegalovirus, *Neisseria gonorrhoeae*, and *Herpesvirus hominis* were also significantly associated with cervical abnormality (L. T. Gutman, in preparation). Antibody prevalence was higher (92%) among *Chlamydia* culture-positive patients than among culture-negative ones (41%), but no paired sera determinations are available for judging sensitivity of isolation by changes in antibody status. Thus, sensitivity of cell culture for determining chlamydial infection in females appears to be unknown at present.

One must also ask whether ability to isolate *C. trachomatis* from genital specimens varies with the immunological type of the infecting strain. No significant difference in rate of isolation by type could be shown for Seattle studies, when the type of isolate obtained in the McCoy cell system was compared with the type of antibody demonstrated by the immunofluorescence technique. The combined data on 118 isolates from 299 patients whose sera showed serological evidence of infection were evaluated. All types, A through K, were represented among the sera, and all but type A were represented among the isolates. Paired sera were not available for all of these patients, and those showing significant changes in antibody titer were distributed among six serotypes. There were, therefore, too few comparisons to evaluate significant differences between types. The overall rate of isolation was 39%, and there was no difference in the distribution of types or of type-specific antibody between the 55% of the patients who were male and the 45% who were female. The lowest rate of isolation, 30%, was found among the 71 patients with broadly reacting sera, not indicative of type. The rates of isolation were highest for types B and I, which were 60% and 56%, respectively, for patients whose sera showed these specificities. However, these rates are not significantly different from the rate of isolation for other types, and it does not appear that inability to isolate certain types was responsible for failure to identify chlamydial etiology. It does appear, therefore, that cell culture may be a relatively sensitive technique for isolation of *Chlamydia* from male NGU patients. Indeed, since we have no serological technique for evaluating the sensitivity of gonococcal isolation methods, perhaps we should speak more confidently of "NCU" than of "NGU."

REFERENCES

1. Gordon FB, Harper IA, Quan AL, et al: Detection of *Chlamydia* (*Bedsonia*) in certain infections of man. I. Laboratory procedures: comparison of yolk sac and cell culture for detection and isolation. J Infect Dis 120:451–462, 1969
2. Darougar S, Dwyer RSC, Treharne JD, et al: A comparison of laboratory methods of diagnosis of chlamydial infection, p 445–460. *In* RL Nichols (ed), Trachoma and related disorders caused by chlamydial agents. Excerpta Medica, Amsterdam, 1971
3. Dunlop EMC, Hare MJ, Darougar S, et al: Detection of *Chlamydia* (*Bedsonia*) in certain infections of man. II. Clinical study of genital tract, eye, rectum, and other sites of recovery of *Chlamydia*. J Infect Dis 120:463–470, 1969
4. Gordon FB, Quan AL: Isolation of the trachoma agent in cell culture. Proc Soc Exp Biol Med 118:354–359, 1965
5. Kuo C-C, Wang S-P, Wentworth BB, et al: Primary isolation of TRIC organisms in HeLa 229 cells treated with DEAE-dextran. J Infect Dis 125:665–668, 1972
6. Blyth WA, Taverne J: Cultivation of TRIC agents: a comparison between the use of BHK-21 and irradiated McCoy cells. J Hyg 72:121–128, 1974
7. Wentworth BB, Alexander ER: Isolation of *Chlamydia trachomatis* by use of 5-iodo-2-deoxyuridine-treated cells. Appl Microbiol 27:912–916, 1974
8. Ford DK, McCandlish L: Isolation of human genital TRIC agents in nongonococcal urethritis and Reiter's disease. Br J Vener Dis 47:196–197, 1971
9. Darougar S, Jones BR, Kinnison JR, et al: Chlamydial infection. Advances in the diagnostic isolation of *Chlamydia*, including TRIC agent, from the eye, genital tract, and rectum. Br J Vener Dis 48:416–420, 1972
10. Dunlop EM, Vaughan-Jackson JD, Darougar S, et al: Chlamydial infection. Incidence in 'non-specific' urethritis. Br J Vener Dis 48:425–428, 1972
11. Oriel JD, Reeve P, Powis P, et al: Chlamydial infection. Isolation of *Chlamydia* from patients with non-specific genital infection. Br J Vener Dis 48:429–436, 1972
12. Richmond SJ, Hilton AL, Clarke SKR: Chlamydial infection. Role of *Chlamydia* subgroup A in non-gonococcal and post-gonococcal urethritis. Br J Vener Dis 48:437–444, 1972
13. Holmes KK, Handsfield HH, Wang S-P, et al: Etiology of nongonococcal urethritis. N Engl J Med 292:1199–1205, 1975
14. Handsfield HH, Alexander ER, Wang S-P, et al: Differences in the therapeutic response of *Chlamydia*-positive and *Chlamydia*-negative forms of nongonococcal urethritis. J Am Vener Dis Assoc 2:5–9, 1976
15. Wang S-P, Grayston JT: Immunologic relationship between genital TRIC, lymphogranuloma venereum, and related organisms in a new microtiter indirect immunofluorescence test. Am J Ophthalmol 70:367–374, 1970
16. Wang S-P, Grayston JT, Alexander ER: Simplified microimmunofluorescence test with trachoma-lymphogranuloma venereum (*Chlamydia trachomatis*) antigens for use as a screening test for antibody. J Clin Microbiol 1:250–255, 1975
17. Oriel JD, Powis PA, Reeve P, et al: Chlamydial infections of the cervix. Br J Vener Dis 50:11–16, 1974

18. Burns DCM, Darougar S, Thin RN, et al: Isolation of *Chlamydia* from women attending a clinic for sexually transmitted disease. Br J Vener Dis 51:314–318, 1975
19. Wentworth BB: Use of gentamicin in the isolation of subgroup A *Chlamydia*. Antimicrob Agents Chemother 3:698–702, 1973
20. Wentworth BB, Bonin P, Holmes KK, et al: Isolation of viruses, bacteria and other organisms from venereal disease clinic patients: methodology and problems associated with multiple isolations. Health Lab Sci 10:75–81, 1973

Chairman's Report and Discussion

DEREK HOBSON

The precision and range of laboratory diagnostic procedures has increased so greatly in the past few years that the diagnosis of active endemic trachoma and of oculogenital trachoma-inclusion conjunctivitis (TRIC) agent infection in temperate countries can now be confirmed in most patients within 2 to 3 days, by direct demonstration of chlamydiae, by isolation in tissue culture, or by finding indirect evidence of infection through specific antibody responses in serum or tears.

By combining these diagnostic approaches, a more exact measure of the true rate of chlamydial infection in a population can now be obtained as an essential base line in evaluating the sensitivity and specificity of future rapid diagnostic procedures and in assessing the results of prophylaxis and chemotherapy.

DIRECT EXAMINATION

Staining of cells scraped from the patient's lesions remains a rapid and useful diagnosis in hyperendemic trachoma, by demonstrating elementary bodies or glycogen matrix in cytoplasmic inclusions. The application of these methods to rapid diagnosis of genital chlamydial infections is less rewarding, mainly because of the difficulties in getting clean enough or large enough samples of infected transitional epithelial cells. The introduction of fluorescein-labeled chlamydial antibodies as a specific stain has improved the precision and sensitivity of these methods. The development of histochemical methods may further improve direct diagnosis from cell scrapings, since the recognition that chlamydiae code for specific enzymes distinct from those of the mammalian cells in which they grow. A useful side-benefit of improving direct methods is that they would be as valuable in detecting early chlamydial growth in tissue culture cells as they are in the patients' own cells, again improving the speed and precision of diagnosis.

An alternative direct approach is to look for free chlamydial elementary bodies shed into the patients' secretions. Electron microscopy is already routine in diagnosing poxvirus and herpesvirus infections and viral hepatitis, so it is logical to explore its role in chlamydial infections, especially since elementary bodies can be precisely identified by specific immunoferritin staining.

ISOLATION OF CHLAMYDIAE

An important milestone in the diagnosis of *Chlamydia trachomatis* infections was the demonstration that these agents grew well in irradiated McCoy cells. Equally important, now, is the increasing evidence that this system is not unique. Tissue culture isolation procedures are now well within the scope of any laboratory equipped for routine virological diagnosis. Replicating cultures of McCoy, HeLa 229, or other cell lines can be used satisfactorily, or McCoy cells can be chemically pretreated to control their replication, which assists in making sufficient nutrients available to the parasite as well as the host cell. In genital infections, all current tissue culture isolation procedures appear to give comparable results in comparable populations, provided time and trouble are taken in standardizing the method itself and in ensuring the quality of the clinical specimens put into it.

In endemic trachoma, difficulties in obtaining clean specimens from active sites of infection and transporting them to distant laboratories make bigger demands on direct examination or serological methods for diagnosis than is the case with genital infections.

IMMUNE RESPONSES TO *C. TRACHOMATIS*

For large-scale diagnostic and epidemiological use, serological methods must be simple, sensitive, and specific. The main drawbacks in the past have been lack of precise information about (i) the antigenic constitution of *C. trachomatis* (TRIC agents) vis à vis lymphogranuloma venereum (LGV) or *C. psittaci* and (ii) the number of TRIC serotypes or their relative frequency in different types of disease in different geographical areas. Practical difficulties in obtaining enough antigen to test, especially from strains freshly isolated from patients, have now largely been overcome by the development of micromethods using purified elementary body suspensions or the actual inclusions in infected tissue culture cells to detect the patient's antibody or to identify the patient's organism by fluorescence methods.

These methods, together with immunochemical and electrophoretic procedures, have divided the TRIC strains of *C. trachomatis* into 12 serotypes sharply demarcated from the three serotypes of LGV agent, and monospecific typing antisera can now be prepared. Types A, B, and C acount for most cases of hyper-endemic trachoma, whereas types D, E, and F are predominant in genital infections. Antibody responses to individual types are sufficiently distinct to be of epidemiological and diagnostic value. A rise in titer, or appearance of immunoglobulin M (IgM) antibody, can usually be found in first-episode infections, and there is good correlation between immunological results and positive tissue culture isolation. Because only a limited number of serotypes of *C. trachomatis* are found in genital infections, it seems justifiable to use only a restricted pool of antigens or even a single type E antigen to screen patients in venereal disease clinics, rather than the full micro-immunofluorescence test. In general, serum antibody frequency and titers are higher in women than in men with genital infection and are higher in genital infections than in endemic trachoma. Specific antibody, presumably IgA, can be found in cervical secretions of many infected women. In endemic trachoma, detection of antibody in tears is a more valuable method of diagnosis than tests on sera. Presumably, these findings reflect differences in the total amount of antigen and its distribution in the body.

The particular problems of sexually transmitted disease clinic populations were discussed here and in the following section, e.g., the need for rapid diagnosis and treatment, the frequent failure of patients to return for serial serum samples, and the high frequency of stable antibody titers from previous infections. In this situation, the detection of antichlamydial IgM antibody, which is often more type specific than IgG responses in these patients, can be of crucial importance.

Much remains to be proven about the exact relevance of antibody to current, as opposed to past, genital infection. Immunological tests on single samples of serum and direct staining procedures do not yet constitute acceptable rapid diagnostic alternatives to tissue culture isolation procedures. However, the most informative results may accrue from a combined diagnostic effort using all these methods. It seems likely that the true incidence of genital infections associated with *C. trachomatis* has already been established by combined clinical and laboratory investigation in America and Europe over the past 10 years. The outstanding problems for the future are thus (i) to determine the etiology of the very common nonspecific sexually transmitted disease which is neither gonococcal nor chlamydial and (ii) to prove that *C. trachomatis* is truly the causal agent of the disease in those patients in whom it is found, and not merely an opportunistic secondary invader.

Session VI

PRACTICAL PROBLEMS IN ISOLATION AND GROWTH OF *CHLAMYDIA TRACHOMATIS* IN TISSUE CULTURE

Quantitative Aspects of the Growth of *Chlamydia trachomatis* in Diagnostic Tissue Culture Procedures

F. WILLIAM A. JOHNSON, DEREK HOBSON, ELISABETH REES, AND I. ANNE TAIT

Tissue culture (TC) techniques for isolating *Chlamydia trachomatis* are of proven value in epidemiological surveys in large research centers. In the next few years, they are certain to be widely adopted for the routine investigation of all patients with suspected sexually transmitted disease. In the scaling-up process, some form of quality control is desirable to ensure that the sensitivity and reproducibility of the present procedures is being maintained at a constant high level and that any future technical modifications, to conform with different laboratory facilities and practice, do not affect these parameters.

Comparison of isolation rates of *C. trachomatis* between successive groups of patients or between patients in different clinics can only give information of limited value. Independent proof of infection (e.g., by direct smear or serology) is rarely feasible in all patients (Schachter and Dawson, p. 337, this volume). Hence, the true base rate of chlamydial infection is unknown, and it is difficult to interpret either concordance or divergence of laboratory results with different patients at different times in different cities.

However, the nature of the current TC procedures provides an inbuilt quantitative means of monitoring their efficiency. Since *C. trachomatis* gives only a single cycle of development (1), the number of inclusions finally detected in TC is directly proportional to the number of viable chlamydial elementary bodies originally inoculated. Thus, a sensitive early warning of any change in laboratory precision can be obtained by determining the overall efficiency of plating, by regularly titrating low-passage laboratory strains of *C. trachomatis*, and by testing repeated swabs at short intervals from known positive, untreated patients.

This paper presents such an investigation of a technique with replicating McCoy cells used for the past 3 years in the diagnosis of chlamydial infections of women in Liverpool (2, 3; Rees et al., p. 67, this volume).

MATERIALS AND METHODS

Details of the preparation of cover slip monolayer cultures of replicating McCoy cells (MCC), media, diluents, and the isolation procedure for clinical specimens have been published (2, 3; Rees et al., p. 67, this volume).

The isolation, propagation, and storage of the laboratory strain SAL of *C. trachomatis* isolated in Liverpool, and methods of investigating its growth under different cultural conditions (pH, incubation temperature, medium, etc.), have previously been described (3). Other technical procedures are described below.

RESULTS

Variation between cover slip cultures. A single batch of MCC was prepared under standard conditions (3). Random samples were stained to confirm that >90% of the cover slip was covered evenly by cells with well-spread cytoplasm. All were then inoculated simultaneously with a single dilution of SAL strain estimated to yield 3,000 inclusions per MCC. After incubation for 48 h at 37°C, all MCC were Giemsa stained, and the total number of inclusions per MCC was counted under ×400 bright-field microscopy. The 48 cover slip counts followed a normal distribution pattern (range, 2,817 to 4,143; mean, 3,200; standard deviation, 325), and 93% of the observed counts were within ± 20% of the mean value.

Effect of different stains and time of examination. A single batch of MCC was inoculated simultaneously with a single dilution of SAL strain estimated to yield 40,000 inclusions per cover slip. All were incubated at 37°C, and randomly selected MCC were removed after 24, 48, or 72 h for staining with iodine, Giemsa, or carmine (4). Total inclusion counts per MCC were calculated from counts of 150 randomly chosen microscopic fields in each of two MCC for each stain and time. The mean size of inclusions was determined by measuring the diameter of 100 randomly selected inclusions in each MCC, with the use of an eye-piece micrometer (Table 1).

Giemsa staining at 48 h revealed the highest inclusion count. Even at 24 h, most of the inclusions were already detectable, but their small size made counting slow and fatiguing. By 72 h, inclusions were big and easy to count, but many were bursting, and presumably other infected McCoy cells had already ruptured and dropped off the glass, since the total count had fallen to 25% of the 48-h value.

Carmine staining was less effective than Giemsa for detecting inclusions at 24 h, but gave similar results at 48 and 72 h. The bright-red inclusions were clearly distinguished against the light-brown staining of the McCoy cells.

TABLE 1. *Comparison of number and size of inclusions of C. trachomatis SAL in simultaneously infected McCoy monolayers stained by different methods after different times of incubation*

Staining procedure	Incubation time (h)	Inclusion data		
		Count[a]	Diameter (μm)[b] Range	Diameter (μm)[b] Mean
Iodine	24	10	4.25–7.25	6
	48	10,250	4–18.75	8.75
	72	8,700	5.5–26.25	11.85
Giemsa	24	44,500	0.87–12	5.25
	48	49,600	7–41.75	12.75
	72	12,120	9.25–38.25	24.27
Carmine	24	11,100	2.0–11.0	5.6
	48	45,100	4.5–46.2	12.85
	72	12,180	8.25–32.75	18.15

[a] Total count per 16-mm cover slip. Mean of two cover slips.
[b] Mean of two diameters of each of 100 inclusions per cover slip.

Iodine resulted in significant underestimation of the number of inclusions present even at 48 h, and seemed incapable of allowing detection of early inclusions or many of the large late inclusions. This may be because of the variability in timing and amount of the glycogen-rich matrix in *C. trachomatis* inclusions in TC (5, 6).

Effect of changing laboratory conditions. Inclusion counts produced by a constant inoculum of SAL strain were determined after various modifications of the standard procedure (3). For each variable factor, two to four MCC were compared with the same number of MCC infected and incubated under standard conditions.

The results summarized in Table 2 indicate the large decrease in the sensitivity of McCoy cells to *C. trachomatis* infection that can result from comparatively small changes in the isolation technique. The only factor which was found to increase the inclusion count significantly was the use of increased force or time of centrifugation. These effects are discussed in greater detail elsewhere (3).

Clinical specimens. *C. trachomatis* was isolated from the cervix of 153 contacts of men with nongonococcal urethritis (NGU) within 2 to 3 days of their first visit to the sexually transmitted disease clinic. On their next visit 7 to 10 days later, a

TABLE 2. *Effect of modifications of the standard tissue culture procedure[a] on the efficiency of plating of C. trachomatis SAL*

Change in procedure	Mean inclusion count per cover slip[b]
None[a]	3,088
Centrifugation	
Nil	62
2,500 × *g*, 15 min	2,120
2,500 × *g*, 3 h	19,830
4,500 × *g*, 1 h	23,400
Incubation temperature	
35°C	2,014
39°C	1,615
pH	
6.6	878
7.8	540
$NaHCO_3$ concentration at pH 7.2	
0.025 M	2,932
0.005 M	405
Medium (+ 10% calf serum, etc.)[c]	
Eagle MEM	1,646
BME	204
McCoy: initial cell count	
5×10^4/ml	633
5×10^5/ml	1,250
McCoy: age of monolayer when infected, 5 days	1,204

[a] In the standard procedure, McCoy cell cultures were prepared from 2×10^5 cells/ml and incubated for 24 h at 37°C. They were then inoculated, centrifuged at 2,500 × *g* for 1 h, and incubated for 48 h at 37°C. The standard medium was medium 199 plus 10% fetal calf serum and 0.02 M $NaHCO_3$.

[b] Mean of 40 cover slips in replicate experiments by standard procedure. Mean of two to four cover slips for each modification.

[c] MEM = minimal essential medium; BME = Eagle basal medium.

second cervical swab was taken and again examined by the standard procedure. There had been no intervening systemic or local therapy. The repeat swab was again positive in 141 (92.5%) of the women; the 8 patients with negative second swabs had inclusion counts of only 2 to 30 per MCC on the first occasion. In a group of 82 NGU contacts with negative primary swabs, the repeat swab was positive in only 2 (2.5%) women, with counts of two and eight inclusions per MCC.

In 98 women with a constant 7-day interval between two positive swabs, inclusion counts were closely similar in most cases on both occasions (Table 3). The overall distribution of counts in the whole group showed little change; only 8% of all swabs yielded fewer than 30 inclusions, whereas 75.4% of primary swabs and 69.8% of second swabs gave 100 or more inclusions per MCC.

DISCUSSION

Current TC techniques for isolating *C. trachomatis* are now well within the scope of any laboratory equipped for routine virological diagnosis. However, comparatively small changes in the detailed procedure or in the staining technique can significantly reduce the number of countable inclusions obtained from a given number of viable elementary bodies (3, 9; Rota, p. 314, this volume) and might thus impair the chance of diagnosing a patient with an early, low-grade, or imperfectly treated chlamydial infection.

The present results suggest that quantitative inclusion counts on all clinical specimens, and on low-passage laboratory strains of *C. trachomatis* at regular intervals, could provide a simple check on the confidence limits of the overall working procedure within the laboratory and, indirectly, on the quality of the specimens reaching the laboratory. Such routine counts may be of future clinical value in estimating the extent and severity of the infection and the possibility of relapse after chemotherapy.

The present methods seem capable of isolating a substantial number of infective particles of *C. trachomatis* from the majority of infected women, and it does not seem that small reductions in the sensitivity of the procedures would be reflected by any significant fall in the total number of new sexually transmitted disease patients found positive. However, a sustained high efficiency of plating is necessary (i) to insure against the marked loss in viability of chlamydia in specimens which need to

TABLE 3. *Distribution of inclusion counts in cervical swabs from 98 untreated women with C. trachomatis infection*

Inclusion counts per cover slip[a]	Percentage of women with stated count in cervical swab taken	
	On 1st visit	7 days later
1–10	6.3	9.9
10–100	17.7	18.8
100–1,000	43.1	36.3
1,000–10,000	18.3	22.6
10,000–100,000	14.0	10.9

[a] Each cover slip was inoculated with one-twelfth of the secretion recovered from each swab.

be sent to distant laboratories and stored overlong before TC inoculation, and (ii) to confirm that chemotherapy has eradicated the patient's infection and not merely depressed it to levels undetectable by the laboratory but still capable of causing clinical relapse at some later date.

The beneficial effect of increasing centrifugation demonstrated here confirms earlier findings (3, 8, 9) and suggests that current techniques may isolate only a fraction of the total number of infective particles shed by the patient. For most laboratories centrifugation of specimens at more than $2{,}500 \times g$ for 1 h is impracticable, but it seems worthwhile to explore further methods of increasing the efficiency of adsorption of *C. trachomatis* to TC cells, to help minimize the effect of the many other variations in laboratory and clinical conditions which can affect the isolation rate of chlamydiae in genital infections.

ACKNOWLEDGMENT

The studies described in this paper were supported by a project grant from the Medical Research Council of Great Britain.

REFERENCES

1. Gordon FB, Dressler HR, Quan AL, et al: Effect of ionizing irradiation on susceptibility of McCoy cell cultures to chlamydia trachomatis. Appl Microbiol 23:123–129, 1972
2. Hobson D, Johnson FWA, Rees E, Tait IA: Simplified method for diagnosis of genital and ocular infections with chlamydia. Lancet 2:555–556, 1974
3. Johnson FWA, Hobson D: Factors affecting the sensitivity of replicating McCoy cells in the isolation and growth of chlamydia A (TRIC agents). J Hyg 76:441–451, 1976
4. Johnson FWA: A comparison of staining techniques for demonstrating group A chlamydia in tissue culture. Med Lab Technol 32:233–238, 1975
5. Gordon FB, Quan AL: Occurrence of glycogen in inclusions of the psittacosis-lymphogranuloma venereum-trachoma agents. J Infect Dis 115:186–196, 1965
6. Croy JR, Kuo CC, Wang SP: Comparative susceptibility of eleven mammalian cell lines to infection with trachoma organisms. J Clin Microbiol 1:434–439, 1975
7. Rota TR, Nichols RL: *Chlamydia trachomatis* in cell culture. Appl Microbiol 26:560–565, 1973
8. Darougar S, Cubitt S, Jones BR: Effect of high speed centrifugation on the sensitivity of irradiated McCoy cell culture for the isolation of chlamydia. Br J Vener Dis 50: 308–312, 1974
9. Reeve P, Taverne J: Strain differences in the behaviour of TRIC agents in cell cultures. Am J Ophthalmol 63:1167–1173, 1967

Trachoma Agent Under Varying Cell Culture Conditions: Infection in McCoy and BHK-21 Cells

TERESA R. ROTA

The infection of cells in culture with ocular strains of *Chlamydia trachomatis* was pioneered by the late Francis Gordon and his associates (1), who brought together some observations from different fields which led to the successful cultivation of slow-growing chlamydiae in cell culture. The basic elements were (i) an increase in the glucose concentration, (ii) centrifugation of the chlamydia onto cell monolayers, and (iii) use of irradiated cells. The effect of glucose had been investigated by Vedros and Gordon (Bacteriol. Proc., p. 134, 1963). The initial observation on the effect of centrifugation had been reported by Weiss and Dressler for the rickettsiae (2). The effect of irradiation was observed by Fernandez (3), who noted the increased susceptibility of irradiated McCoy cells to blue-tongue virus infection. The McCoy cell line was adopted by Gordon et al., and the effects of irradiation were later closely scrutinized for the trachoma organisms in McCoy cells (4). The result of this conjunction of cultural conditions was the opening of a new field of research whose possibilities are still being investigated, and from which much is to be learned about the complicated host-parasite relationships involved in chlamydial infections. Early studies indicated the McCoy cell as a possible host of choice with which to establish knowledge leading to the production of large amounts of trachoma elementary bodies for biochemical studies and for the production of vaccines. These vaccines can readily be tested in highly susceptible animal models available, such as the owl monkey, *Aoutus trivirgatus* (5).

I have studied the effect of varying the environmental conditions on the infectivity of trachoma organisms for irradiated McCoy cells. I also report a comparison of the infectivity of three ocular trachoma strains for irradiated McCoy cells and irradiated BHK-21 Lister cells (6).

The culture conditions were varied by using synthetic media in the absence of serum, by changing the pH, the temperature of centrifugation, and the temperature of incubation, and by testing the effect of diethylaminoethyl-dextran (DEAE-D). These results have been reported in detail elsewhere (7, 8). The trachoma strain used was TRIC/B/SAU/Har-36/OT, which had been propagated serially in irradiated McCoy cells for 28 passages. McCoy cells were used from 7 to 10 days after irradiation with 5,000 rads from a ^{60}Co source. They were planted at 2.3×10^5 cells/ml on 12-mm-diameter cover slips in flat-bottomed tubes.

METHODS

Synthetic media. Three chemically defined synthetic media were used without serum: (i) Eagle minimal essential medium (MEM) with Earle basal salt solution (MEME); (ii) MEM with Hanks balanced salt solution (MEMH); and (iii) a modified MEM obtained in a powdered autoclavable form (Auto Pow; Flow Laboratories).

Preparation of the culture media and pH. Fixed concentrations of $NaHCO_3$ are present in MEME (2,200 mg/liter) and in MEMH (350 mg/liter). MEME is commercially available at pH 7.0, but once opened the pH increases; therefore it needs gassing with 5% CO_2 in air to preserve the pH once a given value has been reached in an opened vial. MEMH is available at pH 7.2 to 7.4. It was used at 7.4. This medium required no gassing. Various amounts of $NaHCO_3$ (from a solution containing 7,500 mg/liter) were added to modified MEM to obtain pH values of 7.0, 7.2, 7.5, 7.6, 7.8, and 8.0. This medium was also gassed in the centrifuged cultures.

All media contained phenol red as pH indicator, and they were supplemented with glucose (30 μmol/ml) and streptomycin (50 μg/ml). The pH values of the media were measured in a pH meter at the start of the experiments and in control cell cultures at the end of the experiments. The pH of the trachoma-inoculated cultures at the end of the experiments was ascertained by comparison with phenol red standards.

Trachoma-inoculated cultures were centrifuged at 800 $\times$ g for 1 h at 20°C and incubated at 34°C for 72 h. Infectivity was assessed by means of iodine staining and was expressed as the geometric mean number of inclusions in the entire cover slip at $\times$320 in six cultures per group per treatment.

RESULTS

Effects of synthetic media and pH. The routine culture medium used up to this time was MEME with 10% horse serum. The geometric mean number of inclusions in this control medium was 1,953. MEME without serum resulted in significantly (analysis of variance, $P < 0.01$) higher inclusion counts at pH 7.0 than at pH 7.4 or 7.7; geometric mean numbers were 2,878, 1,932, and 1,316, respectively. These results indicated that serum was not required for the development of the trachoma inclusions and that the inclusion number increased when the pH was at neutrality or close to it. Modified MEM used at pH 7.0 and 7.2 gave geometric mean numbers of 1,891 and 987, respectively, that is, 1.9-fold more inclusions at pH 7.0 than at 7.2. Modified MEM at pH 7.0 gave 1.5-fold less inclusions than MEME at pH 7.0. MEMH at pH 7.4 gave a geometric mean of 2,179 inclusions, only slightly higher than MEME with or without serum at the same pH. These results are shown in Fig. 1.

Effect of DEAE-D treatment. Pretreatment of irradiated McCoy cells with 20 μg of DEAE-D/ml for 30 min at room temperature followed by two washings in Hanks balanced salt solution (9) significantly increased the number of inclusions in trachoma-inoculated cultures incubated at 34°C for 48 and 72 h. There was a 46-fold increase in inclusions at 48 h and a 40-fold increase at 72 h in noncentrifuged cultures treated with DEAE-D as compared with nontreated, noncentrifuged cultures. Control cultures, centrifuged but not treated with DEAE-D, had 12- to 16-fold greater inclusion counts (at 48 and 72 h) than noncentrifuged DEAE-D-treated cultures. The culture medium used in these early experiments was MEME with 10% horse serum. DEAE-D pretreatment followed by incubation in modified MEM, in the absence of serum, at different pH values resulted in trachoma infectivity that was greater when the pH of the medium was close to neutrality (7.0 and 7.2) than at pH values of 7.5, 7.6, 7.8, and 8.0. The observed differences in inclusion counts at the pH values tested were significant (analysis of variance,

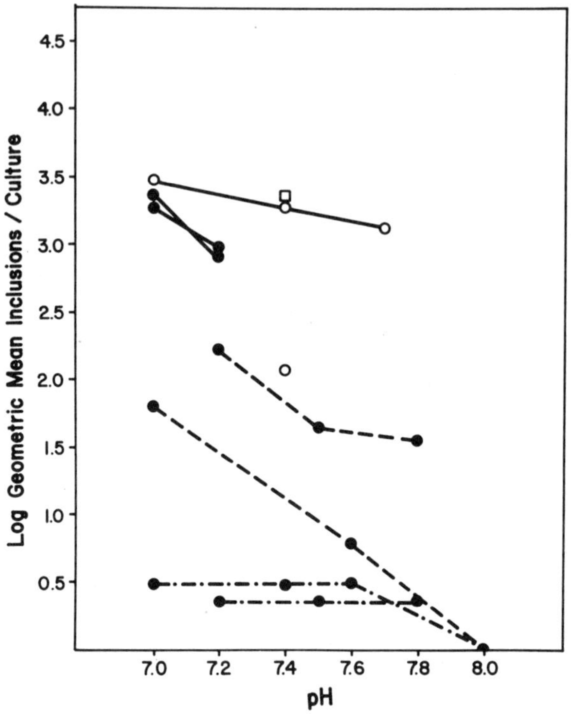

FIG. 1. *Effect of the pH of chemically defined media and effect of DEAE-D treatment on the number of inclusions observed in irradiated McCoy cells. Centrifuged cultures:* ○ —— ○, *MEME;* □, *MEMH;* ● —— ●, *modified MEM. Noncentrifuged, DEAE-D-treated cultures:* ○, *MEME;* ● – – – ●, *modified MEM. Noncentrifuged, non-DEAE-D-treated cultures:* ● –·–·– ●, *modified MEM. Six cultures per group per treatment.*

$P < 0.01$). The results are represented in Fig. 1. Enhancement of the trachoma infection by DEAE-D treatment was 22-fold at pH 7.0 and 73-fold at pH 7.2 over nontreated, noncentrifuged cultures. Centrifuged cultures, not treated with the polycation, had the highest inclusion counts, as reported (7).

Effect of temperature on trachoma infectivity. Figure 2 shows that trachoma-inoculated cultures centrifuged at 20°C and incubated for 48 h had an increase in the number of inclusions at temperatures of incubation above 34°C. There were two peaks of infectivity, at 35 and 37°C. Incubation for 72 h resulted in a decrease in inclusion counts at all temperatures tested, with the exception of 37°C. The increase in inclusions at 37°C may represent a second cycle of infection (4), fragmentation of inclusions within a cell, and/or the appearance of slow-developing inclusions. The reduction in inclusions from 48 to 72 h was due to maturation and rupture of inclusions, which then fall off the glass surface. Culture media were removed from the cultures at 72 h and centrifuged; the pellets were set on slides and stained by Giemsa's method. Cell debris associated with elementary bodies and cells bearing inclusion remnants were observed in these pellets. Inoculated cultures centrifuged

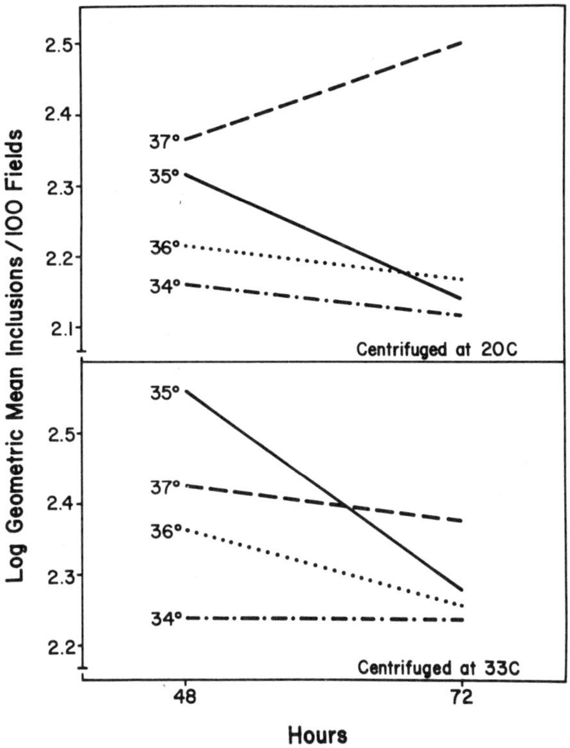

FIG. 2. *Effect of the temperature of centrifugation and the subsequent temperature of incubation on the number of inclusions observed in irradiated McCoy cells in MEME medium. Three cultures per group.*

at 33°C displayed basically the same trends observed when centrifugation was done at 20°C; that is, inclusion numbers increased by 48 h at all temperatures above 34°C, and thereafter the inclusions diminished from 48 to 72 h postinoculation, including those in cultures incubated at 37°C. In general, the inclusion values were higher (up to 1.7-fold) in cultures centrifuged at 33°C than at 20°C. One-way analysis of variance showed significant differences ($P < 0.01$) among the inclusion values observed at a given time after incubation at different temperatures. Two-way analysis of variance based on temperature of centrifugation versus time of incubation showed the temperature effect to be significant ($P = 0.03$) and the incubation time less significant ($P = 0.07$). The effect of centrifugation at 33°C and incubation at temperatures ranging from 35 to 39°C is shown in Fig. 3. Maximal infectivity was reached at 48 h postinoculation, and it was followed by rapid loss of inclusions by 72 h of incubation. Values between 37 and 60% of the inclusions present at 48 h were counted on the cover slips by 72 h.

The temperature of centrifugation (33°C) and the subsequent temperature of incubation had an effect on the number of cells with multiple inclusions. Figure 4 shows that multiply infected cells were more numerous at 35 and 37°C after 48 h of

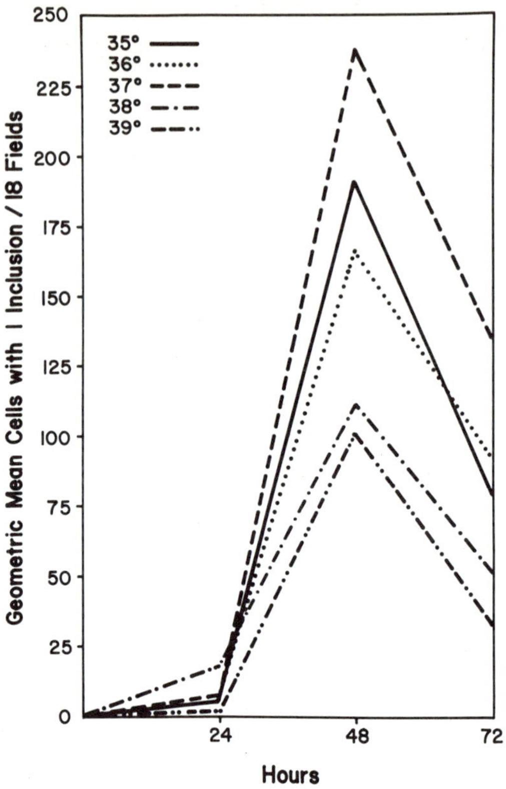

FIG. 3. *Centrifugation at 33°C. Incubation temperatures ranging from 35 to 39°C. Six cultures per group.*

incubation and at 36 and 37°C after 72 h of incubation. Fragmentation, fusion of the inclusions within a cell, and/or maturation and rupture of the inclusions in multiply infected cells are among the factors which contributed to the variations observed after 72 h.

Effect of DEAE-D treatment and the temperature of incubation. Figure 5 shows that the DEAE-D enhancement of the trachoma infectivity for McCoy cells was affected by the temperature of incubation of the cultures. The inclusion counts increased as the temperature was raised above 34°C. The infectivity was higher at 72 h than at 48 h, indicating a slow progression of the development of the inclusions which still remained on the glass surface by 72 h. Non-DEAE-D-treated, noncentrifuged cultures, though minimally infected, also manifested increased inclusion counts as the temperature of incubation increased. The inoculated cultures centrifuged (at 20°C, 800 × *g*, for 1 h) had the highest inclusion numbers, as expected.

Comparison of infectivity of *C. trachomatis* for irradiated McCoy and irradiated BHK-21 Lister cells. Three trachoma strains, TRIC/A/ET/Har-13/OT, TRIC/B/SAU/Har-36/OT, and TRIC/C/SAU/Har-32/OT, belonging to serotypes A, B, and

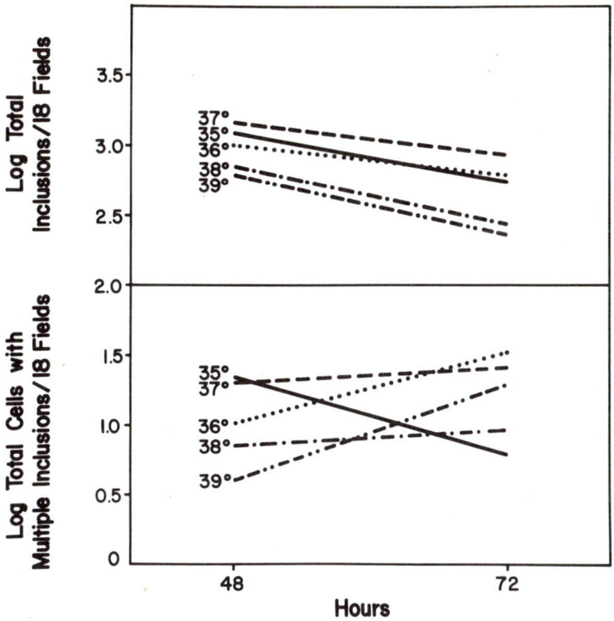

FIG. 4. *Effect of the temperature of centrifugation (33°C) and the subsequent temperature of incubation on the number of inclusions and the number of multiple inclusions per cell observed in irradiated McCoy cells in MEME medium. Six cultures per group.*

C (10), were used in these studies. The strains had been isolated in the yolk sac and later cloned in eggs (11).

BHK-21 Lister cells were used from 3 to 4 days and McCoy cells from 5 to 7 days after irradiation with 5,000 rads. To compensate for differences in cell size, BHK-21 cells were planted at 4.5×10^5/ml and McCoy cells were planted at 2.5×10^5/ml on 12-mm-diameter cover slips in flat-bottomed tubes. The culture medium was modified rubella medium (12) at pH 7.3, supplemented with glucose (30 μmol/ml; Vedros and Gordon, Bacteriol. Proc., p. 134, 1963), glutamine (200 mM), and streptomycin (50 μg/ml).

Titrations of each trachoma strain inoculum were performed by use of 10-fold serial dilutions prepared in the culture medium, which were inoculated into triplicate cultures of each cell type. The cultures were centrifuged at $800 \times g$ for 1 h at 33°C. After incubation at 35°C for 46 h, the cultures were fixed with methanol and stained with iodine. The results (Table 1) are expressed as the geometric mean number of inclusions in 100 fields at $\times 320$. From this value, the apparent titer (inclusion-forming units [IFU] per milliliter) of the original undiluted yolk sac pool in each cell line was calculated.

Inclusion counts reflected the 10-fold difference in the dilutions used. Lower than expected inclusions were attributed to maturation and rupture of inclusions by 46 h and to cell loss, particularly at 2×10^{-4} dilutions.

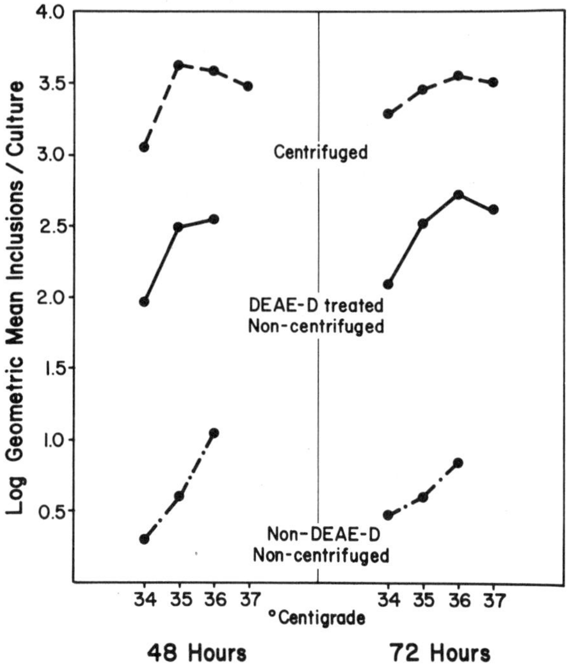

FIG. 5. *Effect of DEAE-D treatment and the subsequent temperature of incubation on the number of inclusions observed in irradiated McCoy cells in MEME medium. Five or six cultures per group per treatment. Cultures centrifuged at 20°C, 800 × g, for 1 h.*

DISCUSSION

In the studies reported here, some variations of the environmental conditions were explored in an effort to obtain increased infectivity of trachoma organisms for cells in culture. The use of chemically defined synthetic media in the absence of serum and varying the pH of the media led to the conclusion that pH values close to neutrality were best when MEME and modified MEM were used. On the other hand, MEMH, which is supplied commercially at pH 7.4 or lower, was as effective as the routine medium previously used (MEME with 10% horse serum gassed with 5% CO_2 in air). MEMH is highly recommended since it requires neither gassing nor additional buffers.

Temperature played a role in the number of inclusions observed and in the enhancing effect of DEAE-D on the trachoma infection of McCoy cells. The role of temperature may be associated with the activation of enzymes within the elementary bodies and/or enzyme systems of the host cell. The observation of two peaks of infectivity at 35 and 37°C suggested the possibility of heterogeneity (as to rate of growth) within the population of elementary bodies. Temperature may act to select elementary bodies which replicate faster at either 35 or 37°C. The experiments reported in this study were performed with an ocular trachoma strain (Har-36) originally cloned in eggs and later propagated serially in irradiated McCoy

TABLE 1. *Comparison of efficiencies of infection of trachoma strains of serotypes A, B, and C in centrifuged[a] cultures of irradiated[b] McCoy and irradiated BHK-21 Lister cells*

Trachoma strain grown in yolk sac	Cell type	Yolk sac dilution	Inclusions (geometric mean number/ 100 fields)	IFU/ml of yolk sac pool	Ratio of IFU/ml, BHK-21 to McCoy
Har-13 serotype A	McCoy	2×10^{-4} 2×10^{-5}	304 31	1.12×10^{7} 1.15×10^{7}	
	BHK-21	2×10^{-4} 2×10^{-5}	958 107	3.54×10^{7} 3.96×10^{7}	3.1 3.44
Har-36 serotype B	McCoy	2×10^{-4} 2×10^{-5}	723[c] 110	2.67×10^{7} 4.07×10^{7}	
	BHK-21	2×10^{-4} 2×10^{-5}	3,854[c] 692	1.43×10^{8} 2.56×10^{8}	5.35 6.29
Har-32 serotype C	McCoy	2×10^{-4} 2×10^{-5}	1,001 97	3.7×10^{7} 3.58×10^{7}	
	BHK-21	2×10^{-4} 2×10^{-5}	2,582[c] 403	9.55×10^{7} 1.48×10^{8}	2.58 4.13

[a] At $800 \times g$, 1 h, 33°C.
[b] With 5,000 rads.
[c] Some mature inclusions had ruptured by 46 h as indicated by cell loss and debris of inclusions left on the cover slips.

cells. As a precautionary measure, trachoma strains are maintained in our laboratory at 35°C, both when grown in the chick embryo yolk sac and when grown in cell cultures. Evidence for slow- and faster-developing elementary bodies has been presented by Becker et al. (13). These authors observed that some elementary bodies began to develop as early as 6 h after infection, whereas others did not initiate enzyme synthesis until 24 h postinoculation. Trachoma organisms of serotypes A, B, and C consistently produced more inclusions in BHK-21 Lister cells than in McCoy cells. The ratios of IFU per milliliter in BHK-21 as compared with McCoy cells were 3.44, 6.29, and 4.13, respectively (Table 1). In conclusion, the BHK-21 cells showed higher susceptibility to infection by the three ocular trachoma strains used than did the McCoy cells.

The work of Blyth and Taverne (14) compared nonirradiated BHK-21 and irradiated McCoy cells. These authors reported that BHK-21 "gradually degenerated" after exposure to 4,000 rads. Degeneration of BHK-21 Lister due to irradiation effect was not observed in my studies, even when the cells received 5,000 rads. Blyth and Taverne tested strains MRC-4, MRC-4 fast, and TW-3 (serotype C), and found ratios of 1.32, 1.85, and 1.51, respectively, more inclusions in nonirradiated BHK-21 than in irradiated McCoy cells. Yolk sac inocula of the trachoma strains used in the present report have not been tested for direct comparison in nonirradiated BHK-21 Lister and irradiated McCoy cells. However, trachoma strain Har-36 (serotype B) propagated serially in irradiated BHK-21 Lister cells consistently was more effective (up to sevenfold) in irradiated than in nonirradiated BHK-21 cells. For this reason, my studies concentrated on the comparison of BHK-21 Lister and McCoy cells after they had been irradiated. Irradiated BHK-21 Lister cells have been used successfully in our laboratory for the serial propagation of trachoma strains and for the isolation of chlamydiae from clinical specimens. The latter are performed by use of the micro-cell culture method reported by McComb and Puzniak (12).

ACKNOWLEDGMENTS

I thank Roger L. Nichols and A. B. MacDonald for support during this investigation, D. McComb for suggestions, and J. Worcester for the statistical treatment of the data. All persons named above are from the Harvard School of Public Health.

This work was supported by Public Health Service grant 9 ROI EY-00812-08 from the National Eye Institute, by the Arabian American Oil Company, and by National Institutes of Health grant RO1-EYO 1568-01.

REFERENCES

1. Gordon FB, Quan AL: Isolation of the trachoma agent in cell culture. Proc Soc Exp Biol Med 118:354–359, 1965
2. Weiss E, Dressler HR: Centrifugation of rickettsiae and viruses onto cells and its effects on infection. Proc Soc Exp Biol Med 103:691–695, 1960
3. Fernandez VM: Irradiation of cells in tissue culture. VII. Studies on the susceptibility of blue-tongue virus on radiation induced giant cells in vitro. Z Zellforsch 50:433–443, 1959
4. Gordon FB, Dressler HR, Quan AL, et al: Effect of ionizing irradiation on susceptibility of McCoy cell cultures to *Chlamydia trachomatis*. Appl Microbiol 23:123–129, 1972
5. Bell SD Jr, Fraser CEO: Experimental trachoma in owl monkeys. Am J Trop Med Hyg 18:568–572, 1969
6. Capstick PB: Growth of baby hamster kidney cells in suspension. Proc R Soc Med 56: 1062–1064, 1963
7. Rota TR, Nichols RL: Infection of cell cultures by trachoma agent. Enhancement by DEAE-D. J Infect Dis 124:419–421, 1971
8. Rota TR, Nichols RL: *Chlamydia trachomatis* in cell culture. I. Comparison of efficiencies of infection in several chemically defined media, at various pH and temperature values and after exposure to diethylaminoethyl-dextran. Appl Microbiol 26:560–565, 1973
9. Harrison MJ: Enhancing effect of DEAE-dextran on inclusion counts of an ovine chlamydia (Bedsonia) in cell culture. Aust J Exp Biol Med Sci 48:207–213, 1970
10. Alexander ER, Wang SP, Grayston JT: Further classification of TRIC agents from ocular trachoma and other sources by mouse toxicity prevention test. Am J Ophthalmol 63:1469–1478, 1967
11. Bell SD Jr, McComb DE, Nichols RL, et al: Studies of trachoma. VII. Isolation of a mixture of Type I and Type II trachoma strains from a child in Saudi Arabia. Am J Trop Med Hyg 19:842–845, 1970
12. McComb DE, Puzniak CI: Micro-cell culture method for isolation of *Chlamydia trachomatis*. Appl Microbiol 28:727–729, 1974
13. Becker Y, Loker H, Sarov I, et al: Studies on the molecular biology of the trachoma agent, p 13–26. *In* RL Nichols (ed), Trachoma and related disorders caused by chlamydial agents. Excerpta Medica, Amsterdam, 1971
14. Blyth WA, Taverne J: Cultivation of TRIC agents: a comparison between the use of BHK-21 and irradiated McCoy cells, J Hyg 72:121–128, 1974

New Simplified Culture Technique for *Chlamydia trachomatis*

K. TORVALD RIPA AND PER-ANDERS MÅRDH

The high incidence of chlamydial oculogenital infections caused by *Chlamydia trachomatis* emphasizes the need for simple diagnostic laboratory facilities (1–6). The techniques for the culture of *C. trachomatis* in current use are laborious and labor-intensive. The development of more easily performed culture procedures is therefore a necessity if the culturing of chlamydia is to be made possible on a large scale.

During the past decade, tissue culture cells have been used for the isolation of *C. trachomatis* (7). A limited number of mammalian cells have proved useful for this purpose (8). The established culture techniques involve the use of untreated McCoy cells (9), as well as McCoy cells pretreated with irradiation (7), 5-iodo-2′-deoxyuridine (IUdR; 10), or cytochalasin B (11). HeLa 229 cells pretreated with diethylaminoethyl (DEAE)-dextran (12) are also used (Table 1).

C. trachomatis is pinocytized by the eukaryotic cell. High-speed centrifugation of chlamydial elementary bodies onto a monolayer of susceptible cells or the use of DEAE-dextran enhances the intracellular uptake of chlamydiae (12, 13). Treatment of the cells by irradiation or with IUdR and cytochalasin B favors the intracellular parasitism by preventing host cell replication and thus making more of the nutrients and precursors in the medium available to the chlamydiae.

In the present study, the aims of pretreatment of eukaryotic cells to be used for the isolation of *C. trachomatis* are discussed. A new, simplified culture technique which uses cycloheximide-treated McCoy cells and which does not involve pretreatment of the cells is described.

MATERIALS AND METHODS

McCoy cells obtained from the Institute of Ophthalmology, University of London, were used. *C. trachomatis* serotypes C (TW-3), E (UW-5), and lymphogranuloma venereum (LGV) III (404) were obtained from S. P. Wang, University of Washington, Seattle.

Yolk sac material harvested from embryonated eggs inoculated when 7 to 11 days old and harvested 7 to 8 days later was used as inoculum. Samples of the yolk sac material were diluted 1:20 in 2-SP (14) and stored at −90°C until used. After thawing, further dilutions were made in 2-SP in order to obtain inclusion counts suitable for the evaluation of the different culture techniques studied.

In addition, 82 male and 135 female urethral and 96 cervical specimens from patients with suspected genital infection were studied. These specimens were collected with cotton-tipped swabs and transported in 2-SP with 3% fetal calf serum, 50 μg of gentamicin per ml (Schering Corp.), and 2.5 μg of amphotericin B per ml (Squibb & Sons).

The tissue culture medium consisted of Earle balanced salt solution (Flow Laboratories) with the addition of amino acids (50×, Flow Laboratories),

TABLE 1. *Different techniques used for the isolation of C. trachomatis in eukaryotic cells*

Authors (year of publication)	Cell type	Treatment of cells	Time lag from start of treatment until inoculation
Gordon and Quan (1965)	McCoy	Irradiation	5–6 days
Kuo et al. (1972)	HeLa 229	DEAE-dextran	30 min
Wentworth and Alexander (1974)	McCoy	IUdR	3 days
Sompolinsky and Richmond (1974)	McCoy	Cytochalasin B	3 days
Hobson et al. (1974)	McCoy	None	—
Ripa and Mårdh (present investigation)	McCoy	Cycloheximide	Direct inoculation[a]

[a] Treatment of cells starts 3 h after the inoculation.

$NaHCO_3$, vitamins (100×, Flow Laboratories), glucose, and 5% fetal bovine serum. Gentamicin and amphotericin B in the same concentrations as used in the transport medium were also added.

Cycloheximide (Acti-dione, The Upjohn Co.), IUdR (Iduridin, from Ferring, Malmö, Sweden; and Stoxil, from Smith, Kline & French Inc., Philadelphia, Pa.), DEAE-dextran, molecular weight 2×10^6 (Pharmacia, Uppsala, Sweden), and gelatina nordica (ACO, Stockholm, Sweden) were used as described below.

McCoy cells were cultured in glass bottles for 72 h. The cells were suspended with the aid of ethylenediaminetetraacetate (Merck), and 1 ml containing 10^5 cells was passed to a flat-bottomed plastic tube containing a cover slip, 13 mm in diameter. The cells were used after 24 h of incubation at 37°C, by which time a confluent cell layer had formed. The diluted yolk sac material and the clinical specimens were inoculated onto the cell cultures. Amounts of 0.5 ml of the specimens in 2-SP were centrifuged at $3,100 \times g$ at 35°C for 60 min. After an additional 2 h, the medium was replaced by 2 ml of fresh tissue culture medium containing 0.5 or 1.0 μg of cycloheximide per ml and gentamicin and amphotericin B as indicated above. The tubes were incubated at 37°C for 3 days, after which the cover slips were stained with iodine for the counting of inclusions.

McCoy cells were pretreated with IUdR according to Wentworth and Alexander (10) by use of Iduridin or Stoxil. Pretreatment of the cells with DEAE-dextran was done according to Kuo et al. (12); alternatively, the cells were treated with gelatina nordica, in a concentration of 0.1 μg/ml, analogous to the DEAE-dextran treatment.

RESULTS

IUdR pretreatment of the McCoy cells by use of the European preparation Iduridin did not affect the inclusion count, compared with untreated cells, unless 800 μg/ml or higher concentrations were used. However, these concentrations also exhibited a cytotoxic effect.

The inclusion counts of trachoma-inclusion conjunctivitis (TRIC)-C, TRIC-E, and LGV III per cover slip when untreated cells, cells treated with cycloheximide in concentrations of 0.5 and 1.0 μg/ml, and cells pretreated with Stoxil (10) were used are shown in Table 2. Cycloheximide used in a concentration of more than 1.0 μg/ml exhibited a cytotoxic effect.

TABLE 2. *Effect of treatment of McCoy cells infected with a 1:4,000 dilution of yolk sac material, infected with TRIC-C, TRIC-E, and LGV III*

Treatment of cells	Concn used (μg/ml)	Inclusions per cover slip[a]		
		TRIC-C	TRIC-E	LGV III
No treatment		3.6×10^1	2.6×10^1	10^4
Cycloheximide[b] ...	0.5	4.8×10^2	8.5×10^2	10^4
	1.0	7.1×10^2	1.2×10^3	10^4
IUdR (Stoxil)[c]	25	1.0×10^2	2.9×10^2	10^4

[a] The cover slips had a diameter of 13 mm.
[b] Cycloheximide present in the medium added to the cells 3 h after inoculation.
[c] Pretreatment of the cells for 3 days.

When cycloheximide was used, no further increase of the inclusion count was noted when the cells were pretreated with DEAE-dextran or gelatina nordica, compared with cells not pretreated with either of these two compounds.

The inclusions formed by TRIC-C (compared with those of TRIC-E) were small when IUdR-treated cells were used. When cycloheximide-treated cells were used, the inclusions of TRIC-C were of the same size as those formed by TRIC-E.

In the series of 313 urethral and cervical specimens, cycloheximide-treated cells gave an isolation rate equal to that of IUdR-treated McCoy cells.

DISCUSSION

The established techniques in use for the isolation of *C. trachomatis*, i.e., the use of untreated McCoy cells (9), McCoy cells pretreated with irradiation (7), IUdR (10) or cytochalasin B (11), or DEAE-dextran treatment of HeLa 229 cells (12), work successfully in many laboratories. However, these methods are hardly suitable for use in large-scale diagnosis of *C. trachomatis* infections, with the possible exception of the use of untreated McCoy cells.

In the search for a simple culture technique for *C. trachomatis* which can work without pretreatment of the tissue culture cells, the intention of such pretreatment was considered. The elementary bodies must come into such close contact with the cell membrane of the eukaryotic cell that they can be pinocytized. The mammalian cell surface has a negative net charge. This is also likely to be true of chlamydial cells, the cell walls of which are reminiscent of those of gram-negative bacteria and are known to be negatively charged. The repelling forces between the elementary bodies and the tissue culture cells can be partly overcome by the use of DEAE-dextran, which is positively charged, or by the centrifugation of the elementary bodies onto the cell layer. DEAE-dextran of the molecular weight recommended by Kuo et al. (12) is, however, no longer commercially available (at least not from Pharmacia, the original manufacturer of dextran). We found gelatina nordica, which is available, to be just as useful as DEAE-dextran when using IUdR-pretreated McCoy cells. However, the effect was marginal in our hands. This collagen seems to neutralize the effect of the repelling forces in a way comparable to that of DEAE-dextran. Irradiation, and treatment with IUdR and cytochalasin, are known to render McCoy cells more vulnerable to chlamydial infection by interfering with cell metabolism.

Cycloheximide, a glutaramide antibiotic, depresses the cell metabolism of eukaryotic cells, but has no such effect on prokaryotic cells (15). This substance has been used in metabolic studies of chlamydiae (16, 17), but, as far as we know, has not been employed in the isolation of chlamydiae from clinical specimens. We found cycloheximide to be of potential value in this connection when used in concentrations not completely inhibiting the metabolism of McCoy cells but only depressing it to such an extent that the intracellular energy parasite will be at an advantage.

The isolation technique described, in which 1 μg of cycloheximide per ml was added to the tissue culture medium after the adsorption phase when changing the medium, gave at least the same number of inclusions as when McCoy cells pretreated with IUdR were used. Our technique makes it possible to prepare the cells the day before use, which also has the advantage that fresher cells can be used. The technique is suitable for routine isolation of *C. trachomatis* on a large scale as it does not involve pretreatment of the cells.

ACKNOWLEDGMENT

This work was supported by grant B76-16X-04509 from the Swedish Medical Research Council.

REFERENCES

1. Grayston JT, Wang SP: New knowledge of chlamydia and the diseases they cause. J Infect Dis 132:87–105, 1975
2. Holmes KK, Handsfield HH, Wang SP, et al: Etiology of nongonococcal urethritis. N Engl J Med 292:1199–1205, 1975
3. Burns DCM, Darougar S, Thin RN, et al: Isolation of *Chlamydia* from women attending a clinic for sexually transmitted disease. Br J Vener Dis 51:314–318, 1975
4. Oriel JD, Reeve P, Thomas BJ, et al: Infection with *Chlamydia* group A in men with urethritis due to *Neisseria gonorrhoeae*. J Infect Dis 131:376–382, 1975
5. Schachter J, Hanna L, Hill EC, et al: Are chlamydial infections the most prevalent venereal disease? J Am Med Assoc 231:1252–1255, 1975
6. Smith TF, Weed LA, Segura JW, et al: Isolation of *Chlamydia* from patients with urethritis. Mayo Clin Proc 50:105–110, 1975
7. Gordon FB, Quan AL: Isolation of the trachoma agent in cell culture. Proc Soc Exp Biol Med 118:354–359, 1965
8. Croy TR, Kuo CC, Wang SP: Comparative susceptibility of eleven mammalian cell lines to infection with trachoma organisms. J Clin Microbiol 1:434–439, 1975
9. Hobson D, Johnson FWA, Rees E, et al: Simplified method for diagnosis of genital and ocular infections with chlamydia. Lancet 2:555–556, 1974
10. Wentworth BB, Alexander ER: Isolation of *Chlamydia trachomatis* by use of 5-iodo-2-deoxyuridine-treated cells. Appl Microbiol 27:912–916, 1974
11. Sompolinsky D, Richmond S: Growth of *Chlamydia trachomatis* in McCoy cells treated with cytochalasin B. Appl Microbiol 28:912–914, 1974
12. Kuo CC, Wang SP, Wentworth BB, et al: Primary isolation of TRIC organisms in HeLa 229 cells treated with DEAE-dextran. J Infect Dis 125:665–668, 1972
13. Darougar S, Cubitt S, Jones BR: Effect of high-speed centrifugation on the sensitivity of irradiated McCoy cell culture for the isolation of *Chlamydia*. Br J Vener Dis 50:308–312, 1974

14. Gordon FB, Dressler HR, Quan AL, et al: Effect on ionizing radiation on susceptibility of McCoy cell cultures to *Chlamydia trachomatis*. Appl Microbiol 23:123–129, 1972
15. Sisler HD, Siegel MR: Cycloheximide and other glutarimide antibiotics, p 283–307. *In* D. Gottlieb and PD Shaw (ed), Antibiotics, vol 1. Springer Verlag, New York, 1967
16. Alexander JJ: Separation of protein synthesis in meningopneumonitis agent from that in L cells by differential susceptibility to cycloheximide. J Bacteriol 95:327–332, 1968
17. Hatch TP: Competition between *Chlamydia psittaci* and L cells for host isoleucine pools: a limiting factor in chlamydial multiplication. Infect Immun 12:211–220, 1975

Growth of Trachoma Organisms in HeLa 229 Cell Culture

CHO-CHOU KUO, SAN-PIN WANG, AND J. THOMAS GRAYSTON

Cell cultures have generally replaced chick embryo yolk sac culture for isolation of trachoma organisms (1–4). However, serial passage and growth of trachoma organisms in large quantity in cell culture have not been reported. In 1966, Jenkin in our laboratory reported that HeLa 229 was a sensitive cell line for the growth of trachoma organisms (5). In his report he also described a method of serial passage of trachoma strains by use of HeLa 229 cells in a monolayer bottle culture. Since then, we have been using HeLa 229 cells for growth of trachoma organisms in large quantity for antigenic analysis (6). We have described a method of isolation of trachoma organisms in HeLa 229 cells pretreated with diethylaminoethyl (DEAE)-dextran (2), and we have compared 11 commonly used mammalian cell lines for their susceptibility to trachoma organisms (7). HeLa 229 was the most sensitive cell line. By comparing the growth of ocular and genital trachoma strains in HeLa 229 cell and the chick embryo yolk sac culture, we found that genital strains definitely grow better in the cell culture than in embryonated hens' eggs (8). Based on these experiences, we have now entirely eliminated chick embryo yolk sac culture from our routine laboratory work. At present, we are using HeLa 229 cell culture for both isolation and growth of trachoma organisms for immunotyping, antigenic analysis, and study of the biology and pathogenesis of the microorganisms. In this report we describe the methods of serial passage and growth of trachoma organisms in HeLa 229 cell culture currently used in our laboratory.

MATERIALS AND METHODS

Clinical Specimens

Conjunctival swab specimens were obtained from trachoma patients and their family members in our trachoma pathogenesis study on Taiwan (Grayston et al., this volume, p. 113).

Chlamydia trachomatis Strains

The *C. trachomatis* strains used in this study were trachoma strains A/G-17/OT, B/TW-5/OT, Ba/Ap-2/OT, C/TW-3/OT, D/UW-3/Cx, E/UW-5/Cx, F/UW-6/Cx, F/UW-54/Cx, G/UW-57/Cx, H/UW-43/Cx, H/UW-4/Cx, I/UW-12/Ur, J/UW-36/Cx, CJ/UW-61/Cx, K/UW-31/Cx, K/UW-38/Cx, K/UW-46/Cx, and K/UW-53/Cx, and lymphogranuloma venereum (LGV) strains L_1/440/Bu, L_1/EA-1/Bu, L_2/434/Bu, L_2/EA-2/Bu, L_2/EA-5/Cx, L_3/404/Bu, L_3/347/Bu, L_3/EA-3/Bu, and L_3/EA-4/Ur. Most of the strains have been described in our previous reports (9–12). LGV strains with EA designation were from Ethiopia and a recent collaborative study with Peter L. Perine.

Cell Line

HeLa 229 cells have been maintained in this laboratory since 1963. The cell line was originally obtained from W. F. Scherer (13), University of Minnesota, in 1961. The cell line, HeLa 229, was an early passage from the original HeLa cell line from G. O. Gey which had been kept frozen since 1956.

Cell Culture Medium

Eagle minimal essential medium with 10% heat-inactivated fetal calf serum (MEM) was used for culture of the stock cell line (without antibiotics) and for cells infected with chlamydiae (with streptomycin and vancomycin or ristocetin each at a 100-μg/ml concentration).

Methods of Cell Culture Growth of Chlamydiae

Culture in vials. The method used has been described (2). The cell culture vial used was a disposable, flat-bottomed, 1-dram glass vial shell (outside diameter, 15 mm; height, 45 mm) containing a round cover slip (diameter, 12 mm) and plugged with a no. 0, nontoxic silicone rubber stopper (Bellco, Vineland, N.J.). Into each vial 1 ml of MEM containing 2×10^5 HeLa cells was placed. The vial was then incubated overnight to obtain a confluent cell monolayer. Before inoculation of organisms or clinical specimen, MEM was removed and replaced with 1 ml of Hanks balanced salt solution (HBSS) containing 30 μg of DEAE-dextran/ml (molecular weight 5×10^5 or 2×10^6; Pharmacia, Uppsala, Sweden), which was allowed to react at room temperature for 20 to 30 min. The monolayer was washed twice with DEAE-dextran if immediate inoculation was desired. The DEAE-dextran was removed and 0.1 ml of specimen was inoculated. At least four vials were regularly inoculated per specimen. The vials were centrifuged at $900 \times g$ for 60 min at 20 to 22°C. After centrifugation, the inoculum was removed, and 1 ml of MEM was added. DEAE-dextran treatment was omitted in LGV culture. The vials were incubated for 3 days at 35°C for trachoma strains and at 37°C for LGV strains. One vial was used for estimation of infectivity. The cover slip was fixed with methanol and stained with May-Grünwald-Giemsa, and the inclusions were counted. A micrometer (7 by 7 mm) was placed in the ocular piece to assist inclusion counting. The cell monolayers of the remaining vials were harvested in SPG (sucrose, 75 g; KH_2PO_4, 0.52 g; Na_2HPO_4, 1.22 g; glutamic acid, 0.72 g; water to 1 liter; pH 7.4 to 7.6) by scraping with a Pasteur pipette. (The amount of SPG used varied according to the infectivity and the amount required for inoculation of new culture vials.) The harvested material was either inoculated into the new culture vials immediately or frozen at −65°C. The number of vials which could be inoculated depended on the infectivity as estimated by inclusion count.

Culture in bottles. The method of Jenkin (5) as described in our previous report (6) was used. The culture vessel used was a disposable 32-ounce (about 960-ml) glass prescription bottle. A 1-day-old cell monolayer was used for inoculation. Before inoculation, MEM was removed and the cell monolayer was washed twice with 5 ml of HBSS containing 30 μg of DEAE-dextran/ml (omitted in LGV culture). A 1-ml amount of inoculum was used. Absorption was carried out at room temperature for 2 to 2.5 h with intermittent agitation of the bottle. After absorption, 80 ml of MEM was added, and the bottle was incubated for 3 days at 35°C with trachoma strains and at 37°C with LGV strains. If infected yolk sac material was used for inoculation, the monolayer was washed twice, 10 ml each, with HBSS after absorption to remove excess yolk sac material. The culture was harvested after 3 days of incubation. The culture medium was discarded. Sterile 4-mm-diameter glass beads were added with 10 ml of HBSS. Cells were rolled off the bottle. The cell suspension was transferred to a 30-ml Nalgene tube (The Nalge Co., Rochester, N.Y.). The bottle was rinsed with HBSS, and the fluid was added to the cell suspension. The cell suspension was sonically treated for 20 s (Biosonik III Sonicator, Bronwill Scientific, Rochester, N.Y.) and then submitted to one cycle of differential centrifugation (500 $\times g$ for 10 min and 30,000 $\times g$ for 20 min). The pellet was resuspended in 2.5 ml of SPG per 32-ounce culture bottle and sonically treated again to obtain a homogeneous suspension. It was inoculated immediately into fresh cell cultures or stored at −65°C. If necessary, the infectivity could be monitored by simultaneous inoculation into the culture vials as described above. The number of elementary body (EB) particles was enumerated by an electron microscopic method (14).

Radioisotope labeling of chlamydial organisms. Chlamydial organisms can be selectively labeled by incorporation of radioactive amino acids in the culture medium in the presence of cycloheximide (15) or emetine (16), which inhibit protein synthesis of the host cells without affecting that of the parasites. We have previously described a method of radioisotope labeling of *C. trachomatis* organisms and the purification of the labeled organisms (18). Bottles of HeLa 229 cell cultures were infected as described above. About 18 h after the infection, MEM was replaced with low-leucine MEM (1/10 concentration of leucine) in which 1 μg of emetine/ml (Sigma Chemical Co., St. Louis, Mo.) was added to inhibit more than 98% of the host protein synthesis without affecting that of the parasite. Two hours after the addition of emetine, 1 μCi each of ^{3}H-labeled amino acids mixture and leucine (30 to 50 Ci/mmol; New England Nuclear Corp., Boston, Mass.) per ml was added. The infected cells were harvested on the third day as described above. The pellets from two culture bottles were resuspended in 6 ml of HBSS, layered on top of 24 ml of 30% sucrose in 0.03 M tris(hydroxymethyl)aminomethane-buffered saline, and centrifuged at 8,000 $\times g$ for 60 min with a swinging-bucket rotor SW25.1 in a Spinco model L ultracentrifuge. Each pellet was resuspended in 1.5 ml of HBSS, layered on a 30-ml linear gradient of 30 to 65% Renografin (Squibb & Sons Inc., Princeton, N.J.), and centrifuged at 45,500 $\times g$ for 90 min. The band of infectious particles at the middle of the gradient was removed. They were resuspended with HBSS and pelleted. The pellet was washed once with HBSS, resuspended in SPG (1 ml per 32-ounce culture bottle), and frozen in small portions at −65°C. The final product was assayed for EB particle count, radioactivity, and infectivity. The EB particles were counted by the electron microscopic method (14). The radioactivity of the labeled organisms was counted with a scintillation counter and expressed as counts per minute (17). The infectivity was assayed by titration in HeLa 229 cell culture and expressed as inclusion-forming units (IFU; 17).

TABLE 1. *Effect of centrifugation and/or DEAE-dextran on inclusion formation of trachoma TW-5 strain in HeLa 229 cell culture*

Treatment	Avg no. of inclusions per 30 fields[a]
Control (no centrifugation and no DEAE-dextran)	7
Centrifugation (900 × *g* for 60 min at 20°C)	33
DEAE-dextran (cell monolayer washed twice with 30 μg of DEAE-dextran/ml)	23
Centrifugation and DEAE-dextran	96

[a] Each result is the average of three experiments. Each experiment was run with quadruplicate cultures.

RESULTS AND DISCUSSION

I. Culture in Vials: Isolation and Immunotyping of the Isolates

Factors which enhance cell culture sensitivity. Weiss and Dressler (18) were the first to apply centrifugal force for the enhancement of chlamydial infectivity in cell culture. We have shown that pretreatment of HeLa 229 cells with DEAE-dextran (30 μg/ml) enhanced the infectivity of trachoma but not LGV organisms (19). Therefore, we omit DEAE-dextran pretreatment when culturing for LGV. An additive enhancing effect was obtained by combining DEAE-dextran pretreatment and centrifugation (2) as shown in Table 1.

We postulated that a more concentrated inoculum would more effectively infect cell cultures. The experiments shown in Table 2 confirm this idea and lead us to recommend a smaller volume of inoculum (0.1 ml) than is the usual practice in other laboratories (1, 20).

We also found that SPG had a deteriorative effect on cells if present in the culture medium during incubation. Therefore, if the inoculum was suspended in SPG it was removed after absorption. 5-Iodo-2′-deoxyuridine (IUdR), which has been shown to enhance the sensitivity of McCoy cells (3), was shown to have no effect on HeLa 229 cells (7). Other factors which have been shown to enhance the sensitivity of cell cultures, such as increasing the centrifugal force when a large volume of inoculum is used (20) and increasing the absorption temperature (17, 20), were not incorporated in our practice because they did not provide sufficient enhancement to make them practical.

TABLE 2. *Effect of inoculum volume on inclusion formation of E/UW-5/Cx strain in HeLa 229 cell culture with DEAE-dextran treatment and centrifugation*

Inoculum vol (ml)	No. of inclusions per 30 fields[a]	
	Expt A	Expt B
0.1	102	298
0.5	61	172
1.0	52	152

[a] Average of three cover slips counted.

We have used the most essential factors for successful cell culture: a sensitive cell line, HeLa 229; centrifugation; DEAE-dextran pretreatment of the cells; and small inoculum volume.

Primary isolation. We have shown that isolation of genital *C. trachomatis* by use of HeLa 229 cells pretreated with DEAE-dextran was as sensitive as the commonly used irradiated McCoy cells (2) or IUdR-pretreated McCoy cells (7). The HeLa 229 cell culture method has also been successful for the isolation of ocular *C. trachomatis*. Conjunctival swab specimens of trachoma patients and their families were shipped in dry ice from Taiwan to Seattle for isolation. As shown in Table 3, 24 isolations were obtained from 48 active trachoma cases, a rate of 50%. The isolation rate was 86% (6 of 7) in the most active cases (trachoma stage II in the World Health Organization classification of trachoma). Of all the 29 isolations obtained in this study, 24 (83%) were obtained in the first passage and 5 (17%) were obtained in the second passage.

Serial passage of isolates. We have estimated the amount of antigen required for successful immunization of a mouse to be one vial monolayer containing more than 20 inclusions per micrometer field at ×400 magnification (22). To achieve this infectivity, an average of five passages (range, three to seven) was required.

Examples of serial passages are presented in Table 4. In doing serial passage, the number of culture vials was not increased until the infectivity reached more than two inclusions per high-power field. Then the number of culture vials could be doubled. When more than 80% (about 40 inclusions per micrometer field) of the cells were infected, the number of culture vials could be multiplied by five with a high degree of infectivity. This high-titer infectivity (more than 80%) was obtained with eight isolates in four to six passages. Serial passages failed in 2 of 29 isolates in which cases only one inclusion was detected in the whole cover slip in the first passage.

Immunotyping of isolates. The cells from six culture vials with infectivity of more than 20 inclusions per micrometer field at ×400 magnification were resuspended in 3 ml of SPG and frozen in two portions. Two mice per specimen were injected intravenously in the tail vein with 0.5 ml on day 0 and day 7. The mice were exsanguinated on day 10. The antibodies were titrated by the micro-immunofluorescence (micro-IF) antibody test (23). With 27 strains, the average titer

TABLE 3. *Isolation of trachoma organisms in HeLa 229 cell culture from the eye of trachoma patients and household contacts in Taiwan*

Trachoma diagnosis[a]	No. tested	No. positive
NP	34	3 (9)[b]
ND	4	1 (25)
Tr I	23	12 (52)
Tr II	7	6 (86)
Tr III	18	6 (33)
Tr IV	16	1 (6)

[a] World Health Organization classification of trachoma: NP = no trachoma; ND = not determined; Tr I, II, and III = active trachoma; Tr IV = inactive healed trachoma.

[b] Numbers in parentheses indicate percent positive.

TABLE 4. *Examples of isolation of trachoma organisms from the eye and serial passage in HeLa 229 cells for preparation of antigens for immunotyping*

Strains	No. of passages in HeLa cells[a]							Mouse antibody titers[b]	Immuno-type
	1	2	3	4	5	6	7		
TW-442	(2)	39	1,020					128	C
TW-460	45	313	540					64	C
TW-440	(0)	13	618	882				256	C
TW-457	9	40	380	286	[80%][c]			128	B
TW-448	(1)	(3)	1	51	380	756		64	D
TW-443	(1)	(1)	1	2	51	202	600	128	C
TW-445	23	94	228	130	218	283	556	128	B

[a] Infectivity after the indicated number of passages in HeLa cells is given as the number of inclusions per 30 fields except for the values in parentheses, which represent the number of inclusions found in the whole cover slip.

[b] From mice immunized with the material harvested from the final HeLa passage shown.

[c] Percentage of cells infected.

of antibodies was 128 and the range was 32 to 256. These titers were high enough to determine the immunotypes of the isolates. Although both the immunizing antigens and the prototype slide antigens were prepared in HeLa 229 cells, immunological interference due to the HeLa material was not observed.

II. Culture in Bottles: Preparation of Antigens in Large Quantity

Technical problems. Serial passage in bottle culture is more difficult than in vial culture because centrifugation cannot be used. More passages will be needed unless a high-titered original inoculum is used. The harvesting procedure is more complicated, increasing the chance of bacterial contamination. The important available factors for enhancement of the infectivity and yield of bottle cultures are DEAE-dextran pretreatment of the cells (19), prolonged absorption (2 to 2.5 h; 5, 6, 17), and sonic disruption of the infected cells on harvesting (5, 6).

Serial passage. We have successfully adapted 17 trachoma strains to infect more than 80% of the cells in bottle culture. These included at least one strain from each of the 12 known immunotypes. The serial passages were started from highly infected yolk sac materials (11 strains), infected McCoy cells (only once with K/UW-31/Cx), or clinical specimens (5 strains). All passages were carried out in bottles with B/TW-5/OT and C/TW-3/OT. These were the first two strains successfully adapted to grow in large quantity in HeLa 229 cells (6) before introduction of DEAE-dextran pretreatment and centrifugation (2). With the remaining strains, serial passage was first done in the culture vials with centrifugation to raise the infectivity titer before switching to bottle culture. The vial culture was switched to bottle culture when the infectivity titer was more than 30 inclusions per high-power field. Fourteen such vials were pooled to infect one 32-ounce bottle. As shown in Table 5, it took an average of six passages (an average of three each in vial and bottle) to obtain more than 80% infectivity in the bottle when infected yolk sacs were used as starting inoculum. With five strains the adaptation was started from clinical specimens. These were recent isolates obtained in HeLa 229 cell culture (2). They were not passed in bottles

TABLE 5. *Growth of trachoma and LGV strains in monolayer culture of HeLa 229 cells in a 32-ounce bottle*

Organism[a]	Starting material	Passage history	No. of strains	Avg no. of passages		Yield of organisms/bottle[b]
				Vials	Bottles	
Trachoma	Infected yolk sacs	Bottle	2	0	7[c]	9.6 × 10^9
		Vial and bottle	10[d]	3	3	
	Clinical specimens	Vial and bottle	5	17[e]	5	
LGV	Infected yolk sacs	Bottle	4	0	2	2.5 × 10^{10}
	Clinical specimens	Vial and bottle	5	4	1	

[a] Trachoma: A/G-17/OT, B/TW-5/OT, Ba/Ap-2/OT, C/TW-3/OT, D/UW-3/Cx, E/UW-5/Cx, F/UW-54/Cx, G/UW-57/Cx, H/UW-43/Cx, H/UW-4/Cx, I/UW-12/Ur, J/UW-36/Cx, CJ/UW-61/Cx, K/UW-31/Cx, K/UW-38/Cx, K/UW-46/Cx, K/UW-53/Cx. LGV: L_1/440/Bu, L_1/EA-1/Bu, L_2/434/Bu, L_2/EA-2/Bu, L_2/EA-5/Cx, L_3/404/Bu, L_3/347/Bu, L_3/EA-3/Bu, L_3/EA-4/Ur.

[b] Average of six trachoma strains and four LGV strains.

[c] Average number of passages required to infect 80 to 100% of the cells in bottle.

[d] Infected McCoy cells were used with one strain (UW-31).

[e] Experiments not designed for growth in large quantities.

until after many passages in vials. Many fewer passages would have been adequate. Most of the strains adapted were prototype strains which we were using for serology and immunotyping. One strain, K/UW-31/Cx, was adapted to grow in bottles of HeLa 229 cells from McCoy cells because it was the original type K isolate (11).

When doing serial passages in bottles, the number of bottles was not increased until 50% of the cells were infected (with some experience inclusions can be directly visualized under an inverted microscope). Increase in the infectivity was not always observed in every passage. In some cases there would be no increase or even a drop of infectivity at the beginning. However, once the infectivity could be raised to more than 50% of the cells, there was a steady increase in the infectivity in the subsequent passages. Once the infectivity reached more than 80%, the number of culture bottles could be multiplied from one to five times and still yield more than 80% infectivity. We found no evidence of differential susceptibility among different immunotypes. Among 18 strains, we failed to raise the infectivity titer by serial passage in bottle culture with only one (F/UW-6/Cx).

Yields of organisms. The yield as estimated from 17 preparations from six different strains was 9.6 × 10^9 organisms per bottle with more than 80% infectivity (Table 5). In comparison, LGV organisms were much easier to grow than trachoma organisms, and the yield was about 2.5 times greater (Table 5). Although the mass of trachoma antigen produced by cell culture methods described in this study is still small compared with bacterial studies, we have successfully used the material for immunotyping of strains (22), antigenic analysis (6, 25), and studies of the pathogenesis (25, 26) and biology (17, 21) of trachoma organisms.

Preparation of antigens for micro-IF test. HeLa 229 bottle culture-grown strains have been used successfully as antigen in the micro-IF test. The organisms from one 32-ounce culture bottle of 100% infectivity were resuspended in 2.5 to 5 ml of phosphate-buffered saline. Homogenized normal yolk sacs (20%, wt/vol) were added to the suspension to make the final concentration of yolk sac 3 to 5%. The purpose of adding the yolk sac is to make the antigens stick to the microscope slide and to facilitate the localization of the antigens. The suspension of antigen was distributed in small portions (0.2 ml/vial) and frozen at −20°C. In routine work one vial of the antigen lasted 2 weeks if 20 slides were prepared daily. The antigen was discarded after 2 weeks of storage at 4°C. At this speed of usage, the antigen prepared from one culture bottle lasted 1 year.

Radioisotope labeling of organisms. Radioisotope labeling was carried out with B/TW-5/OT and L_2/434/Bu strains. Six preparations were made with TW-5. After purification, the average yield was 2.4×10^9 organisms/bottle; the average efficiency of labeling, 38,000 cpm/10^9 organisms; and the average ratio of infectivity as expressed by IFU to physical particles, 1.3. The results of the average of three LGV-434 preparations were as follows: yield, 3.2×10^9 organisms/bottle; labeling efficiency, 33,000 cpm/10^9 organisms; and IFU/physical particle, 1.8. The labeled organisms have been used successfully for studies of chlamydia-HeLa cell interaction (17).

Suspension culture. Although we have grown LGV organisms in mouse L-cell suspension culture (100% infectivity was obtained; 24), we still have not been able to grow trachoma strains in suspension culture. One of the reasons is that those cell lines, e.g., mouse L cells, which can be grown in suspension have a limited susceptibility to the growth of trachoma organisms.

ACKNOWLEDGMENT

This study was supported by U.S. Public Health Service research grant EY 00219 from the National Eye Institute.

REFERENCES

1. Gordon FB, Harper AL, Quan AL, et al: Detection of chlamydia (Bedsonia) in certain infections of man. I. Laboratory procedures: comparison of yolk sac and cell culture for detection and isolation. J Infect Dis 120:451–462, 1969
2. Kuo CC, Wang SP, Wentworth BB, et al: Primary isolation of TRIC organisms in HeLa 229 cells treated with DEAE-dextran. J Infect Dis 125:665–668, 1972
3. Wentworth BB, Alexander ER: Isolation of *Chlamydia trachomatis* by use of 5-iodo-2-deoxyuridine-treated cells. Appl Microbiol 27:912–916, 1974
4. Hobson D, Johnson FWA, Rees E, et al: Simplified method for diagnosis of genital and ocular infection with Chlamydia. Lancet 2:555–556, 1974
5. Jenkin HM: The continuous passage of agents of trachoma in cell culture. I. Characteristics of TW-3 and Bour strains of trachoma cultivated in serial passage in HeLa 229 cells. J Infect Dis 116:390–399, 1966
6. Kuo CC, Kenny GE, Wang SP: Trachoma and psittacosis antigens in agar gel double immunodiffusion, p 113–123. *In* RL Nichols (ed), Trachoma and related disorders caused by chlamydial agents. Excerpta Medica, Amsterdam, 1971

7. Croy TR, Kuo CC, Wang SP: Comparative susceptibility of eleven mammalian cell lines to infection with trachoma organisms. J Clin Microbiol 1:434–439, 1975
8. Kuo CC, Wang SP, Grayston JT: Comparative infectivity of trachoma organisms in HeLa 229 cells and egg culture. Infect Immun 12:1078–1082, 1976
9. Alexander ER, Wang SP, Grayston JT: Further classification of TRIC strains from ocular and other sources by the mouse toxicity prevention test. Am J Ophthalmol 63:1469–1478, 1967
10. Wang SP, Grayston JT, Gale JL: Three new immunologic types of trachoma-inclusion conjunctivities organisms. J Immunol 110:873–879, 1973
11. Kuo CC, Wang SP, Grayston JT, et al: TRIC type K, a new immunologic type of *Chlamydia trachomatis*. J Immunol 113:591–596, 1974
12. Wang SP, Grayston JT: *Chlamydia trachomatis* immunotype J. J Immunol 115: 1711–1716, 1975
13. Scherer WF, Syverton JT, Gey GO: Studies on propagation in vitro of poliomyelitis viruses. IV. Viral multiplication in stable strain of human malignant epithelial cells (strain HeLa) derived from an epidermoid carcinoma of the cervix. J Exp Med 97:695–709, 1953
14. Wang SP, Grayston JT: A potency test for trachoma vaccine utilizing the mouse toxicity prevention test. Am J Ophthalmol 63:1443–1454, 1967
15. Alexander JJ: Separation of protein synthesis in meningopneumonitis agent from that in L cells by differential susceptibility to cycloheximide. J Bacteriol 95:327–332, 1968
16. Becker Y, Asher Y: Synthesis of trachoma agent proteins in emetine-treated cells. J Bacteriol 109:966–970, 1972
17. Kuo CC, Grayston JT: Interaction of *Chlamydia trachomatis* organisms and HeLa 229 cells. Infect Immun 13:1103–1109, 1976
18. Weiss E, Dressler HR: Centrifugation of rickettsia and viruses onto cells and its effect on infection. Proc Soc Exp Biol Med 103:691–695, 1960
19. Kuo CC, Wang SP, Grayston JT: Differentiation of TRIC and LGV organisms based on enhancement of infectivity by DEAE-dextran in cell culture. J Infect Dis 125:313–317, 1972
20. Reeve P, Owen J, Oriel JD: Laboratory procedures for the isolation of *Chlamydia trachomatis* from the human genital tract. J Clin Pathol 28:910–914, 1975
21. Kuo CC, Wang SP, Grayston JT: Effect of polycations, polyanions, and neuraminidase on the infectivity of trachoma-inclusion conjunctivitis and lymphogranuloma venereum organisms in HeLa cells. Infect Immun 8:74–79, 1973
22. Wang SP, Kuo CC, Grayston JT: A simplified method for immunological typing of trachoma-inclusion conjunctivitis-lymphogranuloma venereum organisms. Infect Immun 7:356–360, 1973
23. Wang SP: A micro immunofluorescence method. Study of antibody response of TRIC organisms in mice, p 273–288. *In* RL Nichols (ed), Trachoma and related disorders caused by chlamydial agents. Excerpta Medica, Amsterdam, 1971
24. Caldwell HD, Kuo CC, Kenny GE: Antigenic analysis of chlamydiae by two-dimensional immunoelectrophoresis. I. Antigenic heterogeneity between *C. trachomatis* and *C. psittaci*. J Immunol 115:963–968, 1975
25. Kuo CC, Wang SP, Grayston JT: Studies on delayed hypersensitivity with trachoma organisms, p 158–167. *In* RL Nichols (ed), Trachoma and related disorders caused by chlamydial agents. Excerpta Medica, Amsterdam, 1971
26. Kuo CC, Grayston JT: Studies on delayed hypersensitivity with trachoma organisms. III. Lymphokines. J Immunol 112:540–545, 1974

Comparative Efficiency of Various Diagnostic Methods for Chlamydial Infection

JULIUS SCHACHTER AND CHANDLER R. DAWSON

The purpose of this paper is not to review the literature on diagnostic procedures for chlamydial infections of the eye and genital tract. Rather, it is to review the experiences of one laboratory in applying these various techniques to diagnosing chlamydial infections in different anatomic sites. We have tested more than 30,000 specimens, from 10,000 patients, for chlamydial infection. Over the years, we have used a variety of isolation procedures, and as techniques improved we have compared some of these different procedures before establishing new routines in the laboratory. Thus, we have performed trachoma-inclusion conjunctivitis (TRIC) agent isolation attempts from the eye or genital tract in the yolk sac of embryonated hen eggs (1) and in irradiated (2, 3), 5-iodo-2′-deoxyuridine (IUdR)-treated (4), cytochalasin B-treated (5), or normal McCoy or L929 cells (6). In addition, we have performed cytological evaluations using the Giemsa and fluorescent-antibody stains and have used serological procedures such as the complement fixation (CF) and micro-immunofluorescence (micro-IF; 7) tests on sera collected from these and other patients (8). For the purposes of this paper, we shall present the order of positivity of the various tests performed on the assumption that only positive patients are being tested. The patient is considered positive if any single test detects *current* chlamydial infection. Thus, the perfect test would yield a 100% infection rate. The positive specimen pool for these comparisons has been 60 patients with inclusion blennorrhea, 130 patients with adult inclusion conjunctivitis, 200 men with chlamydial infection of the urethra, and 300 women with chlamydial infection of the cervix.

INCLUSION CONJUNCTIVITIS OF THE NEWBORN

Inclusion blennorrhea is an acute conjunctival infection of the neonate. It is perhaps the easiest chlamydial infection to diagnose. Most tests are positive, and we have found that isolation by tissue culture, fluorescent-antibody procedure, or Giemsa stain is almost always positive in infants with acute and active conjunctivitis. Even in the relatively insensitive yolk sac procedure, 65% of the specimens have yielded isolates (Table 1). The few failures by any single cytological test have usually represented inadequate specimen collection. It is often difficult to obtain an adequate sample of conjunctival cells from infants. We have not performed routine serological tests in this disease. When there has been a reason to draw a blood specimen from the infant, we have found high micro-IF antibody levels but no CF antibodies.

ACUTE ADULT INCLUSION CONJUNCTIVITIS

It is clear that the micro-IF test and tissue culture isolation procedures are the most sensitive indicators of chlamydial infection in adult inclusion conjunctivitis (Table 1). The Giemsa stain is positive in approximately half of the patients, and the

TABLE 1. *Sensitivity of laboratory tests for chlamydial infection*

Test	Anatomic site infected with chlamydiae			
	Conjunctiva			
	Neonates	Adults	Male urethra	Female cervix
Isolation				
Tissue culture	100[a]	95	100	95
Yolk sac	65	38	26	40
Direct staining of cell scrapings				
Giemsa	95	45	15	41
Fluorescent antibody	95	85	60	66
Serum antibody				
Complement fixation	ND	50	15	40
Micro-immunofluorescence	ND	100	90	99

[a] Percentage of specimens yielding a positive result. ND = not done.

agent may be recovered from approximately 38% by the yolk sac procedure. The micro-IF test has always been positive in adults with inclusion conjunctivitis, whereas only 50% of these patients have had CF titers in excess of 1:16 (Table 2).

It should be noted that occasional specimens from the conjunctiva are positive by cytological methods but not by tissue culture isolation. In most of these cases, we have found that the patients have had some exposure to antibiotics. It is a routine finding that 1 or 2 days of chemotherapy will prevent recovery of the organism although the clinical disease is not cured.

INFECTION OF THE MALE URETHRA

The tissue culture isolation procedures have their greatest superiority over other procedures in identifying chlamydial infection of the male urethra (Table 1). In our hands this test is four times more successful than the yolk sac method in obtaining isolates from the male urethra. The Giemsa stain is not useful in diagnosis of urethral infection, being positive in only 15% of patients yielding isolates. If cytological tests are required, fluorescent-antibody procedures are considerably better. Ninety percent of the men tested have had micro-IF antibodies, whereas only 15% have had CF titers equal to or greater than 1:16 (Table 2).

INFECTION OF THE CERVIX

It appears from our results that considerably more organisms are present in cervical specimens than in the urethral swabs. Both egg isolation and Giemsa positivity rates are increased over those observed with the male urethra. However, the most sensitive technique is still tissue culture isolation (Table 1). Fluorescent-antibody techniques are restricted in evaluation of cervical specimens because of the increased fluorescence associated with bacteria and other antigenic material trapped in mucus.

A consistent result throughout our studies has been the higher antibody titers and the higher rates of antibody reactivity observed in women with chlamydial genital

TABLE 2. *Comparison of complement fixation and micro-immunofluorescence tests for chlamydial antibodies in human sera*

Population tested	Percent positive	
	CF[a]	Micro-IF[b]
Screening studies		
"Normal" adults	3	25–40
Pediatric sera	<1	10
Males, sexually active	5–10	15–25
Females, sexually active	10–20	50–70
Prostitutes	30–60	85
Venereal disease	8	60
Proven chlamydial infection (isolation)		
Lymphogranuloma venereum	100	100
Adult inclusion conjunctivitis	50	100
Male urethritis	15	90
Female cervicitis	40	99

[a] Titer ≥ 1:16.
[b] Titer ≥ 1:8.

tract infections than in men with genital tract infections. Of more than 300 women with proven chlamydial infections of the cervix, we have had only 2 who did not have micro-IF antibodies (Table 2). Even the CF test (positive in only 15% of the men) is positive in 40% of the women at levels of 1:16 or greater. It would appear that women are exposed to greater levels of antigen. Simplistically, this may reflect the trapping of antigen in mucous and further processing by host defense mechanisms, in contrast to flushing mechanisms in the male urethra. We have noticed increasing titers in postpartum women. We assume that this is due to increased distribution of antigen after laceration of the cervix but cannot rule out the possibility that the chlamydial infections may occasionally become systemic.

EVALUATION OF SEROLOGY IN THE DIAGNOSIS OF TRIC AGENT INFECTION

Results shown in Table 1 indicate relatively low CF rates but high micro-IF antibody rates in patients with proven chlamydial infection. Despite the very good reactor rates observed with this latter test, we are not hopeful about the use of routine serology with this or any other test in the diagnosis of chlamydial infections. In parallel with the enhanced sensitivity of the test, there is a much higher background rate of seroreactors in different population groups. As can be seen in Table 2, this background reactor rate ranges from a low of 10% in pediatric populations (the stimulus for this antibody response is not known although part of it could be attributed to neonatal infection). In normal populations, micro-IF antibody levels of 25 to 40% are common, and in sexually active individuals they may reach as high as 70% of women who do not have active chlamydial infection (9). In a general venereal disease clinic population, we have found approximately 60% seroreactors. Faced with this background, we consider it unlikely that this test can be used for routine diagnosis of chlamydial infections, although its merits in epidemiological studies are not disputed. The CF test suffers from the same problems of background reactor rates, but is also a relatively insensitive test for any of the chlamydial genital tract infections except lymphogranuloma venereum (LGV). In LGV it has been found positive in all 15 patients from whom we have isolated the organism. The CF titers in LGV tend to be very high, as 12 of the 15 have had CF titers of >1:64, above the titers seen in men with simple uncomplicated urethritis. In parallel, the micro-IF titers in LGV are often 1:2,000 to 1:4,000, which is higher than we usually see in men with urethritis.

The micro-IF procedure allows one to determine the immunoglobulin (Ig) class of reacting antibodies. One could wish that the presence of IgM antichlamydial antibodies would provide a good indication of recent or current chlamydial infection. However, similar to the experience of Philip et al. (10), we have not found it to be a useful diagnostic test. In general, we found IgM antibodies in approximately one-third of patients with active chlamydial infection; in a venereal disease clinic population approximately 20% of the isolate-negative patients may have IgM antibodies by the micro-IF test.

Prior to the introduction of the micro-IF method by Wang and Grayston (7), we used a fluorescent-antibody assay with infected tissue culture cells as the antigen. We still use the technique for standardizing fluorescent-antibody reagents but have discarded it for human serology because of the marked superiority of the micro-IF test.

COMPARISON OF TISSUE CULTURE METHODS

With the introduction of the irradiated cell technique and further modifications of the procedure, we compared the various procedures in the laboratory. This has been done by running series of 100 specimens at a time in irradiated cells, with split specimens tested in parallel by another technique. We found the IUdR-treated cells and the cytochalasin B-treated cells to be essentially similar to the irradiated cells. There were no statistical differences between recovery rates or inclusion counts, although there was a tendency for the IUdR-treated cells to be slightly superior. IUdR-treated cells offer an advantage in terms of simplicity of maintenance procedures within the laboratory and in not requiring irradiation sources. But IUdR is considered to be a dangerous and potentially carcinogenic chemical, and there are restrictions on its handling. Thus, the technicians in a laboratory may prefer to work with irradiated cells rather than IUdR-treated cells. We have consistently found that normal McCoy or L929 cells are inferior to the treated cells for the isolation of chlamydiae. The best result we have ever obtained in a comparative trial yielded a recovery rate of 35% of that obtained with the irradiated cells. Inclusion counts on positive specimens have ranged between 5 and 10% of the inclusion counts in the same specimens inoculated into irradiated cells. In addition, the normal cells have been much more difficult to read because there are many smaller inclusions which are hard to identify with certainty. Therefore, this technique has never been used on a standard basis in our laboratory.

ACKNOWLEDGMENTS

This study was supported by Public Health Service grants EY-00427 from the National Eye Institute, AI-10193 from the National Institute of Allergy and Infectious Diseases, and CA-14427 from the National Cancer Institute.

REFERENCES

1. Schachter J, Rose L, Dawson CR, et al: Comparison of procedures for laboratory diagnosis of oculogenital infections with inclusion conjunctivitis agents. Am J Epidemiol 85:453–458, 1967
2. Gordon FB, Quan AL: Isolation of the trachoma agent in cell culture. Proc Soc Exp Biol Med 118:354–359, 1965
3. Darougar S, Jones BR, Kinnison JR, et al: Chlamydial infection: advances in the diagnostic isolation of *Chlamydia*, including TRIC agent, from the eye, genital tract, and rectum. Br J Vener Dis 48:416–420, 1972
4. Wentworth BB, Alexander ER: Isolation of *Chlamydia trachomatis* by use of 5-iodo-2-deoxyuridine-treated cells. Appl Microbiol 27:912–916, 1974
5. Sompolinsky B, Richmond SJ: Growth of *Chlamydia trachomatis* in McCoy cells treated with cytochalasin. Appl Microbiol 28:912, 1974
6. Schachter J, Hanna L, Hill EC, et al: Are chlamydial infections the most prevalent venereal disease? Am Med Assoc 231:1252–1255, 1975
7. Wang S-P, Grayston JT: Immunologic relationship between genital TRIC, lymphogranuloma venereum, and related organisms in a new microtiter indirect immunofluorescence test. Am J Ophthalmol 70:367–374, 1970
8. Schachter J, Dawson CR, Balas S, et al: Evaluation of laboratory methods for detecting acute TRIC agent infection. Am J Ophthalmol 70:375–380, 1970
9. Schachter J, Hill EC, King EB, et al: Chlamydial infection in women with cervical dysplasia. Am J Obstet Gynecol 123:753–757, 1975
10. Philip RN, Casper EA, Gordon FB, et al: Fluorescent antibody responses to chlamydial infection in patients with lymphogranuloma venereum and urethritis. J Immunol 112:2126–2134, 1974

Chairman's Report and Discussion

JULIUS SCHACHTER

The major emphasis in this session was a discussion of cultural conditions for the isolation and growth of chlamydial agents from human genital tract specimens. Thus, it was shown that wide fluctuations in cultural conditions had a deleterious effect on the growth of the organism. Maximal yield in terms of isolation rates and antigenic mass was obtained by close control of pH (neutrality appears to be optimal) and temperature of centrifugation (33°C being superior to 20°C). The various cell systems being used in different laboratories were discussed. Evidence was presented that irradiated BHK-21 cells might be superior to irradiated McCoy cells for propagation of chlamydiae. Techniques for isolation and larger-scale antigen preparation in which diethylaminoethyl-treated HeLa 229 cells are used were presented. An alternate technique in which cycloheximide treatment of the cells is used was discussed. This technique has the potential advantage of not requiring pretreatment of the cells since this inhibitor of mammalian cell metabolism can be added after the inoculation. It was suggested that serum-free systems could be useful in propagating chlamydiae for immunological studies. The possibility of alternate diagnostic tests in which enzyme probes for enzymes unique to chlamydiae would be used was discussed.

A discussion of the currently available diagnostic tests emphasized that the tissue culture isolation procedures are clearly superior, but there is still a place for cytological studies, particularly in supplying rapid diagnosis of neonatal infections. Reservations were expressed concerning the utility of serology in the diagnosis of chlamydial genital tract infections because of the high background reactor rates in sexually active populations.

The final discussion revolved around the ultimate utility of chlamydial isolation tests. It is recognized that chlamydiae are significant causes of nongonococcal and postgonococcal urethritis, but it is equally clear that such patients routinely receive tetracycline in the absence of a specific etiological diagnosis. A key question in treatment of individual patients arises in dealing with cervicitis in females whose contacts may not be available or who have no known epidemiological association with nongonococcal urethritis. Ultimately, there must be some selection of which patient gets tested for chlamydial infection (considering the current technological restrictions). It is clear that the ability of laboratories to diagnose these infections and to grow chlamydiae has increased at a rate greater than the increase in our knowledge of the clinical and pathological effects of chlamydial infection. The necessity of further studies to identify the effects of chlamydial infection was stressed.

Session VII

UREAPLASMA UREALYTICUM

Biology and Taxonomy of *Ureaplasma urealyticum*

MAURICE C. SHEPARD

For 20 years, *Ureaplasma urealyticum* masqueraded under the trivial names "T-strain mycoplasma" and, more recently, "T mycoplasma." It has now been given the respectability of an official binomial, as proposed by Shepard and others in 1974 (1). At the time of their discovery in 1952 (in urethral exudates from nongonococcal urethritis [NGU] patients), growth of *U. urealyticum* cells in primary agar cultures was so sparse and their size was so small that they were almost unrecognizable. For approximately a year it was undecided whether this was an artifact in the primary cultures or a bizarre colonial manifestation of some previously unknown living agent. The minute colonies of the organism were first briefly described and illustrated in two photomicrographs in 1954 (2), and they were described in greater detail 2 years later (3). The nutritional inadequacies of the earlier agar culture media and the strongly alkaline reaction of such media made progress in the study of this new organism extremely slow and laborious. It was finally learned that most of the established techniques for proper mycoplasmology were inappropriate for the primary isolation, identification, and cultivation of *U. urealyticum* (4). In fact, it was not until the enrichment was changed from human ascitic fluid to horse serum that the first success was achieved in propagating the organism in unlimited serial subcultivation by the agar push-block method.

Attempts at cultivating the organism in fluid cultures were even more disastrous since serial cultivation in broth cultures generally resulted in progressively declining titers and early demise. The discovery by Ford (5) that *U. urealyticum* reached maximal titers after only 16 h of incubation at 36°C resolved much, but not all, of this difficulty. Since those earlier years, steady progress has been made in our knowledge of *U. urealyticum* and in our ability to isolate, identify, and cultivate the organism. The purpose of this report is to summarize briefly what is known about the biological characteristics of *U. urealyticum*.

MORPHOLOGY AND STAINING

The morphology of *U. urealyticum* is basically that of a round or coccobacillary organism, approximately 350 nm in diameter, with a range of 200 to 1,000 nm (1, 6, 7). In Giemsa-stained preparations of exudates from male NGU patients, the organisms are essentially coccoidal in morphology, but many pleomorphic elements can also be seen. Pointed, rickettsia-like organisms in such preparations are very frequently observed. Examinations by electron microscopy suggest that the pointed end is really a small bud (1). As with other species of *Mycoplasma*, short bacillary elements, filaments, annular forms, signet-ring forms, and bipolar elements can also be observed, depending on age of the organisms, strain differences, and methods of examination. *U. urealyticum* is bounded by a single, trilaminar membrane, 10 nm thick, which appears to be covered on the outer surface by a layer of pilus-like projections, 5 to 8 nm long (6, 8, 9). In addition, the organisms contain ribosomes, 12 to 15 nm in size, frequently arranged in regular geometric patterns, and nuclear fibroids and vacuoles (6, 8). The mode of multiplication of the organism is believed to consist basically of a simple budding process, in different directions, producing single organisms, pairs, small aggregates, or multidirectional (filamentous) elements. These may continue to multiply, producing growth aggregations or colonies in fluid and in agar cultures (1, 6, 10). Aggregates of *U. urealyticum* can often be observed in association with infected epithelial cells from the urethral mucosa (11). The organism is also found in very large numbers in many urine mucous threads and free in urethral exudates from NGU patients.

U. urealyticum is gram negative, although it stains weakly by this method (1). In Gram-stained smears of exudate from either NGU patients or certain gonorrhea patients, however, *U. urealyticum* can be recognized in association with certain epithelial cells by the trained observer. The usual method of demonstrating this organism in urethral exudates or in broth cultures is by means of Giemsa-stained smears (11). In our laboratory, briefly air-dried smears are stained for 18 h in a 1:50 dilution of stock Giemsa stain (Azure B type) in pH 7.0 phosphate buffer solution. In such preparations the organisms stain purple or reddish purple in color.

We have recently developed a hematoxylin-Giemsa staining method for demonstration of *U. urealyticum* in urethral exudates, obtained by epithelial scraping, and in fluid cultures of the organism. This staining method is superior to the usual 18-h Giemsa method alone for demonstration of ureaplasmas. Smears are made on specially cleaned glass microscope slides. Commercially available precleaned microscope slides are unsatisfactory for this method, since they frequently have an invisible film which adsorbs stain during a prolonged staining period. Therefore, slides were soaked for 30 min in sulfuric acid-sodium dichromate cleaning solution and rinsed with agitation in running tap water for 15 min; they were then rinsed in deionized water and dried in a glassware drying oven. Freshly made smears were briefly air-dried, fixed for 5 min in ACS methanol, and air-dried. They were then stained for 2 min in regular-strength Gill hematoxylin (Lerner Laboratories, Stamford, Conn.). Hematoxylin-stained smears were rinsed in pH 7.0 phosphate buffer solution (KH_2PO_4, 1.0 g; Na_2HPO_4, 2.0 g; distilled water, 1,000 ml) and, without drying, were placed in *fresh* 1% sodium acetate solution for 2 min. Smears were then transferred immediately (without drying or rinsing) to a Coplin jar containing 40 ml of 1:50 Giemsa stain solution and allowed to stain at room temperature for 18 h. The Coplin jar was tilted forward at an angle of approximately 45°, and the slides were placed smear side downward. The Giemsa stain employed was a 1:50 dilution (in the above pH 7.0 buffer solution) of an 0.8% stock Giemsa solution prepared by use of 3.76 g of Azure B-type Giemsa stain powder (Harleco catalog no. 2281) and 470 ml of a mixture of equal volumes of ACS glycerol and ACS methanol. The dye powder was dissolved in the solvent mixture in a 1-liter Erlenmeyer flask or bottle containing pyrex glass beads. The container was tightly stoppered and secured with tape. The stock stain was allowed to ripen for at least 4 days, with frequent shaking by hand, and was stored at room temperature in a brown glass bottle with a tight screw cap. After the 18-h staining period, the smears were washed in pH 7.0 buffer solution, dried, and examined. *U. urealyticum*, stained by the hematoxylin-Giemsa method described above, stains in a basophilic manner, generally more purple than red-purple. The typical morphology of the organism stained by this hematoxylin-Giemsa method is illustrated in Fig. 1.

CULTURAL CHARACTERISTICS

On horse serum agar media, such as our A5H standard agar medium (12), colonies of *U. urealyticum* are generally small in size (20 to 30 μm) and normally grow without zones of surface growth. In the manner of all *Mycoplasma* colonies, these colonies stain blue on standard agar media by Dienes stain. The existence of unusual, very small (10 to 20 μm), refractile colonies of *U. urealyticum* in primary cultures of certain urine specimens has been reported (13). These colonies often resemble chemical deposits and can easily be overlooked. In uncrowded conditions, and depending also on strain differences, agar volume, inoculum size, and other factors (14), agar colonies may attain diameters of 30 to over 50 μm. They are circular in shape and are characterized by an irregular border and a granular structure. Larger colonies are often multilobate and exhibit a characteristic "cauliflower head" colony (1). Under special conditions of cultivation, such as increased carbon dioxide, increased agar hardness, and increased cation content

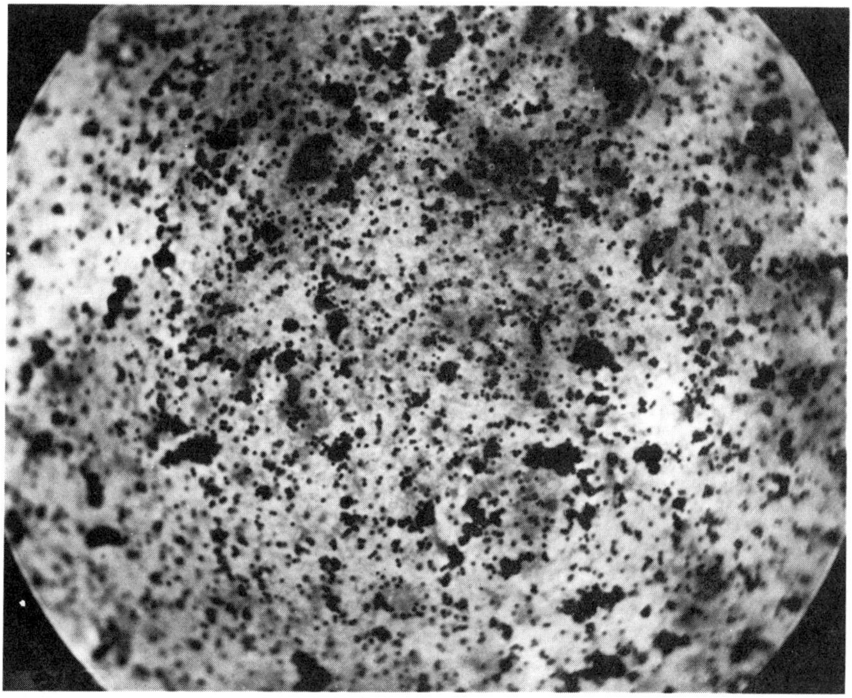

FIG. 1. *Ureaplasma urealyticum from a 17-h culture in fluid medium U17B stained by the hematoxylin-Giemsa method of Shepard. The culture was centrifuged and washed twice in phosphate-buffered saline. The final pellet of organisms was resuspended in Tyrode's pH 7.0 buffer (containing 0.8% formaldehyde) to a final concentration of 80-fold over the original volume of culture* (×2,000).

(14, 15), or incorporation of HEPES (*N*-2-hydroxyethylpiperazine-*N*′-2-ethanesulfonic acid) buffer (16), for example, zones of surface growth may develop in agar cultures. Under such conditions, *U. urealyticum* colonies mimic those of classical, large-colony *Mycoplasma* species with which they could be confused. However, possible misinterpretation is avoided by the application of the direct test for urease (17). The entire colony, including the zone of surface growth, produces the specific manganese reaction product and identifies the colony as that of *U. urealyticum*.

In primary agar cultures of exudates from NGU patients, colonies of *U. urealyticum* are frequently associated with urethral epithelial cells. Outgrowth of multiple colonies suggests that they originated from multiple foci of infection of such epithelial cells.

A differential agar medium known as medium A7 has recently been developed (12) which incorporates the biochemical principle of the direct test for urease. Colonies of *U. urealyticum* are positively identified on this differential medium by virtue of a deep golden-brown manganese reaction product which precipitates within and on the surface of the colonies.

REQUIREMENTS FOR GROWTH

Very satisfactory growth of *U. urealyticum* can be obtained with media prepared from commercially available nutrient agars and broths, especially BBL Trypticase soy broth and Difco PPLO broth. The previously recommended solidifying agent (Oxoid Ionagar no. 2) is no longer available, and in its place we recommend a bacteriological-grade agar (Grand Island Biological Co. catalog no. 1400000) available from GIBCO Diagnostics, Madison, Wis. Our current standard agar is medium A5H (12). *U. urealyticum* grows best in media adjusted to pH 6.0 ± 0.5 (18). The organisms are microaerophilic, and best growth is obtained in a gaseous mixture containing 5 to 15% carbon dioxide in nitrogen or air (10). In a medium of pH 6.0, aerobic growth can be obtained, but growth of colonies is not luxuriant. Excellent growth is, however, obtained in agar cultures of pH 6.0 under anaerobic conditions (1).

Native Protein Requirement

The native protein requirement is best served by enrichment with 20% unheated normal horse serum, which, among other things, furnishes cholesterol and urea. Members of genus *Ureaplasma* are unique among the *Mycoplasmatales* in that they possess a urease or urease-like enzyme system and actively hydrolyze urea, with the production of ammonia and carbon dioxide (19–23). Urea is believed to be required for best growth and multiplication of freshly isolated strains of *U. urealyticum* in primary cultures, although an absolute requirement for urea has not been established and many questions remain to be answered (24–27). We (M. C. Shepard and C. D. Lunceford, unpublished data) and other investigators believe that the minimal urea requirement for multiplication of *U. urealyticum* lies between 2.5 and 10 μg per ml (26). Different investigators have provided evidence that other compounds can be substituted for urea, such as putrescine and allantoin (24, 25).

Action on Carbohydrates

The usual carbohydrates are not fermented by *U. urealyticum* (as determined by conventional indicator color change methods in carbohydrate-containing growth media), and the organisms lack hexokinase activity (28).

Sterol Requirement

Cholesterol is required for growth by *U. urealyticum* and is generally furnished by the horse serum enrichment of the medium (9, 29). Ureaplasmas are sensitive to digitonin (9, 28).

Amino Acids

As far as we know, the usual amino acids are not utilized as such. L-Histidine has been reported to be active in urea broth cultures of *U. urealyticum*. However, results of experiments performed with ^{14}C-labeled urea demonstrated that the action of L-histidine was to retard the rise in pH of the medium (30). L-Cysteine has been found to result in improved growth of the organisms in both agar and fluid cultures in the presence of this thiol (12, 31). In our laboratory, broth culture titers are at least 10-fold higher in the presence of L-cysteine HCl than in its absence. The amino acids of the urea cycle, ornithine, citrulline, and arginine, are not hydrolyzed.

Fatty Acids

Palmitic acid and oleic acid appear to be incorporated by *U. urealyticum*, but the purpose and fate of these fatty acids in its metabolism are unknown (P. F. Smith, personal communication).

Yeast Extract

The primary value of yeast extract as a supplement to media for *U. urealyticum* is as an aid to distinguishing this organism from classical, large-colony mycoplasmas in mixed mycoplasma cultures. Classical mycoplasmas require yeast extract for proper surface zone development, and, if this factor is lacking in agar media, the classical colonies may fail to produce surface zones and may be confused with colonies of *U. urealyticum*. The usual 10% concentration of yeast extract is inhibitory for some strains of *U. urealyticum*. We have, therefore, recommended inclusion of only 1% supplementation of yeast extract in the medium (1, 10). In some deficient media, yeast extract may be of benefit by virtue of its high magnesium content.

BIOLOGICAL PROPERTIES

The only enzyme systems known for this organism are urease, phosphatase, and aminopeptidase (21, 32, 33). *U. urealyticum* is catalase negative and fails to hydrolyze esculin. Lecithinase activity and serum digestion are negative. Similarly, methylene blue and tetrazolium are not reduced. Peroxide has not been demonstrated (28). A soluble beta-hemolysin is produced which completely lyses guinea pig and rabbit erythrocytes under both aerobic and anaerobic conditions (1, 28). An alpha-prime type of hemolysis has also been reported (34). Hemadsorption of guinea pig erythrocytes by *U. urealyticum* has also been reported, but this was produced only by serotype III organisms (28). *U. urealyticum* also adsorbs to HeLa cells from cell suspensions placed on agar. Attachment to the cells was tenacious and resisted vigorous washing. The specificity of the HeLa cell adsorption phenomenon was suggested by inhibition of cell attachment by specific antiserum (16).

Cell Proteins

Analysis of 12 different strains of *U. urealyticum* by means of electrophoretic analysis of cell proteins has revealed close similarities and has suggested that all of the various strains of *U. urealyticum* tested belong to a single, common group or species (35).

Membranes

Purified membranes from seven strains of *U. urealyticum* were found to contain between 1 and 7% carbohydrates by dry weight. Mannose, galactose, and glucose were identified, with glucose in the largest concentration (range: 1.64 to 5.45% of membrane dry weight, depending upon the strain) (36). In a subsequent study, the same seven *U. urealyticum* strains were examined for amino acid and protein content. The amino acid composition was similar to that of other biological membranes. Seven to 10 protein bands were detected by isoelectric focusing, and they occurred mostly in the pH 4.0 to 7.0 range of the gel. One strain had a distinctive band at pH 3.0, whereas another strain had a prominent band at pH 9.0 (37).

Phosphatase

Phosphatase has been found in over 50 different strains of *U. urealyticum* so far examined (32).

Aminopeptidase

All eight serotypes of *U. urealyticum* were capable of releasing alanine from L-alanyl-L-alanyl-L-alanine. Serotypes VII and VIII were shown to release free amino acids from all the unsubstituted peptides investigated and from leucine amide. The discovery of aminopeptidase activity in *U. urealyticum* adds to our limited knowledge of the biochemical activity of this organism (33).

DNA Base Composition

The deoxyribonucleic acid (DNA) base composition of eight strains of *U. urealyticum* in moles percent guanine plus cytosine (G + C) ranged from 27.7 to 28.5 (38). This low G + C content is similar to that found for other mycoplasmas, which ranged from 26 to 32% (39).

DNA Genome Size

The calculated genome size of DNA from two strains of *U. urealyticum* was found to be 4.7×10^8 and 4.4×10^8 daltons (40). All eight serotypes of *U. urealyticum* were similarly examined, and the calculated DNA genome size was found to range from 4.1×10^8 to 4.8×10^8 daltons (40, 41).

TOXIC PRODUCTS

Ammonia

Although ammonia per se is generally not considered among the toxic products of microorganisms, there is nevertheless evidence that ammonia is capable of damaging tissues and tissue cells in vitro and in vivo (42, 43). Since *U. urealyticum* is a very actively urealytic organism, rather large quantities of ammonia may be produced in small foci of growth at the cellular level on the urethral mucosa or within infected mucous glands and ducts of Littré, for example. It is noteworthy that the organism is able to hydrolyze urea to ammonia within a few seconds of contact with substrate. It is possible that such strong local accumulations of ammonia might predict a pathogenic role for *U. urealyticum* in NGU.

Toxic Factor

Evidence has been presented that a toxic product is produced by *U. urealyticum* during active growth and that the accumulation of such a toxic product in cultures may explain sudden restriction of multiplication (44). It is equally possible that such a toxic substance, elaborated by *U. urealyticum* in the lumen of the male urethra, may also contribute to its pathogenic potential.

ACTION OF ANTIMICROBIAL AGENTS

Susceptibility to Antibiotics

U. urealyticum is inhibited by the following antibiotics, listed in order of approximate decreasing effectiveness: the tetracyclines (including the semisynthetic tetracyclines), erythromycin, streptomycin, chloramphenicol, gentamicin, kanamycin, and partially by clindamycin (5, 28, 45–47).

Resistance to Antibiotics and Antimicrobial Agents

U. urealyticum is resistant in vitro to the inhibitory properties of the following agents: the penicillins (including the semisynthetic penicillins), sulfadiazine, rifampin (Shepard and Lunceford, unpublished data), cephaloridine (controversial) (45, 47, 48), aurothiomalate (47, 49), and lincomycin (45, 50, 51). Lincomycin is selectively inhibitory to *M. hominis* in genital tract cultures and is, therefore, useful in the separation and purification of *U. urealyticum* from mixed *Mycoplasma* species (52). In addition, the organism is unaffected by the folic acid antagonist trimethoprim lactate at concentrations which are completely inhibitory to most strains of *Proteus* (5 μg/ml) in an agar system of pH 6.0 (Shepard and Lunceford, unpublished data).

Selective Growth Inhibition by Antimicrobial Agents

Growth of *U. urealyticum* in vitro is selectively inhibited by the following agents: thallium acetate (4, 19, 47, 53), erythromycin (46, 47, 52), 5-iodo-2′-deoxyuridine (4), hydroxyurea (4), acetohydroxamic acid (Shepard and Lunceford, unpublished data), and sorbyl, benzoyl, and 3-aminobenzoyl hydroxamic acids (49). We have found 1.0 mM concentrations of acetohydroxamic acid to be completely inhibitory to *U. urealyticum* multiplication in agar cultures and in U9B urease color test medium. Other investigators have found somewhat different results with acetohydroxamic acid inhibition of growth (25). Erythromycin selectively inhibits *U. urealyticum* growth, and its action is opposite to that of lincomycin described above (52). The human classical mycoplasmas are generally unaffected by these selective agents at concentrations which are completely inhibitory to *U. urealyticum*. *M. pneumoniae*, like *U. urealyticum*, is sensitive to erythromycin (8). The selective inhibition of *U. urealyticum* in agar cultures by both erythromycin and thallium acetate is significantly enhanced at an alkaline reaction of pH 7.5 or greater (4).

Ether Sensitivity

U. urealyticum treated for 18 h at 4°C in a liquid medium containing 20% diethyl ether is killed by this treatment, as are three classical *Mycoplasma* species, *M. hominis*, *M. pneumoniae*, and *M. bovigenitalium* (47).

Serology

U. urealyticum is antigenically distinct from all other recognized mycoplasmas and represents a serologically heterogeneous group. Six different serotypes were originally identified and were shown to be antigenically distinct (54) by means of a metabolism-inhibition test (20, 55). The existence of a seventh serotype was subsequently confirmed (56), and an eighth serotype was included later (57). The validity of these eight serotypes was established by comparative serology, employing four different serological methods (56). The growth inhibition test was subsequently further modified for application to the serological study of *U. urealyticum* serotypes (57).

An immune-inactivation test for *U. urealyticum* was developed which employed a complement-dependent, mycoplasmacidal reaction (58). By use of this test, five serogroups on the basis of shared common antigens and 11 serotypes within these five groups (59) were demonstrated. The serological heterogeneity of *U. urealyticum* was more precisely studied, and each of the proposed 11 serotypes was further characterized by the presence of both major and minor antigens (60). It has been firmly established that at least eight different serotypes of *U. urealyticum* exist on the basis of the classification proposed by Black (56). According to the findings of Lin and Kass, at least 11 serotypes of this organism can be recognized (60).

Habitat

U. urealyticum is found chiefly in the male and female genitourinary tracts and occasionally also in the pharynx and the rectum (10, 54, 55, 61). The organism was first isolated in 1952 from male patients with primary and recurrent NGU and is associated with this disease, which is referred to as the sixth venereal disease (62), in from 60 to 93% of over 3,000 NGU patients we have studied to date. A recent study of *U. urealyticum*-positive NGU patients in relapse after inadequate treatment showed a close quantitative relationship of this organism to the clinical course of the disease (63, 64). This is the most meaningful evidence to date supporting a pathogenic role of *U. urealyticum* as one of the major etiological agents of NGU. The organism can also be isolated from the genitourinary tract of apparently healthy adult carriers or individuals harboring inapparent infections (10, 65). However, the isolation rate of *U. urealyticum* from NGU patients reported in a recent study (64) was 151 of 219 men (69%), whereas the isolation rate of the organism from adult male controls, matched for age, sexual activity, and occupation, was 111 of 385 men (28.8%). Thus, the incidence of *U. urealyticum* in the NGU patients was 2.4 times greater than in the matched control group.

The organism has also been reported to colonize the genital tract and pharynx of newborn infants (66, 67). The isolation rates of *U. urealyticum* from asymptomatic carriers (controls) and from NGU patients reported by other investigators, however, have varied widely (61). The organism has been associated with reproductive failure in females (68–71) and with low birth weight of newborn infants (66, 72), and it has been isolated from cases of acute salpingitis (73). *U. urealyticum* has also been isolated from patients with puerperal sepsis (74) and amnionitis with mycoplasmemia (75). *U. urealyticum* (strain T960) produced vesical calculi and bladder stones in male rats after bladder or renal inoculation of broth-grown organisms (76).

Although beyond the scope of this brief review, it should be emphasized that species of *Ureaplasma* have also been isolated from lower animals, and the importance of pathogenic ureaplasmas in certain animal infections should be indicated. Ureaplasmas have been isolated from the genital tracts of cattle (47, 77, 78), the genital tracts of dogs (79), the throats of squirrel monkeys (79), the genital tracts of female talapoin, patas, and macaque monkeys (80), and the oropharynges of cats (81). A "San Angelo" strain of bovine *Ureaplasma* was isolated from pneumonic lung lesions of feedlot cattle during an epizootic of pneumonia. This strain was unrelated serologically to two human strains of *U. urealyticum* and to two bovine strains of urinary tract origin (82). The isolation of *Ureaplasma* strains from sheep and goats was first reported in Texas in the United States in 1975 (83), and the lamb isolates (and one goat isolate) were serologically distinct entities. Pathogenic bovine ureaplasmas, isolated from pneumonic calf lungs, produced pneumonia in experimentally inoculated calves (84), and these same pneumonic calf lung *Ureaplasma* strains also produced clinical mastitis in cows by intramammary inoculation (85). The mouse has recently been suggested as an experimental animal model for studying *Ureaplasma* infections (mouse mastitis) due to either bovine or human strains (86).

TAXONOMY

U. urealyticum is distinguished by its ability to hydrolyze urea and its possession of a urease or urease-like enzyme system. This singular property significantly sets *U. urealyticum* apart from all other members of the *Mycoplasmatales*. On the basis of this distinguishing property, the new, separate genus *Ureaplasma* was proposed in the family *Mycoplasmataceae* in which to classify species isolated from humans and lower animals (1). A single human species was proposed for the present, namely, *Ureaplasma urealyticum*. This species consists of at least eight different serotypes. The type strain is *U. urealyticum* 960-(CX8), serotype VIII, ATCC no. 27618.

ISOLATION AND IDENTIFICATION

The recent development of improved standard agar medium for primary isolation and cultivation of *U. urealyticum* (12), together with differential methods for identification of the organism based on demonstration of urease (23), has greatly simplified the reliability of positive identification of *U. urealyticum*. The three differential methods now available for identification are the U9B urease color test fluid medium (87), the direct spot test for urease in colonies of *U. urealyticum* (17), and a differential agar medium (12).

REFERENCES

1. Shepard MC, Lunceford CD, Ford DK, et al: *Ureaplasma urealyticum* gen.nov., sp.nov.: proposed nomenclature for the human T (T-strain) mycoplasmas. Int J Syst Bacteriol 24:160–171, 1974
2. Shepard MC: The recovery of pleuropneumonia-like organisms from Negro men with and without nongonococcal urethritis. Am J Syph Gonorrhea Vener Dis 38:113–124, 1954
3. Shepard MC: T-form colonies of pleuropneumonialike organisms. J Bacteriol 71:362–369, 1956
4. Shepard MC: Cultivation and properties of T-strains of mycoplasma associated with nongonococcal urethritis. Ann NY Acad Sci 143:505–514, 1967
5. Ford DK: Culture of human genital "T-strain" pleuropneumonia-like organisms. J Bacteriol 84:1028–1034, 1962
6. Whitescarver J, Furness G: T-mycoplasmas: a study of the morphology, ultrastructure and mode of division of some human strains. J Med Microbiol 8:349–355, 1975
7. Furness G: T-mycoplasmas: growth patterns and physical characteristics of some human strains. J Infect Dis 132:592–596, 1975
8. Black FT, Birch-Andersen A, Freundt EA: Morphology and ultrastructure of human T mycoplasmas. J Bacteriol 111:254–259, 1972
9. Rottem S, Pfendt EA, Hayflick L: Sterol requirements of T-strain mycoplasmas. J Bacteriol 105:323–330, 1971
10. Shepard MC: Fundamental biology of the T-strains, p 49–65. *In* L Hayflick (ed), The Mycoplasmatales and the L-phase of bacteria. Appleton-Century-Crofts, New York 1969
11. Shepard MC: Visualization and morphology of pleuropneumonialike organisms in clinical material. J Bacteriol 73:162–171, 1957
12. Shepard MC, Lunceford CD: A differential agar medium (A7) for identification of *Ureaplasma urealyticum* (human T mycoplasmas) in primary cultures of clinical material. J Clin Microbiol 3:613–625, 1976
13. Shepard MC, Lunceford CD: Unusual colonies of *Ureaplasma urealyticum* (T mycoplasmas) in primary agar cultures of certain urine specimens. J Clin Microbiol 2:456–458, 1975
14. Lee Y-H, Donner A, Bailey PE, et al: Effect of agar volume, inoculum size, and Hepes buffer on the size of T-mycoplasmal colonies. J Lab Clin Med 84:766–770, 1974

15. Furness G: T-mycoplasmas: some factors affecting their growth, colonial morphology, and assay on agar. J Infect Dis 128:703–709, 1973
16. Manchee RJ, Taylor-Robinson D: Enhanced growth of T-strain mycoplasmas with *N*-2-hydroxyethylpiperazine-*N'*-2-ethanesulfonic acid buffer. J Bacteriol 100:78–85, 1969
17. Shepard MC, Howard DR: Identification of "T" mycoplasmas in primary agar cultures by means of a direct test for urease. Ann NY Acad Sci 174:809–819, 1970
18. Shepard MC, Lunceford CD: Effect of pH on human mycoplasma strains. J Bacteriol 89:265–270, 1965
19. Shepard MC: Human mycoplasma infections. Health Lab Sci 3:163–169, 1966
20. Purcell RH, Taylor-Robinson D, Wong D, et al: Color test for the measurement of antibody to T-strain mycoplasmas. J Bacteriol 92:6–12, 1966
21. Shepard MC, Lunceford CD: Occurrence of urease in T strains of mycoplasma. J Bacteriol 93:1513–1520, 1967
22. Ford DK, MacDonald J: Influence of urea on the growth of T-strain mycoplasmas. J Bacteriol 93:1509–1512, 1967
23. Shepard MC: Differential methods for identification of T-mycoplasmas based on demonstration of urease. J Infect Dis 127(Suppl):S22-S25, 1973
24. Masover GK, Hayflick L: Growth of T-strain mycoplasmas in media without added urea. Ann NY Acad Sci 225:118–130, 1973
25. Masover GK, Benson RJ, Hayflick L: Growth of T-strain mycoplasmas in medium without added urea: effect of trace amounts of urea and of a urease inhibitor. J Bacteriol 117:765–774, 1974
26. Furness G, Coles RS: The urea requirement and urease production of some human species of T-mycoplasmas. Proc Soc Exp Biol Med 150:807–809, 1975
27. Masover GK, Sawyer JE, Hayflick L: Urea-hydrolyzing activity of a T-strain mycoplasma: *Ureaplasma urealyticum*. J Bacteriol 125:581–587, 1976
28. Black FT: Biological and physical properties of human T-mycoplasmas. Ann NY Acad Sci 225:131–143, 1973
29. Romano N, Smith PF, Mayberry WR: The lipids of a T-strain of mycoplasma. J Bacteriol 109:565–569, 1972
30. Ajello F, Romano N: Effect of L-histidine on the survival of a T-strain of mycoplasma. Appl Microbiol 29:293–294, 1975
31. Tull AH, Blair EB, Fishman DL, et al: Significant enhancement of T-strain mycoplasmal growth on supplemented Shepard's A6 differential agar. J Clin Microbiol 1:234–236, 1975
32. Black FT: Phosphatase activity in T-mycoplasmas. Int J Syst Bacteriol 23:65–66, 1973
33. Vinther O, Black FT: Aminopeptidase activity of *Ureaplasma urealyticum*. Acta Pathol Microbiol Scand Sect B 82:917–918, 1974
34. Manchee RJ, Taylor-Robinson D: Lysis and protection of erythrocytes by T-mycoplasmas. J Med Microbiol 3:539–546, 1970
35. Razin S, Valdesuso J, Purcell RH, et al: Electrophoretic analysis of cell proteins of T-strain mycoplasmas isolated from man. J Bacteriol 103:702–706, 1970
36. Whitescarver J, Castillo F, Furness G: The preparation of membranes of some human T-mycoplasmas and the analysis of their carbohydrate content. Proc Soc Exp Biol Med 150:20–22, 1975
37. Whitescarver J, Trocola M, Campana T, et al: A study of the amino acids and proteins of some human T-mycoplasma membranes. Proc Soc Exp Biol Med 151:68–71, 1976
38. Bak AL, Black FT: DNA base composition of human T strain mycoplasmas. Nature (London) 219:1044–1045, 1968
39. McGee ZA, Rogul M, Wittler RG: Molecular genetic studies of relationships among mycoplasma, L-forms and bacteria. Ann NY Acad Sci 143:21–30, 1967

40. Black FT, Christiansen C, Askaa G: Genome size and base composition of deoxyribonucleic acid from eight human T-mycoplasmas. Int J Syst Bacteriol 22:241–242, 1972
41. Bak AL, Black FT, Christiansen C, et al: Genome size of mycoplasmal DNA. Nature (London) 224:1209–1210, 1969
42. MacLaren DM: The significance of urease in proteus pyelonephritis: a histological and biochemical study. J Pathol 97:43–49, 1969
43. Visek WJ: Urease: its significance in birds and mammals. Agric Sci Rev Coop State Res Serv US Dep Agric 8:9–23, 1970
44. Furness G: T-mycoplasmas: Their growth and production of a toxic substance in broth. J Infect Dis 127:9–16, 1973
45. Braun P, Klein JO, Kass EH: Susceptibility of genital mycoplasmas to antimicrobial agents. Appl Microbiol 19:62–70, 1970
46. Shepard MC, Lunceford CD, Baker RL: T-strain mycoplasma. Selective inhibition by erythromycin in vitro. Br J Vener Dis 42:21–24, 1966
47. Taylor-Robinson D, Williams MH, Haig DA: The isolation and comparative biological and physical characteristics of T-mycoplasmas of cattle. J Gen Microbiol 54:33–46, 1968
48. Csonka GW, Williams REO, Corse J: T strain mycoplasma in nongonococcal urethritis. Ann NY Acad Sci 143:794–798, 1967
49. Ford DK: Inhibition of growth of T-strain mycoplasmas by hydroxamic acids and by aurothiomalate. Antimicrob Agents Chemother 2:340–343, 1972
50. Csonka GW, Corse J: Selective inhibition in vitro of *Mycoplasma hominis* by lincomycin. Br J Vener Dis 46:203–204, 1970
51. Shipley AS, Bowman J, O'Connor JJ: T-strain mycoplasmas in nonspecific urethritis. Med J Aust 1:794–796, 1968
52. Braun P, Klein JO, Lee Y-H, et al: Methodologic investigations and prevalence of genital mycoplasmas in pregnancy. J Infect Dis 121:391–400, 1970
53. Lee Y-H, Bailey PE, McCormack WM: T mycoplasmas from urine and vaginal specimens: decreased rates of isolation and growth in the presence of thallium acetate. J Infect Dis 125:318–321, 1972
54. Ford DK: Relationships between mycoplasma and the etiology of nongonococcal urethritis and Reiter's syndrome. Ann NY Acad Sci 143:501–504, 1967
55. Purcell RH, Wong D, Chanock RM, et al: Significance of antibody to mycoplasma as measured by metabolic-inhibition techniques. Ann NY Acad Sci 143:664–675, 1967
56. Black FT: Serological methods for classification of human T mycoplasmas. Int Congr Infect Dis, 5th, Vienna, Austria 1:407–411, 1970
57. Black FT: Modifications of the growth inhibition test and its application to human T-mycoplasmas. Appl Microbiol 25:528–533, 1973
58. Lin J-S, Kass EH: Immune inactivation of T-strain mycoplasmas. J Infect Dis 122:93–95, 1970
59. Lin J-S L, Kendrick MI, Kass EH: Serological typing of human genital T-mycoplasmas by a complement-dependent mycoplasmacidal test. J Infect Dis 126:658–663, 1972
60. Lin J-S, Kass EH: Serotypic heterogeneity in isolates of human genital T-mycoplasmas. Infect Immun 7:499–500, 1973
61. McCormack WM, Braun P, Lee Y-H, et al: The genital mycoplasmas. N Engl J Med 288:78–89, 1973
62. Shepard MC, Alexander CE Jr, Lunceford CD, et al: Possible role of T-strain mycoplasma in nongonococcal urethritis: a sixth venereal disease? J Am Med Assoc 188:729–735, 1964
63. Shepard MC: T-strain mycoplasmas as a cause of urethritis. Epidemic venereal disease. Proc 2nd Int Vener Dis Symp, St. Louis, Mo., 1972, p 83–87

64. Shepard MC: Quantitative relationship of *Ureaplasma urealyticum* to the clinical course of nongonococcal urethritis in the human male. Les Colloques de l'Institut National de la Santé et de la Recherche Médicale. Les Mycoplasmes (INSERM) 33:375–380, 1974
65. Shepard MC: Nongonococcal urethritis associated with human strains of "T" mycoplasmas. J Am Med Assoc 211:1335–1340, 1970
66. Klein JO, Buckland D, Finland M: Colonization of newborn infants by mycoplasmas. N Engl J Med 280:1025–1030, 1969
67. Foy HM, Kenny GE, Levinsohn EM, et al: Acquisition of mycoplasmata and T-strains during infancy. J Infect Dis 121:579–487, 1970
68. Kundsin RB, Driscoll SG, Ming PL: A strain of mycoplasma associated with human reproductive failure. Science 157:1573–1574, 1967
69. Caspi E, Solomon F, Sompolinski D: Early abortion and mycoplasma infection. Isr J Med Sci 8:122–127, 1972
70. Gnarpe H, Friberg J: Mycoplasma and human reproductive failure. I. The occurrence of different mycoplasmas in couples with reproductive failure. Am J Obstet Gynecol 114:727–731, 1972
71. Friberg J, Gnarpe H: Pregnancies in "infertile" couples treated with doxycycline for T-mycoplasmas. Am J Obstet Gynecol 116:23–26, 1973
72. Braun P, Lee Y-H, Klein, JO, et al: Birth weight and genital mycoplasmas in pregnancy. N Engl J Med 284:167–171, 1971
73. Mårdh P-A, Westrom L: T-mycoplasmas in the genitourinary tract of the female. Acta Pathol Microbiol Scand 78:367–374, 1970
74. Sompolinsky D, Solomon F, Leiba H, et al: Puerperal sepsis due to T-strain mycoplasma. Isr J Med Sci 7:745–748, 1971
75. Caspi E, Herczeg E, Solomon F, et al: Amnionitis and T-strain mycoplasmemia. Am J Obstet Gynecol 111:1102–1106, 1971
76. Friedlander AM, Braude AI: Production of bladder stones by human T mycoplasmas. Nature (London) 247:67–69, 1974
77. Taylor-Robinson D, Haig DA, Williams MH: Bovine T-strain mycoplasma. Ann NY Acad Sci 143:517–518, 1967
78. Taylor-Robinson D, Thomas M, Dawson PL: The isolation of T-mycoplasmas from the urogenital tract of bulls. J Med Microbiol 2:527–533, 1969
79. Taylor-Robinson D, Martin-Bourgon C, Watanabe T, et al: Isolation of T-mycoplasmas from dogs and squirrel monkeys: biological and serological comparison with those isolated from man and cattle. J Gen Microbiol 68:97–107, 1971
80. Kundsin RB, Rowell T, Shepard MC, et al: T-strain mycoplasmas and reproductive failure in monkeys. Lab Anim Sci 25:221–224, 1975
81. Tan RJS, Markham JG: Feline T-strain mycoplasmas. Jpn J Exp Med 41:247–248, 1971
82. Livingston CW Jr: Isolation of T-strain of mycoplasma from Texas feedlot cattle. Am J Vet Res 33:1925–1929, 1972
83. Livingston CW, Gauer BB: Isolation of T-strain mycoplasma from sheep and goats in Texas. Am J Vet Res 36:313–314, 1975
84. Gourlay RN, Thomas LH: The experimental production of pneumonia in calves by the endobronchial inoculation of T-mycoplasmas. J Comp Pathol 80:585–594, 1970
85. Gourlay RN, Howard CJ, Brownlie J: The production of mastitis in cows by the intramammary inoculation of T-mycoplasmas. J Hyg 70:511–521, 1972
86. Howard CJ, Anderson JC, Gourlay RN, et al: Production of mastitis in mice with human and bovine ureaplasmas (T-mycoplasmas). J Med Microbiol 8:523–529, 1975
87. Shepard MC, Lunceford CD: Urease color test medium U-9 for the detection and identification of "T" mycoplasmas in clinical material. Appl Microbiol 20:539–543, 1970

Physiology of Ureaplasmas

SHMUEL RAZIN, GERALD K. MASOVER, AND LEONARD HAYFLICK

In gross morphology and ultrastructure, the ureaplasmas resemble the classical mycoplasmas, as evidenced by electron microscopy (1) and by phase-contrast microscopy of broth cultures (Razin et al., in preparation). The ureaplasmas share with the classical mycoplasmas the "fried egg" colony shape when growth conditions are adequate, such as under a CO_2 atmosphere (Razin et al., in preparation). The ureaplasmas also resemble the classical mycoplasmas in their genome size (2), in their low guanine-plus-cytosine base ratio (3), and in their requirement for cholesterol (4). Nevertheless, they differ from the mycoplasmas in at least two basic properties: (i) their inability to grow in vitro, in the conventional mycoplasma media, to titers higher than 10^6 to 10^7 colony-forming units (CFU) per ml, and (ii) their ability to hydrolyze urea.

Why do the ureaplasmas stop growing once they reach a titer of 10^6 to 10^7 CFU/ml? Comparison of the growth curve of a ureaplasma with that of a classical mycoplasma, such as *Mycoplasma hominis*, reveals that the initial part of the curve, representing the lag and the early logarithmic phase, is essentially similar in both the mycoplasma and ureaplasma cultures, but the logarithmic phase in ureaplasmas stops rather abruptly once the titer reaches 10^6 to 10^7 CFU/ml, whereas in *M. hominis* it continues to rise to 10^9 CFU/ml and higher. Death of the ureaplasmas might then be rapid when the medium is poorly buffered or slower when the medium is well buffered. As a result of this peculiar behavior, the cell yields in ureaplasma cultures are extremely low. Theoretically, on the basis of the growth curve, we would expect about one-thousandth of the yield of a classical mycoplasma, but in practice we may obtain much less. This is because the fast-growing mycoplasmas, such as *M. hominis* or *Acholeplasma laidlawii*, produce filaments, long chains of cocci and large aggregates of cells, each yielding one colony only. In this case the CFU values are actually much lower than the actual number of cells. However, with the ureaplasmas, most of the cells appear either singly or in pairs, so that the CFU values reflect much better the real number of cells. Thus, the difference in cell yields between the ureaplasmas and the classical mycoplasmas is even larger than that depicted by the growth curves.

How much cell yield can one expect from a ureaplasma culture? A well-growing classical mycoplasma would produce about 100 to 150 mg of cell protein per liter of medium, under optimal growth conditions (5). Accordingly, 1 liter of a ureaplasma culture should yield not more than one-thousandth of this yield, that is, only 0.1 mg of cell protein, which is insufficient for the analysis of many of the biochemical activities of the cells. Nevertheless, the pellet obtained after centrifugation of a ureaplasma culture is sometimes sizable, in particular when 10 to 20% horse serum is used to supplement the medium. This pellet consists mostly of precipitated serum proteins (6, 7). We have tried to overcome this problem by reducing the serum content of the medium from 20% to 5% and by filtering the complete uninoculated medium through a 0.22-μm membrane filter. Even with these precautions, the pellet, which was much smaller, still consisted mostly of medium components rather than cells. Judging from the ratio of cytoplasm to the nonsoluble "membrane" material, obtained by treating the pellet with digitonin to lyse the cells, the pellet consisted of about 5 to 10% membranes and the rest was precipitated medium components. Hence, the recent reports of Whitescarver et al. (8, 9), claiming the successful chemical characterization of ureaplasma membranes, should be taken with reservation, as their experimental procedures leave little doubt that their "membrane" material consisted largely of precipitated serum proteins and other components of the growth medium.

In spite of these technical difficulties, we were successful in collecting some data on the enzymatic capacities of *Ureaplasma urealyticum* (strain 960) and their localization in the cells. The easiest task was the localization and characterization of the urease activity, which distinguishes the ureaplasmas from all other mycoplasmas. This hydrolytic enzyme is very active, as are the urease activities in other microorganisms and plants (10). Moreover, we have developed a very sensitive radioassay to quantitate urease activity. In this assay the amount of [^{14}C]urea remaining in the reaction vessel or evolved as radioactive CO_2 is measured. Our results show that the ureaplasma urease, like that of other organisms, is located in the cytoplasm and is not membrane bound. We could demonstrate this point by two independent techniques for cell lysis: by osmotic lysis (5) and by the digitonin procedure (11). Both techniques appear to be adequate for the isolation of ureaplasma membranes free from cytoplasm. Since urease is such an active enzyme, its presence in membrane pellets can be taken to indicate their contamination with intact cells or cytoplasm, and thus serves as a convenient and effective criterion for membrane separation from cytoplasm. The recent work carried out by Masover et al. (12) provided a partial characterization of the urease of *U. urealyticum*. It appears to resemble other ureases in being inhibited by sulfhydryl-blocking reagents, such as *p*-chloromercuribenzoate, as well as by some of the urea analogues. The enzyme is thermolabile and is not inhibited by ethylenediaminetetraacetate.

We tried also to localize two other key enzymatic activities in the ureaplasma cell: (i) adenosine triphosphatase (ATPase), which is ubiquitous and of outstanding importance in energy utilization and transport mechanisms, and (ii) reduced nicotinamide adenine dinucleotide (NADH) oxidase activity, which in the classical mycoplasmas represents the truncate respiratory pathway, responsible for the transfer of electrons from NADH to oxygen or alternate electron acceptors. We found that the ATPase in *U. urealyticum* is localized in the membrane, as in other mycoplasmas and in microorganisms in general. However, the activity of the NADH oxidase was too low to be reliably measured on the minute quantities of cell material available to us. Thus, we are unable to say at present whether the NADH oxidase in ureaplasmas is localized in the membrane, as in *A. laidlawii*, or in the cytoplasm, as in the sterol-requiring mycoplasmas and spiroplasmas (13; Razin and Greenstein, unpublished data).

This brings us to a very important question: how do the ureaplasmas obtain their energy? It appears that they are unable to attack carbohydrates. F. T. Black (Ph.D. thesis, Univ. of Aarhus, Aarhus, Denmark, 1974) could not detect hexokinase in ureaplasmas by the sensitive radioassay method of Cirillo and Razin (14). Hence, it appears that ureaplasmas do not possess the glycolytic pathway. However, unlike the nonfermentative mycoplasmas which are able to derive adenosine triphosphate from the degradation of arginine, the ureaplasmas have so far been shown to lack the arginine dihydrolase system (15). Therefore, we still do not know the energy source for ureaplasmas. Degradation of amino acids might provide energy, and indeed Masover and Hayflick (16) found more ammonia in urea-containing ureaplasma cultures than would be accounted for by the hydrolysis of urea alone. This suggests deamination and perhaps oxidation of amino acids or similar compounds (17). However, we have no direct evidence to support this suggestion. It should be noted that some classical mycoplasmas, such as *M. agalactiae, M. bovigenitalium,* and *M. verecundum*, are not capable of glycolysis or arginine breakdown; hence, the ureaplasmas are not unique in this respect. Rather, it reflects our ignorance of their basic physiological properties.

Lipid biosynthesis has been extensively studied in mycoplasmas. The classical mycoplasmas have been shown to be restricted in their fatty acid biosynthetic ability. Thus, all the classical sterol-requiring mycoplasmas and spiroplasmas tested so far are totally incapable of fatty acid synthesis, and the acholeplasmas can synthesize only the saturated fatty acids (18, 19). Recent studies in our laboratory in Jerusalem, Israel (Romano, Rottem, and Razin, in preparation), have indicated that a strain of *U. urealyticum* (serotype VI) is capable of synthesis of both saturated and unsaturated fatty acids. The ureaplasma was grown with the lipid precursor [^{14}C]acetate. The cell lipids were extracted, and their fatty acids were methylated, separated into saturated and unsaturated fatty acid fractions, and analyzed by gas-liquid chromatography. The radioactivity peaks eluted from the column corresponded to myristic, palmitic, and stearic acids in the saturated fatty acid fraction, and to $C_{14:1}$ or $C_{16:1}$ in the unsaturated fraction. If this finding is confirmed and extended to other *U. urealyticum* strains, it will mean that ureaplasmas, unlike other mycoplasmas, are capable of synthesizing the major portion of the hydrophobic lipid bilayer of their membrane, a property of great physiological as well as phylogenetic significance.

The aspect of ureaplasma physiology that has attracted most attention so far is that concerning the role of urea and urease activity in ureaplasma growth. The first workers to show that urea promotes growth of ureaplasmas were Shepard and Lunceford (20) and Ford and MacDonald (21). They found that dialysis of the serum component of the growth medium abolished its ability to support ureaplasma growth. This was restored by the addition of urea. Moreover, they could show that the ureaplasmas, unlike all other mycoplasmas, were capable of hydrolysis of urea to ammonia and CO_2. Consequently, these workers investigated whether there is any correlation between the urease activity and promotion of ureaplasma growth by urea. The experiments of Ford et al. (22), employing [^{14}C]urea, indicated that the CO_2 liberated by urease action apparently is not assimilated by the organisms and thus cannot be regarded as the growth-promoting factor. We have confirmed this finding by showing that no radioactivity from [^{14}C]urea was associated with the cell pellet harvested from 3 liters of medium supplemented with high-specific-activity [^{14}C]urea. Perhaps a better argument against CO_2 being the active growth factor is the failure of CO_2 by itself to replace urea in the growth of ureaplasma. One would then consider the second degradation product of urea, ammonia. The most cited hypothesis for the biological role of urease is that the ammonia liberated by its action provides the microorganism or plant with ammonia used for assimilation (23). Unfortunately, experiments with [^{15}N]urea are complicated by many technical difficulties so that we cannot test directly whether the ammonia derived from urea is incorporated into the ureaplasma cells. As with CO_2, all attempts to replace urea with ammonia failed.

We approached this problem by utilizing an observation made in our laboratory that an atmosphere of 100% CO_2 markedly improves the growth of ureaplasmas on solid media. The colonies obtained under CO_2 are very large (up to 300 μm in diameter) and have the characteristic "fried egg" shape. The reason for growth improvement under CO_2 appears to be very simple: CO_2 is an excellent buffer at the acid pH range, between 6.0 and 6.5, found by Shepard and Lunceford (24) to be optimal for ureaplasma growth, so that this optimal pH can be maintained in the presence of excessive amounts of ammonia derived from urea hydrolysis. Figure 1 shows that under CO_2 the pH does not rise above 6.5, but, on transfer of the culture to N_2 or air, the pH rises rapidly, leading to more rapid death of the organisms. Figure 1 also shows that the decline phase is somewhat longer under CO_2. However, the titer of the viable organisms never exceeded 5×10^7 50% color-changing units (CCU_{50}), even under CO_2.

We tested, therefore, whether the addition of more urea would improve growth. The ability to keep the pH almost constant under CO_2 enabled us to add substantial quantities of urea in portions not toxic by themselves. Five consecutive additions of urea to give 0.01 M, resulting in the accumulation of more than 2,000 μg of ammonia/ml, did not enhance or inhibit growth. It thus appears that neither CO_2 nor ammonia is the factor limiting the growth of the ureaplasmas. Thus, it is not the exhaustion of the urea supply which acts as a barrier to growth. Furthermore, the accumulation of ammonia per se does not inhibit growth as long as the medium is kept at the acid pH range. The accumulation of ammonia in the absence of a CO_2 atmosphere might inhibit growth indirectly by raising the pH and activating a toxic factor, though the possibility that the ammonium ion becomes toxic at high pH cannot be ruled out. It is also possible that high pH by itself retards growth.

Another interesting point shown in Fig. 1 is that urease activity could be detected after all the organisms had died. This means that at least part of the ureaplasma cells do not lyse after death, since lysis would result in the loss of the soluble cytoplasmic enzyme into the growth medium.

So far, it appears that the degradation products of urea resulting from urease activity cannot account for the growth-promoting effects of urea. We should perhaps consider the possibility that the urea molecule by itself is the active factor, probably affecting the cell membrane, ion balance, etc. Complicating the picture even further is the finding of Masover and Hayflick (17) that putrescine can replace urea after the adaptation of the organisms to this amine. Experiments were carried out recently in our laboratory in which the number of colonies and their size on agar media supplemented with 5% dialyzed calf serum or horse serum, and various combinations of urea and putrescine, were determined. Dialyzed calf serum by itself did not support growth, confirming the findings of Shepard and Lunceford (20). However, the addition of 0.01 M urea alone supported very poor growth, and its combination with 0.01 M putrescine gave good growth. Putrescine by itself could not support growth. The beneficial effect of the mixture of putrescine and urea is also reflected by the larger colony size in medium containing horse serum and by a faster growth rate in liquid media. Although we have no explanation for the

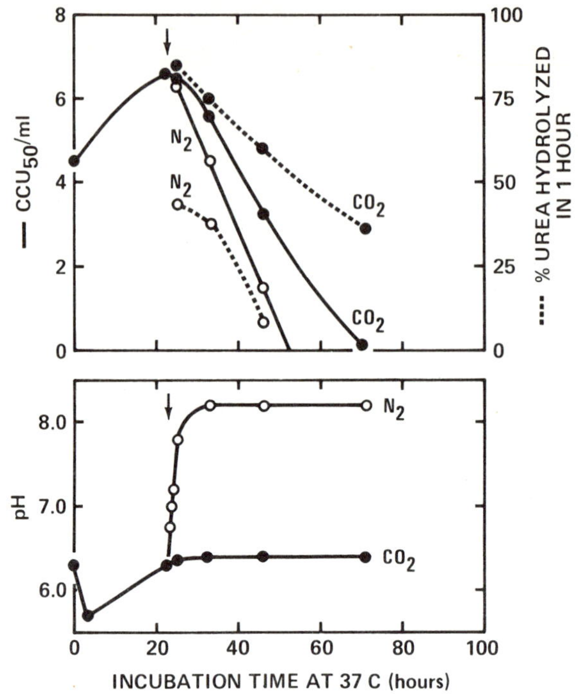

FIG. 1. *Effect of changing the atmosphere on the pH, viability, and urease activity of a U. urealyticum culture grown with 0.01 M urea and 0.01 M putrescine. At the time indicated by the arrow, the culture grown under 100% CO_2 was divided into two equal portions; one was transferred to 100% N_2, and the other remained under 100% CO_2. At time intervals indicated in the figure, samples were taken and tested for urease activity by adding [^{14}C]urea and determining the residual radioactivity in the medium after 1 h.*

beneficial effect of putrescine, we would recommend the routine addition of a mixture of 0.01 M urea plus 0.01 M putrescine to all ureaplasma media.

Coming back to our major issue of why ureaplasmas stop growing once their culture reaches the titer of 10^6 to 10^7 CFU/ml, our findings appear to rule out the possibility that this is due to the exhaustion of urea. The possibility that the rise in pH due to liberated ammonia is causing the low titer can be ruled out by our findings that under CO_2 this does not happen and yet the titer remains low. Furness (25) suggested that the abrupt cessation of ureaplasma growth is caused by the accumulation of a toxic factor, which is catalase resistant, thermostable, and apparently responsible for the failure of spent cultures to support growth of ureaplasmas, even when supplemented with fresh serum and urea. The toxic factor is only slowly dialyzable and its identity is still unknown. The hypothesis of Furness should be investigated further.

ACKNOWLEDGMENT

This work was supported by Public Health Service grant AI11805 from the National Institute of Allergy and Infectious Diseases.

REFERENCES

1. Black FT, Birch-Andersen A, Freundt EA: Morphology and ultrastructure of human T-mycoplasmas. J Bacteriol 111:254–259, 1972
2. Bak AL, Black FT, Christiansen C, et al: Genome size of mycoplasmal DNA. Nature (London) 224:1209–1210, 1969
3. Bak AL, Black FT: DNA base composition of human T-strain mycoplasmas. Nature (London) 219:1044–1045, 1968
4. Rottem S, Pfendt EA, Hayflick L: Sterol requirements of T-strain mycoplasmas. J Bacteriol 105:323–330, 1971
5. Razin S, Rottem S: Isolation, solubilization and reconstitution of mycoplasma membranes, p 459–468. *In* S Fleischer and L Packer (ed), Methods in enzymology, vol. 32B. Academic Press Inc., New York, 1974
6. Razin S, Valdesuso J, Purcell RH, et al: Electrophoretic analysis of cell proteins of T-strain mycoplasmas isolated from man. J Bacteriol 103:702–706, 1970
7. Rosendal S: Analysis of the electrophoretic pattern of mycoplasma proteins for the identification of canine mycoplasma strains. Acta Pathol Microbiol Scand 81:273–281, 1973
8. Whitescarver J, Castillo F, Furness G: The preparation of membranes of some human T-mycoplasmas and analysis of their carbohydrate content. Proc Soc Exp Biol Med 150:20–22, 1975
9. Whitescarver J, Trocola M, Campana T, et al: A study of amino acids and proteins of some human T-mycoplasma membranes. Proc Soc Exp Biol Med 151:68–71, 1976
10. Sumner JB: Urease, p 873–892. *In* JB Sumner and K Myrbach (ed), The enzymes, vol 1, part 2. Academic Press Inc., New York, 1951
11. Rottem S, Razin S: Isolation of mycoplasma membranes by digitonin. J Bacteriol 110:699–705, 1972
12. Masover GK, Sawyer JE, Hayflick L: Urea-hydrolyzing activity of a T-strain mycoplasma: *Ureaplasma urealyticum*. J Bacteriol 125:581–587, 1976
13. Pollack JD: Localization of reduced nicotinamide adenine dinucleotide oxidase activity in *Acholeplasma* and *Mycoplasma* species. Int J Syst Bacteriol 25:108–113, 1975
14. Cirillo VP, Razin S: Distribution of a phosphoenolpyruvate-dependent sugar phosphotransferase system in mycoplasmas. J Bacteriol 113:212–217, 1973
15. Woodson BA, McCarty KS, Shepard MC: Arginine metabolism in mycoplasma and infected strain L-929 fibroblasts. Arch Biochem Biophys 109:364–371, 1965
16. Masover GK, Hayflick L: Dialysis culture of T-strain mycoplasmas. J Bacteriol 118:46–52, 1974
17. Masover GK, Hayflick L: The growth of T-strain mycoplasmas in media without added urea. Ann NY Acad Sci 225:118–130, 1973
18. Razin S: Physiology of mycoplasmas. Adv Microb Physiol 10:1–80, 1973
19. Freeman BA, Sissenstein R, McManus TT, et al: Lipid composition and lipid metabolism of *Spiroplasma citri*. J Bacteriol 125:946–954, 1976
20. Shepard MC, Lunceford CD: Occurrence of urease in T strains of mycoplasma. J Bacteriol 93:1513–1520, 1967
21. Ford DK, MacDonald J: Influence of urea on the growth of T-strain mycoplasmas. J Bacteriol 93:1509–1512, 1967
22. Ford DK, McCandlish KL, Gronlund AF: Metabolism of ^{14}C-urea by T-strain mycoplasma. J Bacteriol 102:605–606, 1970
23. Seneca H, Peer P, Nally R: Microbial urease. Nature (London) 193:1106–1107, 1962
24. Shepard MC, Lunceford CD: Effect of pH on human *Mycoplasma* strains. J Bacteriol 89:265–270, 1965
25. Furness G: T-mycoplasmas: their growth and production of a toxic substance in broth. J Infect Dis 127:9–16, 1973

Interaction of *Ureaplasma urealyticum* with Eucaryotic Cells In Vitro

GERALD K. MASOVER, MARINA PALANT, ZENAIDA ZERRUDO, AND LEONARD HAYFLICK

Mycoplasma contamination of cells grown in culture is a widespread and serious problem for those who use cell culture in biomedical research (1, 2). The problem is of such importance that many of the recent microbiology, mycoplasma, and cell culture meetings have addressed this topic in special sessions. However, I know of only a single report which describes a *Ureaplasma* contaminant in a cell culture (3). This is unusual in view of the fact that ureaplasmas resemble other mycoplasmas in most major properties. In addition, ureaplasmas occur frequently in bovine as well as human hosts (4), and, as a result, there would seem to be ample opportunity for these microorganisms to contaminate cell cultures either via the human cell culturist or from bovine cell culture medium supplements. However, ureaplasmas either do not infect cells for some as yet unknown reason or perhaps do infect cells but are not detected. In any case, if infection of cells by ureaplasmas can be established or if a cytopathic effect can be demonstrated, then the possibility of *Ureaplasma* contamination of cell cultures must be considered. Mazzali and Taylor-Robinson (5) studied the growth of ureaplasmas in cell cultures and found that they grow when first inoculated but do not persist beyond a few days and cause only a minimal cytopathic effect.

It is of more interest from our point of view that an effect, however inapparent, of ureaplasmas on cells in culture, or perhaps the failure of these organisms to cause an effect under some circumstances, might yield useful insights into ureaplasma-host cell interactions in vivo. This, of course, would provide a useful model for possible pathogenicity of ureaplasmas. With this in mind, we studied the effects of *Ureaplasma* on normal human cells in vitro.

We assessed the effect of *U. urealyticum* on WI-38 cells first by observing cell multiplication and plating efficiency. In typical growth curve experiments, hemocytometer counts of cell nuclei were made (by the method of Sanford et al. [6]) on days 1, 3, and 5 after the cells were seeded and infected. Cell culture medium was routinely changed on day 3. Experiments in which a multiplicity of about 3.0 CCU_{50} (50% color change units) of *Ureaplasma* per cell were used to infect WI-38 cells growing in Eagle basal medium (BME) supplemented with 10% unheated calf serum showed a modest effect, in that the cells did grow but at a slower rate. After 5 days, the number of cells in infected cultures was 10 to 25% of the number of cells in uninfected control cultures. Infection of WI-38 cells with similar multiplicities of *Mycoplasma hominis* or *Acholeplasma laidlawii* caused either complete cessation of multiplication or death of the fibroblasts. In some experiments, unheated horse serum was used instead of calf serum and the medium was not changed on day 3. Horse serum, which is considered a superior enrichment for ureaplasmas, did not enhance the cytotoxic effect. It did slow WI-38 cell growth and was, in fact, found to be unsuitable for long-term experiments because it is toxic to the WI-38 cells even when mixed with various proportions of calf serum.

In dose-response experiments in which either the same or different numbers of WI-38 cells were seeded in each flask, it was found that a multiplicity of infection (MOI) greater than 1.0 was needed to cause a reduction of WI-38 cell multiplication. In these experiments, the survival of cells in infected cultures was about the same as in the previous experiments.

An experiment in which a high initial number of WI-38 cells (10^6 cells per T-25 flask) was seeded was repeated with both an MOI of 40 and a lower MOI (<1.0). There was no cytotoxic effect in either case. This is an important result for two reasons. First, it suggests that the reduced multiplication of WI-38 cells is not due to some toxic substance elaborated by the ureaplasmas, because if that were the case a decrease in cell number would be expected, and none was observed. Second, this result at least suggests that the cells must be multiplying in order to be sensitive to the cytotoxic effect of the ureaplasmas. This point was confirmed in experiments in which we used lower numbers of WI-38 cells and the usual multiplicity of about 3 CCU_{50} of *Ureaplasma* per WI-38 cell, but medium with 2% serum or no serum at all. The low serum retards or prevents WI-38 cell multiplication, and it also retarded the cytotoxic effect.

We also found that, if these growth experiments were initiated in medium containing chlortetracycline instead of penicillin, the cytotoxic effect of the ureaplasmas was prevented. We used similar growth experiments to see whether the reduction in cell multiplication was a result of nutrient depletion. We found that addition of fourfold-concentrated BME vitamins plus fourfold-concentrated essential and nonessential amino acids did not prevent the cytotoxic effect. Use of Ham's F-12 medium containing nucleic acid precursors or addition of 10 μg of each nucleic acid precursor per ml to the usual BME medium did not prevent the cytotoxic effect on multiplication.

Thus, ureaplasmas caused a reproducible dose-dependent cytotoxic effect on WI-38 cell multiplication which appeared to require that the host cell be growing. The effect was sensitive to chlortetracycline but not to penicillin; it was not related to depletion of the usual vitamins, amino acids, or nucleic acid precursors from the cell growth medium; and it was probably not the result of a cytotoxic substance elaborated by the ureaplasmas.

If the WI-38 cells were subcultivated by trypsinization several days after infection, the cytotoxic effect of cell multiplication either was not observed or was very small. The growth reduction was not eliminated by changing the medium without subcultivation. It was possible to demonstrate this effect very dramatically in plating efficiency experiments, and one such experiment is shown in Fig. 1. In this experiment, about 1,000 cells were seeded per 60-mm petri dish (triplicates), and half of the dishes were infected. The medium was changed on days 5, 10, and 15, and the cells were fixed and stained on day 20. The relative plating efficiency of the cells in infected cultures was 37%, and, in addition, there were not only fewer colonies but fewer cells per colony. Thus, the cytotoxic effect on growth was shown to persist for at least 20 days through three medium changes without subcultivation.

These observations were confirmed and extended by autoradiographic experiments in which both early and late effects of ureaplasmas could be observed. Short (4 h) exposure to tritiated thymidine resulted in reduced numbers of labeled nuclei in cells infected 1 day or 30 days prior to addition of the label. There was no

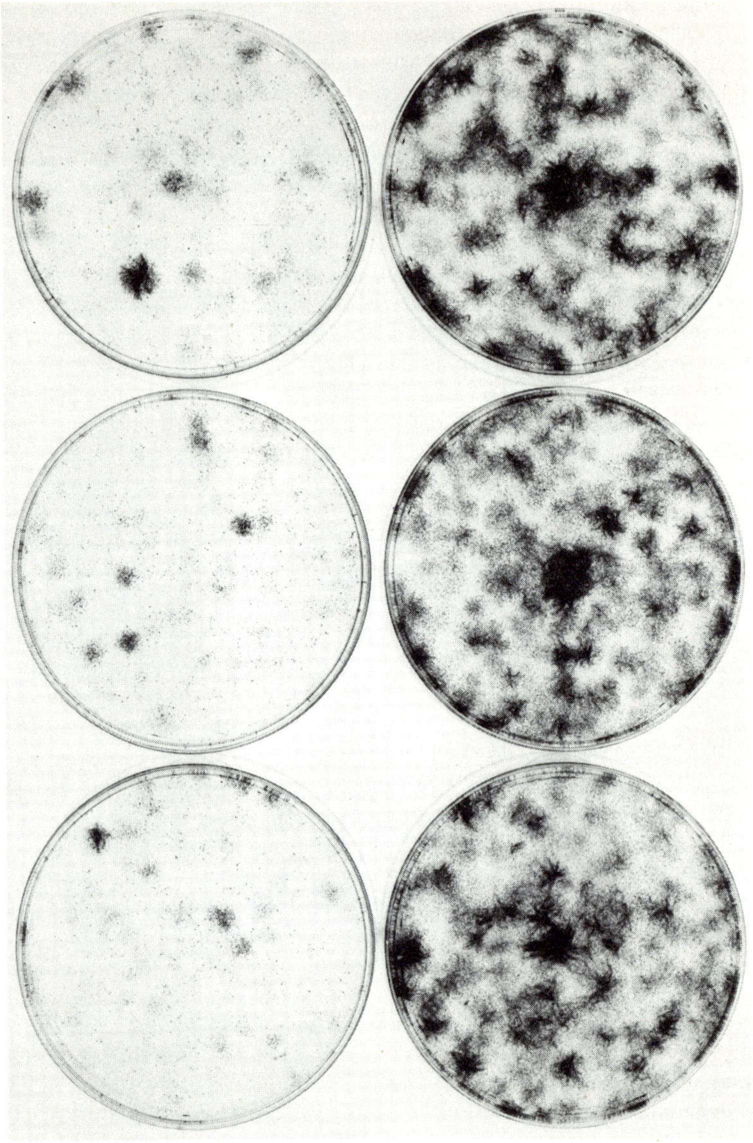

FIG. 1. *Effect of ureaplasma infection on the plating efficiency of WI-38 cells. Petri dishes (60 mm) were seeded with 1,000 cells/dish in 5 ml of complete cell culture medium. Three replicate infected cultures (left) received 0.1 ml of growing Ureaplasma urealyticum broth culture to give a multiplicity of infection of 3.0 ureaplasmas per WI-38 cell. The three replicate control cultures (right) received 0.1 ml of sterile mycoplasma medium. Culture medium was changed on days 5, 10, and 15. Cultures were fixed and stained with Giemsa stain on day 20.*

apparent cytoplasmic labeling at this low MOI, and the number of nuclear grains was about the same for both infected and uninfected cells which were labeled for the shorter period. When the exposure to label was increased to 48 h, both infected and uninfected cells gave the same high labeling index, indicating that the infected cells do synthesize deoxyribonucleic acid, but either at a slower rate or, more likely, after a block in the cell cycle has been overcome. This latter result also argues against the possibility that the thymidine is altered or not absorbed by infected cells because localization of the label in the cell nucleus does occur after a longer pulse.

Autoradiographic experiments were done also on WI-38 cells infected with 20 CCU_{50} of *Ureaplasma* per cell and scored after each of three subcultivations by trypsinization. Results for uninfected controls were the same as for the previous experiment, with the exception that the effect of a 24-h labeling period as well as the 4- and 48-h periods was observed. This was important to note because the WI-38 cell doubling time is about 24 h for our conditions, and, if the cell cycle was proceeding normally, we could be justified in expecting the maximal number of labeled nuclei within 24 h. This was observed. The effect of the higher than usual MOI was apparent at the first subcultivation, and it was seen that, as these cells were subcultivated by trypsinization, the numbers of labeled nuclei approached control levels. We were unable to demonstrate the characteristic *Ureaplasma* urease activity during these experiments. This experiment indicates again that although a high MOI caused a lack of deoxyribonucleic acid synthesis during 48 h in 76% of the cells at the first subcultivation, the cells were not killed. On subcultivation of these same cells, the labeling index increased. This was taken as evidence that either the ureaplasmas or some effect resulting from their association with the cells was removable by trypsin. The effect of trypsin is of interest for reasons discussed below.

The effects of ureaplasmas on WI-38 cell morphology correspond to reported effects of other mycoplasmas on cells in culture. In Giemsa-stained preparations of infected WI-38 cells, vacuolated cytoplasms, cytoplasmic protrusions, and cytoplasmic inclusions could be seen. Similar observation were possible by phase-contrast microscopy.

We were able to conclude from these experiments that ureaplasmas *do* interact with WI-38 cells in culture to produce a modest but reproducible effect. The fact that the infection did not persist is of interest and, when taken with our newer data, is a very useful finding. We recently found that a more persistent infection of these cells could be established in five of six trials in which an MOI between 1 and 100 was used, but only if the ureaplasmas were pretreated with trypsin. In several separate experiments we observed that purified trypsin does not affect ureaplasma viability during the 30-min treatment period usually used. In our opinion, this is a very important observation with relevance to mechanisms involved in mycoplasma-host cell interactions in general.

The fact that trypsin pretreatment of *Ureaplasma* enables a persistent infection to occur suggests that a close membrane-to-membrane association is required for a productive infection to occur and that this is made possible by removal of trypsin-sensitive material from the ureaplasmas. This is consistent with our finding that ureaplasmas grown in vitro have much more protein associated with them than would be expected to be found as cellular material. This result of trypsinization of the ureaplasmas is not inconsistent with the observation that untreated ureaplasmas can be removed from their host cells by trypsinization. It suggests, in fact, that the ureaplasmas are able to produce a cytotoxic effect on cell multiplication despite what is probably a tenuous association with the cell.

In one experiment a WI-38 culture was infected with trypsin-treated ureaplasmas, and the cells were allowed to remain confluent without subcultivation for more than 60 days. The medium was changed every 3 to 4 days, and the old medium was tested for ureaplasmas and found to be positive at titers of 10^2 to 10^4 per ml of supernatant cell culture medium throughout the experiment. It was negative after 64 days, but the cells were still viable as indicated by their ability to undergo two subsequent subcultivations before the experiment was terminated. This clearly indicates that the ureaplasmas were multiplying and being released into the medium over a considerable period of time, contrary to what was found with untrypsinized ureaplasmas. In this case, the confluent WI-38 cells are not multiplying. Thus, if a close ureaplasma-cell association is accomplished, then host cell multiplication is not required for the ureaplasmas to multiply. On the other hand, during the 60 days before the culture medium became ureaplasma negative, there was little or no obvious cytopathic effect.

This type of demonstrable cell association without discernible cytopathic effect was recently reported with *M. hyopneumoniae* in primary cultures of pig kidney (7), whereas the same cultures infected with *M. hyorhinis*, another common porcine mycoplasma, showed considerable cytopathic effects. There is a certain amount of appeal in a model for disease in which a microorganism is able quietly to coexist with a host cell, producing no cytopathic effect until the host cell starts to multiply, perhaps under hormonal influence, at which time a cytopathic effect is exerted and expressed as clinical disease. It seems reasonable to us that subtle effects on the host cell rather than overt cytocidal effects in vitro might parallel the more important effects of infectious agents in vivo in the production of disease. Whether or not the subtle effects we have observed thus far in vitro are relevant to disease remains to be proven, and the mechanisms involved must be elucidated.

Another recent development in our efforts to create a reasonable in vitro model to study the role of ureaplasmas in disease has been our development of epithelioid cells, which we think might serve as a more likely target for a mycoplasma. We now have an epithelioid cell derived from the synovium of a strain 13 guinea pig about 1 year ago. The epithelioid character of these cells has been maintained for more than 80 population doublings so far. They now have an increasing proportion of binucleate and multinucleate cells, but their chromosomes are still characteristic of guinea pig chromosomes in morphology and number. Experiments are in progress to test these cells for malignancy by injecting them back into strain 13 animals.

It is our opinion that tissue culture techniques can be used to advantage in the study of urethritis and all manner of pathology which might relate to mycoplasma-cell interactions, provided, of course, that we pursue both the basic biology of the infectious agent, in this case *Ureaplasma*, and an appropriate host cell, which we hope can be developed.

ACKNOWLEDGMENT

This work was supported by Public Health Service grant AI11805 from the National Institute of Allergy and Infectious Diseases.

REFERENCES

1. Hayflick L: Tissue cultures and mycoplasmas. Tex Rep Biol Med 23(Suppl)1:285–303, 1965
2. Barile MF, Hopps HE, Grabowski MW, et al: The identification and sources of mycoplasmas isolated from contaminated cell cultures. Ann NY Acad Sci 225:251–264, 1973
3. Sethi KK: On the incidence of mycoplasma contamination in cell cultures. Zentralbl Bakteriol Parasitenkd Infektionskr Hyg Abt I Orig A 219:550–554, 1972
4. Taylor-Robinson D, Furr PM: The distribution of T-mycoplasmas within and among various animal species. Ann NY Acad Sci 225:108–117, 1973
5. Mazzali R, Taylor-Robinson D: The behaviour of T-mycoplasmas in tissue culture. J Med Microbiol 4:125–138, 1971
6. Sanford KK, Earle WR, Evans VJ, et al: The measurement of proliferation in tissue cultures by enumeration of cell nuclei. J Natl Cancer Inst 11:773–795, 1951
7. Pijoan C: The effects of *Mycoplasma hyopneumoniae* on pig kidney primary tissue culture. Br Vet J 131:586–594, 1975

Ureaplasma urealyticum: Serotyping and the Human Humoral Response

JUEY-SHIN L. LIN

Ureaplasma urealyticum and *Mycoplasma hominis* are the principal mycoplasmas which have been isolated from the human genital tract. The mycoplasmas are acquired primarily through sexual contact, although other factors undoubtedly operate. Many sexually experienced men and women are colonized with the mycoplasmas, but the etiological significance of these agents in human diseases is not yet clearly defined (1, 2). The role of genital mycoplasmas in human disease is discussed in detail by McCormack (p. 98, this volume).

Immunological studies of *U. urealyticum* have been particularly difficult because pure antigens (or specific antisera) have been difficult to prepare. The mycoplasmas require rich and complex media for growth, and they do not yield sufficient antigenic mass for many studies; the maximal number of colony-forming units (CFU) in broth cultures of *U. urealyticum* is usually 100- to 1,000-fold lower than that found for the classic large-colony-forming mycoplasmas. Therefore, growth patterns and metabolic characteristics of the mycoplasmas were utilized to define the specificity of serological tests. Because of the ability of ureaplasmas to hydrolyze urea (3), metabolic inhibition (MI) became convenient for their study. The release of ammonia from the metabolite increases the pH of the growth medium as the mycoplasmas multiply. Specific inhibition of the mycoplasmas can be detected by delay in or prevention of increases in pH.

In 1966, Purcell et al. (4) first utilized the MI test to determine serum antibody titers in a population composed of different age groups. MI antibody titers to a strain of *U. urealyticum* were demonstrated in a large proportion of adults but were infrequent in children. Later, in 1967, the same workers showed that there were at least six distinct serotypes of the mycoplasmas and that paired sera from patients with nongonococcal urethritis (NGU) did not show any serological responses when the six reference strains were used as antigens in the MI test (5). Ford (6) also observed that there were at least six serologically distinct subgroups of the mycoplasmas by MI; oral strains were serologically similar to the genital ones, and strains isolated from patients with NGU were serologically diverse. Most recently, Black used four different serological tests (7–9) to validate and establish eight serotypes of *U. urealyticum* which are now recognized as standard reference strains (3). Growth inhibition on agar plates and indirect immunofluorescence (IMF) with unfixed colonies were specific tests that were suitable for serotyping of clinical isolates. The indirect hemagglutination (IHA) and MI reactions were more sensitive but less specific than growth inhibition or IMF and were better for detecting humoral responses than for initial serotyping. Paired sera from patients with NGU did not show significant serological responses by MI and IHA using all eight reference strains as antigens.

We have developed a complement-dependent mycoplasmacidal test and have used it to show that there are at least 11 serological subgroups of *U. urealyticum* (10,

11). Strains isolated from pregnant women who delivered infants with low birth weight and strains isolated from patients with NGU were serologically diverse. Thus, published studies have repeatedly demonstrated that *U. urealyticum* is serologically heterogeneous, consisting of at least 11 subgroups, and that strains of *U. urealyticum* isolated from patients with NGU are serologically diverse. Bovine *Ureaplasma* (12) and other mycoplasmas such as *M. pulmonis* (13), *M. hominis* (14), and *M. hyorhinis* (15) have also been shown to be serologically heterogeneous.

Our interest in the genital mycoplasmas originated from studies from our laboratory (16, 17) which demonstrated that colonizations of pregnant women with these agents were associated with delivery of infants of low birth weight. Serological studies were initiated to develop sensitive and reliable methods for serotyping the mycoplasmas and for detecting humoral responses in patients. Complement-dependent immune inhibition of *U. urealyticum* provided a specific and sensitive test (10), which can be adapted to a microtiter system for seroepidemiological surveys (11).

In the present mycoplasmacidal test, twofold serial dilutions of sera are incubated at 37°C for 1 h with the mycoplasmas and with complement (fresh guinea pig serum) in a diluent. After incubation to permit killing of the mycoplasmas by antibody, growth medium is added to allow the surviving organisms to grow. The highest serum dilution which prevented the changes in pH that occur after 4 to 5 days of incubation is recorded as the mycoplasmacidal antibody titer. The sensitivity and specificity of the mycoplasmacidal test are affected by the concentration of antigen. If more than 10^2 50% color change units (CCU_{50}) (11) of mycoplasma are used, little or no antibody is detectable. Therefore, to detect humoral responses in humans, low concentrations of antigen must be used.

While comparing the serotype behavior of reference strains, we suspected and confirmed that initial genital cultures might be serotypically heterogeneous (18). Freshly obtained vaginal cultures often contained more than one serotype of *U. urealyticum*. Two well-isolated colonies of *U. urealyticum* from each culture were picked and grown separately in broth. Each strain was serotyped with the 11 reference antisera. Examination by this colony-picking procedure of 30 cultures obtained from pregnant women who had attended the prenatal clinic at Boston City Hospital (17) showed that 4 of them contained two different serotypes of *U. urealyticum*. Of 54 cultures obtained from college students, 3 showed serotype heterogeneity. It is of interest that the cultures with more than one serogroup were obtained from sexually active people (Table 1). One of eight cultures studied from patients with salpingitis (supplied by K. K. Holmes) also contained more than one serogroup. A similar study with *M. hominis* showed that vaginal cultures from pregnant women were serologically homogeneous. However, cultures obtained at different times from the same patient may contain a different serotype of *M. hominis* from that originally found (19).

Since sexual activity had been shown to be one of the important determinants of colonization with the genital mycoplasmas, and the number of different partners seems to be epidemiologically significant in determining the rate of colonization (20), it is not surprising that several serotypes of *U. urealyticum* are present in human genital tracts. The reasons why *M. hominis* tends to be found in pure culture are less evident.

TABLE 1. *Presence of mixed serotypes of U. urealyticum in vaginal cultures*

Source	No. tested	No. showing mixed serotypes
Pregnant women	30	4
College students[a]		
Sexually active	27	3
Sexually nonactive	27	0
Patients with salpingitis	8	1

[a] Sexually active, those who had ≥7 partners in past. Sexually nonactive, those who had ≤1 partners in past.

The serotype distribution of *U. urealyticum* isolated from various human sources is summarized in Table 2. Strains isolated from men with urethritis, from women with salpingitis, and from pregnant women who delivered low-birth-weight or stillborn infants have all been serologically diverse. These results suggest that the pathogenicity of the mycoplasmas, if any, may not be directly associated with particular serotypes. On the other hand, the diversity could be due to the presence of multiple serotypes in some individuals.

We have also used our reference sera to serotype Black's eight strains (types I through VIII) which were officially recognized as reference serotypes of *U. urealyticum* (3). These eight strains were also serologically distinct by the mycoplasmacidal test; types II through VIII were assigned as our serogroups 5, 9, 4, 7, 8, 6, and 1, respectively. Type I showed multiple cross-reacting patterns and could not be assigned.

Because of the presence of multiple serological subgroups in *U. urealyticum* and because of the presence of multiple serotypes in some individuals, we have studied how to detect the most antibody titers and titer changes in human sera. Paired sera

TABLE 2. *Serotype distribution of U. urealyticum from various sources*

Serogroup	Pregnant women[a]		Newborn infants[b]		Men with NGU	Women		Total
	N.B.W.	L.B.W.	L.B.W.	S.B.		Salpingitis	College	
1	2	2			1			5
2	1	2	2		2			7
3		5	4	2	1		1	13
4	2	1			3	1		7
5	4	7			2	5	22	40
6		3						3
7	4	5			1	1	5	16
8	6	7	1				4	18
9	11	4	3			2	7	27
10	7	4		1	1	1		14
11	5	4	1	1		1	17	29
Untypable		3		1			5	9
Total	42	47	11	5	11	11	61	188

[a] Pregnant women who delivered normal-birth-weight (N.B.W.) and low-birth-weight (L.B.W.) infants.

[b] Infants of low birth weight (L.B.W., ≤2,500 g) or still birth (S.B.).

TABLE 3. *Postpartum mycoplasmacidal titer changes to U. urealyticum detected by serum donors' isolates and 11 reference strains*

Titer changes	No. detected by:	
	Own isolates	Reference strains
Rise ≥ 4×	4	9
Fall ≥ 4×	1	1
No change (≥1:8)	2	7
No change (<1:8)	22	12
Total	29	29

were obtained from pregnant women (21, 22). First sera were taken immediately after delivery and second sera were taken 6 to 12 weeks later. Vaginal swabs were obtained at the former time for cultivation of the genital mycoplasmas. The mycoplasmacidal titers of paired sera were determined with the serum donor's genital isolates and 11 reference strains used as antigens. The results showed that all 11 reference strains were needed to detect most titer rises and that there was no advantage in using the donor's own strains. The serum donors' isolates detected 4 significant titer rises out of 29 paired sera, whereas reference strains picked up 9 such rises (Table 3). Other data also indicate that the reference strains detected more and higher titers (Lin et al., unpublished data). A similar study with *M. hominis* showed that only three of seven reference strains were required to detect most antibody titers and titer rises (Lin et al., in preparation).

Using 11 reference strains of *U. urealyticum* and 3 of *M. hominis* as antigens, we have surveyed postpartum humoral responses to both genital mycoplasmas. More postpartum titer rises to *U. urealyticum* were observed in women who had positive cultures of *U. urealyticum* at delivery (16 of 57) than in those who had negative cultures (3 of 24, not significant). When postpartum humoral responses were studied in patients who were colonized with *M. hominis*, titer rises of fourfold or more were significantly associated with the presence of the mycoplasma in vaginal cultures and with low serum mycoplasmacidal titers (<1:8) at delivery (Lin et al., in preparation). As summarized in Table 4, significantly more patients had postpartum

TABLE 4. *Postpartum humoral responses to the genital mycoplasma*[a] *detected by the mycoplasmacidal (MC) tests*

Species	Fraction with ≥4× rise in MC titers	
	Mycoplasma in blood	*Mycoplasma* in vagina
U. urealyticum	1/11 (9%)	16/57 (28%)
M. hominis	9/9 (100%)	19/32 (60%)
χ^2	13 (corrected)	8.3
P	<0.001	<0.01

[a] Pregnant women who had the mycoplasmas isolated from blood or vaginal cultures immediately after delivery.

TABLE 5. *Serum mycoplasmacidal (MC) titers to the genital mycoplasmas in pregnancy*

Species	No. tested	No. with MC titers of:				Fraction with titer changes[a]
		≥1:4	≥1:8	≥1:16	≥1:32	
U. urealyticum	117	92 (77.6%)	22 (18.8%)	5 (4.3%)	4 (3.4%)	13/104 (12.5%)
M. hominis	113	83 (73.5%)	47 (41.6%)	26 (22.8%)	21 (18.6%)	32/104 (30.8%)
P			<0.001	<0.001	<0.001	<0.01

[a] Significant titer changes of fourfold or more were observed between the first visit and the delivery.

humoral responses to *M. hominis* than to *U. urealyticum*; only 1 of 11 of those who had *U. urealyticum* isolated from postpartum blood (21) had significant rises to *U. urealyticum*, whereas all 9 of those who had *M. hominis* in postpartum blood cultures had significant postpartum rises in titer. Of those who were colonized with each mycoplasma, 16 rises to *U. urealyticum* were observed in 57 patients and 19 rises to *M. hominis* were observed in 32 patients ($\chi^2 = 8.3, P < 0.01$). Most, if not all, of the rises were observed among those who had low titers at the first sera (<1:8), and higher rises were observed to *M. hominis* (usually ≥1:32 in second sera) than to *U. urealyticum* (usually ≤1:32 in second sera).

In recent studies, we have followed 117 pregnant women who had normal deliveries at the Boston City Hospital. All the women studied had positive cultures for either or both genital mycoplasmas at the first prenatal visit, and most had at least three serum samples collected between the first visit and delivery (17). Mycoplasmacidal titers to both mycoplasmas were determined for each serum sample (Table 5). Of this selected group, 95% had *U. urealyticum* and 62% had *M. hominis* isolated from their vaginal cultures. About 78% had mycoplasmacidal titers of ≥1:4 to *U. urealyticum* and 74% to *M. hominis*, indicating that the mycoplasmacidal test for *U. urealyticum* is as sensitive as the test for *M. hominis*. However, significantly more women had higher titers to *M. hominis*: only 19% of them had titers of ≥1:8 to *U. urealyticum* whereas 42% of them had such titers to *M. hominis* ($\chi^2 = 14$, $P < 0.001$). Similar differences were observed for titers of ≥1:16 ($\chi^2 = 17$, $P < 0.001$) and for titers of ≥1:32 ($\chi^2 = 35$, $P < 0.001$). Significant titer changes of fourfold or more were observed to *U. urealyticum* in 13% of women and to *M. hominis* in 31% of women who were studied ($\chi^2 = 9$, $P < 0.01$). Thus, *M. hominis* was more invasive and/or induced stronger humoral responses than *U. urealyticum* in these patients.

In summary, useful and sensitive tests have continued to be developed for the study of the *U. urealyticum*. By application of these procedures, during the next few years it should be possible to assemble the necessary seroepidemiological information to assess the role of these microorganisms in disease.

ACKNOWLEDGMENTS

This study was supported by Public Health Service research grant HD-03693 from the National Institute of Child Health and Human Development.

The assistance given by Edward H. Kass, William M. McCormack, Mary I. Kendrick, Katalin M. Radnay, and Susan Alpert is gratefully acknowledged.

REFERENCES

1. McCormack WM, Braun P, Lee Y-H, et al: The genital mycoplasmas. N Engl J Med 288:78–89, 1973
2. Freundt EA: Present status of the medical importance of mycoplasmas. Pathol Microbiol 40:155–187, 1974
3. Shepard MC, Lunceford CD, Ford DK, et al: Ureaplasma urealyticum gen. nov., sp. nov.: proposed nomenclature for the human T (T-strain) mycoplasmas. Int J Syst Bacteriol 24:160–171, 1974
4. Purcell RH, Taylor-Robinson D, Wong D, et al: Color test for the measurement of antibody to T-strain mycoplasmas. J Bacteriol 92:6–12, 1966
5. Purcell RH, Wong D, Chanock RM, et al: Significance of antibody to mycoplasmas as measured by metabolic inhibition techniques. Ann NY Acad Sci 143:664–675, 1967
6. Ford DK: Relationships between mycoplasma and the etiology of nongonococcal urethritis and Reiter's syndrome. Ann NY Acad Sci 143:501–504, 1967
7. Black FT: Serological methods for classification of human T-mycoplasmas. Int Congr Infect Dis, 5th, Vienna, Austria 1:407–411, 1970
8. Black FT: Modification of the growth inhibition test and its application to human T-mycoplasmas. Appl Microbiol 25:528–533, 1973
9. Black FT, Krogsgaard-Jensen A: Application of indirect immunofluorescence, indirect haemagglutination and polyacrylamide-gel electrophoresis to human T-mycoplasmas. Acta Pathol Microbiol Scand Sect B 82:345–353, 1974
10. Lin J-S, Kass EH: Immune inactivation of T-strain mycoplasmas. J Infect Dis 122:93–95, 1970
11. Lin J-S, Kendrick MI, Kass EH: Serological typing of human genital T-mycoplasmas by a complement-dependent mycoplasmacidal test. J Infect Dis 126:658–663, 1972
12. Howard CJ, Gourlay RN: Serological comparison of bovine T-mycoplasmas. J Gen Microbiol 79:129–134, 1973
13. Forshaw KA, Fallon RJ: Serological heterogeneity of *Mycoplasma pulmonis*. J Gen Microbiol 72:501–510, 1972
14. Lin J-S, Kass EH: Serological reactions of *Mycoplasma hominis*: differences among mycoplasmacidal, metabolic inhibition, and growth agglutination tests. Infect Immun 10:535–540, 1974
15. Gois M, Kuksa F, Franz J, et al: The antigenic differentiation of seven strains of *Mycoplasma hyorhinis* by growth inhibition, metabolism-inhibition, latex agglutination, and polyacrylamide-gel electrophoresis tests. J Med Microbiol 7:105–115, 1974
16. Klein JO, Buckland D, Finland M: Colonization of newborn infants by mycoplasmas. N Engl J Med 280:1025–1030, 1969
17. Braun P, Lee Y-H, Klein JO, et al: Birth weight and genital mycoplasmas in pregnancy. N Engl J Med 284:169–171, 1971
18. Lin J-S, Kass EH: Serotypic heterogeneity in isolates of human genital T-mycoplasmas. Infect Immun 7:499–500, 1973
19. Lin J-SL, Radnay K, Kendrick MI, et al: Serological studies on *Mycoplasma hominis*: serotypic distribution of recent clinical isolates. Scand J. Infect Dis 8:45–48, 1976
20. McCormack WM, Almedia PC, Bailey PE, et al: Sexual activity and vaginal colonization with genital mycoplasmas. J Am Med Assoc 221:1375–1377, 1972
21. McCormack WM, Rosner B, Lee Y-H, et al: Isolation of genital mycoplasmas from blood obtained shortly after vaginal delivery. Lancet i:596–599, 1975
22. McCormack WM, Lee Y-H, Lin J-S, et al: Genital mycoplasma in postpartum fever. J Infect Dis 127:193–196, 1973

Antigenic Analysis of the *Mycoplasmatales*

GEORGE E. KENNY

Antigenic analysis of the *Mycoplasmatales* has proven difficult because of the small amounts of material which can be obtained (5 to 20 mg of organisms per liter of culture). The prospects of antigenic analysis of *Ureaplasma urealyticum* appear even more difficult because present media permit the production of approximately 10-fold less antigen.

U. urealyticum, or the T strains, as they were previously called, is a species which presently includes all of the human T strains. The organisms are similar in that they form small colonies and produce ammonia from urea; they differ in that they show a bewildering number of serological types (1–3). I would like to address this presentation to several issues for which a general discussion of the antigenic structure of the *Mycoplasmatales* might be helpful. (i) Does *U. urealyticum* have any relationship to the organisms in the *Mycoplasmatales* or *Mycoplasmataceae*? (ii) Are the types of *U. urealyticum* closely related to each other or are they very distant? (iii) What immunological methods might be most suitable for antigenic analysis of *U. urealyticum*?

TAXONOMY

The taxonomy of the *Mycoplasmatales* is evolving rapidly, not only because many new species have been discovered in the past 10 years but also because extensive surveys and comparisons have been made of groups of species. In contrast, *U. urealyticum* has only recently been classified (4), and as a consequence few relationships are clearly understood (Table 1). Three genera are presently recognized among the *Mycoplasmatales*: *Mycoplasma*, *Acholeplasma*, and *Ureaplasma*. Separate families have been established for genus *Mycoplasma* and genus *Acholeplasma*.

SEROLOGICAL TESTS

A variety of serological tests can be employed with the *Mycoplasmatales*. Each of these tests has varying sensitivities for measurement of antigen and antibody, and some information is available as to the antigenic site measured (Table 2). Accordingly, a discussion of the sensitivity, analytical abilities, and specificities of each serological test is relevant to their usefulness for *U. urealyticum* serology. The inhibition of formation of colonies on agar by disks soaked in antibody has been used as a major taxonomic tool (5). This method is the least sensitive for measurement of antibody, but it requires only a small number of organisms. Powerful antisera are required, and suitable antibodies only arise late in immunization (6). This test has difficulties for *U. urealyticum*. In the metabolic inhibition test, antibody prevents the formation of some normal metabolic product (acid from glucose, ammonia from arginine or urea) (2). Only small amounts of antibody are required, and the role of complement in this

TABLE 1. *Taxonomy of the Mycoplasmatales*

Taxa	Identification
Mycoplasmatales	Small membrane-bounded procaryotic cells
Mycoplasmataceae	Require cholesterol
Mycoplasma	Utilize glucose or arginine
Mycoplasma species (40 +)	Serology—growth inhibition, fluorescent antibody, metabolic inhibition
Subspecies	Serology—metabolic inhibition, analytical serology
Ureaplasma	Utilize urea
U. urealyticum	Human isolate
Types	Serology—metabolic inhibition, complement-mediated killing
Acholeplasmataceae	Do not require cholesterol
Acholeplasma species	Serology—growth inhibition, metabolic inhibition

reaction has been controversial. The complement-mediated killing reaction is a rapid reaction (7, 8) and is highly sensitive for measurement of antibody. Each of these tests apparently measures antibodies against membrane components. Both metabolic inhibition testing and complement-mediated killing measure antibody against surface components; this localization of the antigenic site is not so clear with growth inhibition. The fluorescent-antibody method also appears to measure antibody against surface components; however, this may depend upon methods of fixation. All the preceding tests are species specific in that they readily detect antibodies and antigens in a given species, but serological reactivity is seldom detected against related species. Antigenic

TABLE 2. *Specificity and sensitivity of serological tests for antigenic analysis of the Mycoplasmatales*

Test	Specificity	Sensitivity		Analytical[a] abilities	Antigens[b]
		Antibody	Antigen		
Growth inhibition (Clyde)	Species	+	++++	+	M?
Metabolic inhibition	Species/subspecies	+++	++++	+	MS
Fluorescent antibody	Species/subspecies	++++	++++	+	MS?
Complement-mediated killing	Species/subspecies	++++	++++	+	MS
Complement fixation	Species/genus/subspecies	++	++	+	MC
Gel diffusion	Species/genus/subspecies	++	++	++[c]	MC
Two-dimensional immunoelectrophoresis	Species/genus/subspecies, negatively charged antigens	++	+++	++++[c]	MCS

[a] Analytical abilities for directly differentiating antigens.
[b] M = membrane; S = surface; C = cytoplasm (soluble fraction).
[c] Broadly reactive antisera required for maximal analytical results.

variation at the subspecies level can readily be observed by all tests except the growth inhibition technique on agar.

Complement fixation is an important method for measurement of both soluble and insoluble antigens. Both membrane and cytoplasmic antigens are readily detected. The two analytical methods are gel diffusion and two-dimensional immunoelectrophoresis (9). Both methods permit the simultaneous detection of a number of antigenic components, including both membrane and soluble (cytoplasmic) components. However, antigens need to be soluble or to be made soluble by nonionic detergents. Additionally, for two-dimensional immunoelectrophoresis, antigens need to be negatively charged at pH 8.6 (carbamylation of antibodies permits the detection of positively charged antigens) (10). The superior analytical abilities of two-dimensional immunoelectrophoresis arise from its quantitative abilities since not only can antigens be detected, but also their relative (or absolute) quantities can be measured (11). Thus, data obtained for serological relationships may be sharply different depending upon the method employed, and a direct comparison of data from one serological test with that of another may not be possible because of differences in sensitivity or specificity.

GEL DIFFUSION STUDIES

The question of whether *U. urealyticum* might be immunologically related to other genera in the *Mycoplasmatales* can be answered in part by an inspection of the serological relationships observed between species in the *Mycoplasmatales* as observed by gel diffusion (Table 3). When *Mycoplasma* or *Acholeplasma* species are classified by agar inhibition, fluorescent-antibody, or metabolic inhibition testing, the results are highly specific and provide no information for possible immunological relationships of the species. However, when the species are examined by a more analytical method such as gel diffusion, strong evidence for antigenic heterogeneity emerges. The organisms can be divided into six or seven groups which are essentially unrelated.

The methods used for gel diffusion were identical to those previously reported (12). Briefly, serological test antigens were grown in soy peptone fresh yeast dialysate broth (13) supplemented with 10% agamma (gamma globulin-depleted) horse serum, whereas immunogens were cultivated in the same medium supplemented with either 10% agamma calf serum or 10% agamma rabbit serum. Serological test antigens contained 5 to 15 mg of protein per ml after concentration and washing. Two separate serological surveys were completed with two different lots of antigen. Antisera from two different rabbits were used for each organism. The number of precipitin lines reported is the least number which consistently appeared in the experiments.

CROSS-RELATIONSHIPS

The type species of the genus *Mycoplasma* is *Mycoplasma mycoides* var. *mycoides*, an organism which is prohibited in the United States. Consequently, taxonomic studies of the genus are extremely difficult because antigens and antisera have been unavailable. Recently, antiserum was made available by the Reference Reagents Branch of the National Institutes of Health. This antiserum

TABLE 3. *Serological comparison of 23 species in the Mycoplasmatales by double immunodiffusion*

Antigen	Antisera[a]																									Strain	Serol. group[b]
	M	C	1	2	3	4	5	6	7	8	9	10	11	12	13	14	15	16	17	18	19	20	21	22	23		
1 Bovine group 7[c] . . .	5[d]	5	*8*	3						1		2														ATCC 19884	1
2 *M. putrefaciens* . . .	3	2	3	*6*						1																ATCC 15718	1
3 *A. laidlawii*					*7*	4	1																			ATCC 14089	2
4 *A. granularum*					4	*5*																				Friend	2
5 *A. axanthum*					2	2	*6*																			S-743	2
6 *M. bovigenitalium* .								*7*										1								ATCC 19852	3 (?)
7 *M. gallisepticum* . .									*5*																	ATCC 15302	4
8 *M. pneumoniae* . . .							1			*6*																AP-164	5
9 *M. anatis*											*8*	3			3		2	1								1340	6
10 *M. bovismastitidis* .		1						2				*6*			1	1	1	1	1							ATCC 25025	6 (?)
11 *M. canis*											4		*5*	3	3		1	1								ATCC 19525	6
12 *M. edwardii*													3	*4*	4	2	2	1	2							ATCC 23462	6
13 *M. felis*											3			3	*5*	1										B2	6
14 *M. fermentans*											2	1	1		2	*6*										PG 18	6
15 *M. hyorhinis*					1						1	1				3	*8*	3								ATCC 17981	6
16 *M. pulmonis*					1						2	1	1		1		3	*9*	3							63	6
17 *M. neurolyticum* . .											1			2	1		2		*7*	1						ATCC 15049	6
18 *M. synoviae*											1	1	1	1	2		1			*3*						WVU 1853	6
19 *M. arginini*																					*9*	4	1	2	2	G-230	7
20 *M. gateae*																					3	*11*	4	4	2	Siam	7
21 *M. gallinarum*																						1	*8*	2	1	ATCC 15315	7
22 *M. hominis*																						3		*6*		ATCC 14027	7
23 *M. salivarium*																					2			1	*9*	ATCC 23064	7

[a] M = antisera to *M. mycoides* var. *mycoides*, C = antisera to *M. mycoides* var. *capri* (both from the National Institutes of Health Reference Reagents Branch). Numbers of antisera correspond to antigen numbers.

[b] Serological group numbers are provisional and not intended as a final designation.

[c] The strain used for these experiments was obtained by cloning ATCC 19884 (*M. bovirhinis*). The clone was a contaminant and was related if not identical to bovine group 7 as represented by strain N-29B.

[d] Number of precipitin lines; homologous reactions are in italics.

was prepared by Dr. Freundt in Denmark and forms precipitin lines readily with bovine group 7 organisms and *M. putrefaciens*. The bovine group 7 organism was obtained by cloning ATCC strain 19884 (*M. bovirhinis*). This strain was recognized as a mixed culture of *M. bovirhinis* and bovine group 7 (J. Tully and D. G. ff Edward, personal communications). The clone was closely serologically related if not identical to authentic bovine group 7 organisms as represented by strain N-29B (received from J. Tully). Consequently, an apology is required for the previous report (14) where this clone was listed as *M. bovirhinis*, a species which is not related to group 1 organisms. Group 2 organisms consist of *Acholeplasma* species. The status of *M. bovigenitalium* is uncertain. It shows a one-way cross-reaction with *M. bovimastitidis* and possibly may be related to group 6 organisms. Both *M. gallisepticum* and *M. pneumoniae* are unique organisms; no related strains have been observed in these studies. Their serological difference is not surprising since *M. pneumoniae* has a high guanine-plus-cytosine (G + C) content in its deoxyribonucleic acid (DNA), i.e., 40% in contrast to 25 to 32% for all other species, and *M. gallisepticum* has a unique morphology. The largest group of organisms is group 6, which comprises 10 glycolytic species which are loosely related. These organisms have of 25 to 28% G + C. The arginine-utilizing organisms (group 7), also with 28% G + C, are closely serologically related. More detailed studies by two-dimensional immunoelectrophoresis indicate that these organisms have clear evidence of common antigens (9, 11).

SIGNIFICANCE OF SEROLOGICAL HETEROGENEITY

The fact that the *Mycoplasmatales* contain six or seven highly distinct groups, when measured by serological tests which detect 5 to 10 antigens, has considerable biological significance. It is clear that groups of organisms which share no common antigens are distant from each other taxonomically and phylogenetically. The question is, how distant? In my opinion, the failure to cross-react suggests that the groups of species should be classified in separate genera and probably distinct families, to recognize the phylogenetic differences these data imply. Thus, at least four additional genera need to be proposed to encompass the heterogeneity shown in Table 3. Should reclassification be done now? My answer is, no. Further serological studies need to be done to clearly identify common and unique antigens; such studies are in progress with the arginine-utilizing strains (9, 11; A. G. Alexander and G. E. Kenny, Abstr. Annu. Meet. Am. Soc. Microbiol. 1975, D36, p. 57). Additionally, chemical and physical studies (15) will be required to verify these differences. It should be noted that the serological differences observed in these studies may serve as a useful guide.

CHEMICAL HETEROGENEITY OF THE ANTIGENS OF THE *MYCOPLASMATALES*

Studies on the chemical nature of the antigens of various *Mycoplasma* and *Acholeplasma* species have been limited (16). Where data are available, a striking heterogeneity is observed (Table 4). All organisms must have protein antigens; however, only certain species contain other antigens such as glycolipids and

TABLE 4. *Relationship of analytical serology to taxonomy of the Mycoplasmatales*

Taxa	G + C content of DNA (%)	Genome size	Analytical serology[a]
Acholeplasmataceae			
Acholeplasma	30–32[b]	10^9	One group[c]
Mycoplasmataceae			
Mycoplasma			
(18 species tested)	25–40	5×10^8	Seven groups
Ureaplasma			
Types 1–	28	5×10^8	?

[a] Groups are defined by lack of cross-reactivity under conditions where to 6 to 10 antigens are measured.

[b] Data limited to some species of genus.

[c] Preliminary data by gel diffusion.

polysaccharides. Evidence for glycolipid antigens has recently been obtained for *U. urealyticum* (17), but it is not known whether humans respond to this component. Accordingly, the heterogeneity observed in both the comparative serology and the chemical nature of the antigens suggests that no species in the *Mycoplasmatales* can be clearly used as a model for serology of *U. urealyticum* (Table 5). It appears unlikely that *U. urealyticum* will have common antigens with all species in the *Mycoplasmatales*. Further, it is not necessary that *U. urealyticum* have any common antigens with any known *Mycoplasma* or *Acholeplasma* species. No clear indication of the heterogeneity of *U. urealyticum* types emerges from this comparison with the *Mycoplasmatales*. The organisms could be serologically very close and hence a species. Alternatively, they could be a closely related group of species (similar to *Acholeplasma* and the arginine-utilizing species), or they could be as heterogeneous as the glycolytic species in group 6.

ACKNOWLEDGMENTS

This study was supported by Public Health Service Research grant AI 06720 from the National Institute of Allergy and Infectious Diseases.

I acknowledge the expert technical assistance of E. Mattson and J. Russell.

TABLE 5. *Characteristics of major antigens of the Mycoplasmatales*

Species	Antigen[a]		
	Protein	Glycolipid	Polysaccharide
M. pneumoniae	+*	++*	+*
M. fermentans	+	++	?
M. pulmonis	++	−	+
M. hominis	+*	−	−
A. laidlawii	+	+	+
U. urealyticum	+ (?)	+	?

[a] Symbols: ++, major antigen; +, active; −, not detected, ?, unknown; *, react with human antisera.

REFERENCES

1. Ford DK: Relationships between mycoplasma and the etiology of non-gonococcal urethritis and Reiter's syndrome. Ann NY Acad Sci 143:501–504, 1967
2. Purcell RH, Chanock RM, Taylor-Robinson D: Serology of the mycoplasmas of man, p 221–264. *In* L Hayflick (ed), The Mycoplasmatales and L-phase of bacteria. Appleton, New York, 1969
3. Lin JS, Kass EH: Serotypic heterogeneity in isolates of human genital T-mycoplasmas. Infect Immun 7:499–500, 1973
4. Shepard MC, Lunceford CD, Ford DK, et al: *Ureaplasma urealyticum* gen. nov. sp. nov: proposed nomenclature for the human T (T-strain) mycoplasmas. Int J Syst Bacteriol 24:160–171, 1974
5. Clyde WA: Mycoplasma species identification based upon growth inhibition by specific antisera. J Immunol 92:958–965, 1964
6. Kenny GE: Immunogenicity of *Mycoplasma pneumoniae*. Infect Immun 3:510–515, 1971
7. Barker LF, Patt JK: Role of complement in immune inactivation of *Mycoplasma gallisepticum*. J Bacteriol 94:403–408, 1967
8. Gale JL, Kenny GE: Complement dependent killing of *Mycoplasma pneumoniae* by antibody: kinetics of the reaction. J Immunol 104:1175–1183, 1970
9. Thirkill CE, Kenny GE: Serological comparison of five arginine-utilizing *Mycoplasma* species by two-dimensional immunoelectrophoresis. Infect Immun 10:624–632, 1974
10. Weeke B: Carbamylated human immunoglobulins tested by electrophoresis in agarose and antibody containing agarose. Scand J Clin Lab Invest 21:351–354, 1968
11. Thirkill CE, Kenny GE: Antigenic analysis of three strains of *Mycoplasma arginini* by two-dimensional immunoelectrophoresis. J Immunol 114:1107–1111, 1975
12. Kenny GE: Serological comparison of ten glycolytic *Mycoplasma* species. J Bacteriol 98:1044–1055, 1969
13. Kenny GE: Heat-lability and organic solvent-solubility of *Mycoplasma* antigens. Ann NY Acad Sci 143:676–681, 1967
14. Kenny GE: Serologic heterogeneity of the Mycoplasmatales. J Infect Dis 127: S2–S5, 1973
15. Neimark HC: Division of mycoplasmas into subgroups. J Gen Microbiol 63: 249–263, 1970
16. Kenny GE: The antigens of the Mycoplasmatales and chlamydiae, p 449–478. *In* M Sela (ed), The antigens, vol 3. Academic Press Inc., New York, 1975
17. Romano N, Scarlata G: Serological activity of lipids of a T strain of *Mycoplasma*. Infect Immun 9:1062–1065, 1974

Chairman's Report and Discussion

DAVID TAYLOR-ROBINSON

M. C. Shepard described the biological features of *Ureaplasma urealyticum*, namely, small colony size, rapid growth in fluid medium, preference for an acid pH, and selective inhibition by thallium acetate, erythromycin, 5-iodo-2′-deoxyuridine, hydroxyurea, acetohydroxamic acid, and sulfhydryl inhibitors. It was pointed out that the single most characteristic feature was the ability to hydrolyze urea, and on this basis a separate genus in the family *Mycoplasmataceae* had been proposed, containing a single species, *U. urealyticum,* within which there were at least eight serotypes. In the discussion, the question of the relationship between ureaplasmas isolated from different animal species arose. Shepard suggested that it might be possible to designate other *Ureaplasma* species according to the animal in which they were most prevalent, but it was agreed that this would be possible only if the characteristics of those apparently belonging to one animal species were sufficiently different from those belonging to another; insufficient work had been done to know whether this was so.

S. Razin, G. K. Masover, and L. Hayflick in discussing the physiology of ureaplasmas noted that they resemble classical mycoplasmas in morphology, genome size, low guanine-plus-cytosine content, and requirement for cholesterol, but differ in their ability to hydrolyze urea and their failure to grow to more than 10^7 colony-forming units per ml. Since repeated additions of urea did not increase the titer, the failure to grow to higher titers was not the result of the exhaustion of urea or lack of its degradation products, CO_2 and ammonia. An atmosphere of 100% CO_2 improved growth, apparently by keeping the pH low. Urease was localized in the cytoplasm, and adenosine triphosphatase was localized in the membrane. One strain had been found to synthesize both saturated and unsaturated fatty acids from acetate, differing in this respect from all mycoplasmas tested so far. In discussion, H. M. Jenkin wondered whether ammonia could be removed from the system by adding, for example, acetic acid so that ammonium acetate would be volatilized. It was pointed out that a system had been used in which ammonia had been kept at a low level (1). This had resulted in an extension of the growth cycle without an increase in the titer of viable organisms. D. K. Ford suggested that urea might be broken down so rapidly that there would never be a sufficient amount available for growth. Dr. Razin did not believe this was likely, but agreed that the function of urea was unknown. The addition of putrescine to the medium resulted in much better growth. G. Deslile then reported that she had looked at the cytoplasm fractions of seven *Ureaplasma* strains, so far untyped, which had been isolated from the genital tract of patients and had discovered four urease isoenzymes. They were produced from 8 to 48 h of the growth cycle, did not change with different serum supplements, and were not present in uninoculated media.

Dr. Masover and colleagues discussed the interaction of *U. urealyticum* with eucaryotic cells in vitro. Infection of WI-38 cells with serotype 960 resulted in reduced rates of multiplication, reduced plating efficiency, and morphological changes. These effects were inhibited by chlortetracycline and did not occur in cells which were not growing. The organisms persisted for only a short time and could not be recovered after subcultivation—probably being removed from the cell surface by trypsinization. However, pretreatment of the ureaplasmas with trypsin resulted in infection of cells for more than 60 days. It seemed that such treatment removed a surface protein from the ureaplasmas, allowing closer contact with the cell membrane so that infection resulted. D. Taylor-Robinson pointed out that there was a paradox between persistent ureaplasma infection in Fallopian tube organ cultures and infection which did not persist in cell cultures unless the ureaplasmas were specially treated. Dr. Masover agreed that this underscored the general lack of knowledge of cell-ureaplasma interrelationships and that more information was needed if we were to validly compare in vitro and in vivo systems, particularly if the former were to be used as a model for urethritis. In contrast to successful infection of the WI-38 cells, he and his colleagues had failed to set up a persistent infection in an epithelioid cell line they had developed from guinea pig synovium. Shepard thought it was important to develop cell lines that were susceptible to ureaplasmas. He described a prostatic carcinoma cell line which exhibited cytopathic changes after ureaplasma infection; after Gram staining, individual cells were seen to contain cytoplasmic inclusions which resembled chlamydial inclusions. However, in response to questions from M. F. Barile, Shepard said that the inclusions had not been identified by immunofluorescent-antibody staining. In addition, he thought that the key to successful infection of a cell line was to include a small amount of urea in the nutrient medium. However, although Masover had not added urea, he had added horse serum, which had not helped much.

J. A. Robertson then described evidence she had obtained for the existence of extramembranous carbohydrate on ureaplasma strains 960 and 681/74, the latter having been isolated from a male with nongonococcal urethritis. The ruthenium red technique had been used to show an outer layer of polyanions about 20 to 30 nm in width. It seemed from further studies using concanavalin A that part of this outer layer was adherent carbohydrate material from the medium. Nevertheless, this was not uninteresting because ureaplasmas might adhere to similar layers in genital tract mucosa, and it was tempting to speculate that the extramembranous layer might be a marker of ureaplasma pathogenicity.

J.-S. L. Lin discussed the serotyping of *U. urealyticum* and the human humoral response to it, using a complement-dependent mycoplasmacidal test. Strains isolated from various groups of persons were serotypically diverse, at least 11 serotypes being recognized, and isolates occasionally contained more than one serotype. Studies on sera taken from women immediately after delivery and 6 to 12 weeks later showed that antibody titer rises were most frequently detected when all 11 serotypes were used and when the women had ureaplasma-positive cultures. However, more women had antibody responses to *Mycoplasma hominis* than to *U. urealyticum*. Dr. Lin said that isolations had been made in liquid medium and then the typing had been done. J. G. Tully suggested, therefore, that the number of reported serotypes in a mixed culture was the minimum that might exist and more were likely to be found by using the epi-immunofluorescence technique with colonies on agar. Furthermore, he thought that low antibody titers were to be expected, particularly in the absence of blood invasion. Dr. Ford also commented on the low antibody titers in human sera which were seen even with Dr. Lin's test. He had not found the method to give results which were really different from those obtained with the conventional metabolism-inhibition technique in which he used complement. Taylor-Robinson wondered whether the inhibitory effect of tetracycline in serum had posed a problem by being confused with antibody. Lin said that there was no confusion because antibody required complement to be active. G. E. Kenny suggested that any problem could be overcome by making antibiotic-resistant mutants, although Dr. Taylor-Robinson thought it would be easier to select naturally occurring ones than to produce them in vitro. He also thought that concurrent use of the ureaplasma serotype numbering system proposed by Black and that proposed by Lin was causing confusion and suggested that there should be amicable agreement on the use of a common system.

Dr. Kenny presented a paper on the antigenic analysis of the *Mycoplasmatales*. He pointed out that serological tests, in particular gel diffusion and two-dimensional immunoelectrophoresis, had shown that the *Mycoplasmatales* contain six or seven highly distinct groups. He thought that groups of organisms which shared no common antigens were distinct from each other taxonomically and phylogenetically and that the groups of species should be classified in separate genera and perhaps distinct families. However, he did not think the time was ripe for reclassification, and little could be said about *U. urealyticum* since the diverse antigenic structure of organisms in the *Mycoplasmataceae* meant that it would be difficult to select a model system for it. However, despite the difficulty of producing ureaplasma antigens in any quantity, the strains appear to possess lipid antigens and show striking serological variability. Dr. Razin commended the two-dimensional immunoelectrophoresis technique in this kind of work and pointed out that taxonomy was an artificial creation for man's use. However, he agreed that it was necessary to move cautiously before attempting to make any changes. Dr. Lin mentioned that each of the seven strains of *M. hominis* which she had studied (2) had its own specific membrane protein antigen. It was also agreed that there was a need for the ready availability of antisera to each of the known serotypes of *U. urealyticum*. Production, it was suggested, might be supported by the Reference Reagents Branch of the National Institutes of Health. The antisera were needed for the purpose of serotyping strains isolated from patients with nongonococcal urethritis and other conditions. Finally, Dr. Shepard briefly described a quantitative method he had developed to establish the closeness of the association of ureaplasmas with the clinical course of nongonococcal urethritis (3). Patients were treated with a suboptimal dose of tetracycline. Of 23 men that were followed successfully, 12 were cured completely, 2 failed to respond, and 9 relapsed after several days. Those regarded as cured remained asymptomatic over 30 days. In the case of relapse, the organisms disappeared from the urine but were found again after about 14 days, just before the disease became reestablished. Dr. Taylor-Robinson commended this novel approach to the problem but thought that the significance of the data would be enhanced by seeking chlamydiae at the same time.

REFERENCES

1. Hendley JO, Allred EN: Prolonged stationary-phase growth of T-strain mycoplasmas in broth culture. Infect Immun 5:164–168, 1972

2. Lin J-SL, Alpert S, Radnay KM: Combined type-specific antisera in the identification of *Mycoplasma hominis*. J Infect Dis 131:727–730, 1975
3. Shepard MC: Quantitative relationship of *Ureaplasma urealyticum* to the clinical course of nongonococcal urethritis in the human male. INSERM 33:375–379, 1974

Author Index

Subject Index